LIGHT AND LIFE IN THE SEA

Edited by

PETER J. HERRING

Institute of Oceanographic Sciences, Deacon Laboratory

ANTHONY K. CAMPBELL

Department of Medical Biochemistry, University of Wales College of Medicine

MICHAEL WHITFIELD

Director, Marine Biological Association

LINDA MADDOCK

Marine Biological Association

A volume arising from the Symposium on Light and Life in the Sea, organised by
the Marine Biological Association of the United Kingdom and held at the
Polytechnic South West, Plymouth, on 10-11 April 1989

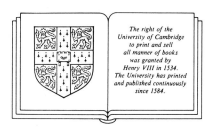

The right of the
University of Cambridge
to print and sell
all manner of books
was granted by
Henry VIII in 1534.
The University has printed
and published continuously
since 1584.

CAMBRIDGE UNIVERSITY PRESS

Cambridge

New York   Port Chester   Melbourne   Sydney

Published by the Press Syndicate of the University of Cambridge
The Pitt Building, Trumpington Street, Cambridge CB2 1RP
40 West 20th Street, New York, NY 10011, USA
10 Stamford Road, Oakleigh, Melbourne 3166, Australia

First published 1990

Printed in Great Britain by University Press, Cambridge

British Library Cataloguing in publication data available

Library of Congress cataloguing in publication data available

ISBN 0 521 39207 1 hard covers

Frontispiece; Light on the sea. Photograph by David Nicholson

Contents

Contributors

J. Aiken. Plymouth Marine Laboratory, Prospect Place, West Hoe, Plymouth, PL1 3DH

R.S. Batty. Dunstaffnage Marine Laboratory, PO Box 3, Oban, Argyll, PA34 4AD

J.H.S. Blaxter. Dunstaffnage Marine Laboratory, PO Box 3, Oban, Argyll, PA34 4AD

I.E. Bellan. Plymouth Marine Laboratory, Prospect Place, West Hoe, Plymouth, PL1 3DH

C. Brownlee. Marine Biological Association, The Laboratory, Citadel Hill, Plymouth, PL1 2PB

Anthony K. Campbell. Department of Medical Biochemistry, University of Wales College of Medicine, Heath Park, Cardiff, CF4 4XN

E.J. Denton, FRS. Marine Biological Association, The Laboratory, Citadel Hill, Plymouth, PL1 2PB

M.J. Dring. Department of Biology, Queen's University, Belfast, N. Ireland, BT7 1NN

Peter J. Herring. Institute of Oceanographic Sciences Deacon Laboratory, Brook Road, Wormley, Godalming, Surrey, GU8 5UB

Ian R. Joint. Plymouth Marine Laboratory, Prospect Place, West Hoe, Plymouth, PL1 3DH

Michael F. Land, FRS. School of Biological Sciences, University of Sussex, Falmer, Brighton, BN1 9QG

Frank McCapra. School of Chemistry & Molecular Sciences, University of Sussex, Falmer, Brighton, Sussex, BN1 9QJ

Linda Maddock. Marine Biological Association, The Laboratory, Citadel Hill, Plymouth, PL1 2PB

H.R. Matthews. Physiological Laboratory, University of Cambridge, Downing Street, Cambridge, CB2 3EG

J.C. Partridge. Department of Zoology, University of Bristol, Woodland Road, Bristol, BS8 1UG

P.L. Pascoe. Plymouth Marine Laboratory, Citadel Hill, Plymouth, PL1 2PB

Trevor Platt. Biological Oceanography Division, Bedford Institute of Oceanography, PO Box 1006, Dartmouth, Nova Scotia, Canada, B2Y 4A2

Geoffrey W. Potts. Marine Biological Association, The Laboratory, Citadel Hill, Plymouth, PL1 2PB

I.S. Robinson. Department of Oceanography, University of Southampton, University Road, Southampton, SO9 5NH

Helen R. Saibil. Department of Crystallography, University of London Birkbeck College, Malet Street, London, WC1E 7HX

Shubha Sathyendranath. Oceanography Department, Dalhousie University, Halifax, Nova Scotia, Canada, B3H 4J1

Paul Tett. School of Ocean Sciences, University College of North Wales, Menai Bridge, Gwynedd, LL59 5EY

M. Whitfield. Marine Biological Association, The Laboratory, Citadel Hill, Plymouth, PL1 2PB

Introduction

M. WHITFIELD

Marine Biological Association, The Laboratory, Citadel Hill, Plymouth PL1 2PB

Without light there would be no life in the sea. Since the seas were the cradle for the evolution of all life forms, the theme of this book is central to our understanding of the interaction between living organisms and their environment. To express the breadth of research in this area, leading experts in topics as diverse as satellite imagery and molecular biology have contributed to this collection of essays on 'Light and Life in the Sea' so that it is unique in both its scope and its authority. The intention is not to provide a comprehensive textbook but rather to present the reader with a sampler of the exciting research that is underway and to provide an introduction to its broad compass.

The book opens with a series of chapters on the physics of light in the sea. Here the spectral characteristics of the incoming radiation are described and their modification by absorption, scattering, fluorescence and bioluminescence is explained. Microscopic drifting plants, the phytoplankton, intervene and modify the characteristics of light reflected and radiated from the surface layers of the ocean. The use and verification of remote sensing techniques for determining the distribution and productivity of phytoplankton in the oceans is described. The harvesting of light for photosynthesis provides the main topic for the following section of the book. The mean depth of the oceans is nearly four kilometres but photosynthesis is restricted to a thin layer, at most a few hundred metres thick. The factors controlling the depth of this shallow photic zone are described and the photosynthetic strategies for organisms ranging from bacteria to seaweeds are explained and the importance of light in controlling the development of plant cells is discussed.

The second half of the book explores the exploitation of light in the sea by marine animals, often at great depths. As well as providing the basic energy source for life in the sea, light is used by animals to provide information about their environment, through vision. The optics of the eyes of marine animals show great diversity but, where a particular structure provides optimal sensitivity or resolution it may evolve by a number of independent routes. A section on vision considers the physiology of vision and the functioning of the photoreceptors of vertebrates and of invertebrates. The following section explores the way in which animals use the visual information and the effects that this has on their behaviour. The impact on the vertical migration and schooling of fish

Herring, P.J., Campbell, A.K., Whitfield, M. & Maddock, L. (ed.). *Light and Life in the Sea.*
© 1990. Cambridge University Press.

and on the diurnal rhythms of animals in reef communities is described. The exploitation of such behavioural responses by man is discussed in an essay considering the use of light lures on fishing gear. Where sunlight can penetrate the oceans this is utilised, but at night or at greater depths the organisms must generate their own light. The final section of the book therefore considers the use of this biologically generated light (bioluminescence) for communication between organisms. The physiology, evolution and biochemistry of bioluminescence are described and examples are given of the exploitation of biolumines- cent proteins for medical research.

The individual essays were originally presented as invited papers at the first Annual Scientific Meeting of the Marine Biological Association which was held at the Polytechnic South West in Plymouth in April 1989. The speakers, and their topics, were selected to provide a balanced view of the exploitation of light by life in the sea. The invited papers were augmented at the meeting by a diverse range of submitted papers and posters. The result was a meeting of extraordinary breadth that was greeted with great enthusiasm by those who participated, not least because it provided an opportunity for experts studying different aspects of the impact of light on life in the sea to understand the context of their investigations. It is our intention that this series of essays should provide an equally bracing stimulus to a much wider audience.

In closing this introduction I would like to express my thanks to all those who have contributed to the production of this book. Tony Campbell and Peter Herring developed the theme of the meeting and were responsible for assembling such an excellent and diverse set of authors. John Green, through his careful organisation and infinite patience, untiringly helped by Jane and Rebecca Green, ensured that the meeting itself ran smoothly so that we could all concentrate on enjoying the delightful variety of ideas that we hope this volume will convey to its readers. Finally, Linda Maddock carried out most of the detailed editorial work, ably assisted by Elwyn Harris, Jinny Rutherford and Robert Maddock, and shepherded the individual contributions into a well-organised volume.

Chapter 1

THE LIGHT FIELD IN THE OCEAN: ITS MODIFICATION AND EXPLOITATION BY THE PELAGIC BIOTA

SHUBHA SATHYENDRANATH

Oceanography Department, Dalhousie University, Halifax, Nova Scotia, Canada B3H 4J1

AND TREVOR PLATT

Biological Oceanography Division, Bedford Institute of Oceanography, PO Box 1006, Dartmouth, Nova Scotia, Canada B2Y 4A2

INTRODUCTION

Life in the sea is intimately linked to the quality and quantity of light available there. Primary production, upon which the entire trophic economy of the sea depends, is controlled by the spectral and angular distribution of underwater light. At the same time, the marine organisms themselves modify the submarine light field by absorption and scattering, and through the processes of fluorescence and bioluminescence. The light produced underwater by marine organisms through fluorescence and bioluminescence is but a minute fraction of the sunlight reaching the sea surface, and is therefore safely neglected in most energy budget calculations for the pelagic ocean. On the other hand, it carries the spectral signature of the organisms that generate it, and thus has potential significance as a diagnostic marker for them. Variations in the pigmented biota usually dominate variations in ocean colour as seen by an observer outside the ocean, a phenomenon which is now being exploited for estimation, by remote sensing, of phytoplankton biomass and primary production.

Here, we describe the light field at the surface and in the interior of the oceans. Light emission within the sea by fluorescence is briefly discussed. We describe how the expected rate of photosynthesis depends on the characteristics of the submarine light field, and how this information can be used in remote sensing applications.

Our discussion is confined to the visible spectrum. Light penetration in water, marine photosynthesis, fluorescence by phytoplankton, and bioluminescence all show strong wavelength dependence, and within the spectral range of interest it is therefore important to consider the spectral distribution, rather than just the total flux available. The terms and notations used in this paper are described in the Glossary in Appendix 1.

Herring, P.J., Campbell, A.K., Whitfield, M. & Maddock, L. (ed.). *Light and Life in the Sea.*
© 1990. Cambridge University Press.

LIGHT REACHING THE SEA SURFACE

The extra-terrestrial solar irradiance spectrum is now fairly well known. Changes in the distance between earth and sun introduce a seasonal variation of just ±3.4% around the mean extra-terrestrial irradiance. If, however, we compute the flux per unit area on the surface of the earth (neglecting atmospheric effects), it is seen to vary markedly with latitude and with season, the seasonal variation being most pronounced at the poles and weakest at the equator. The daily path of the sun across the sky, which determines both the day length and the daily maximum in solar flux per unit surface area, is also a function of latitude and season. In addition to these effects, which can be computed on purely geometrical considerations, we also have to take into account the consequences of variable transmission through the atmosphere. Atmospheric transmission changes not only the quantity and spectral quality of light reaching the sea surface, but also its angular distribution: part of the light reaching the sea surface arrives from the direction of the sun (direct sunlight), and the remainder (diffuse sky light) has a complex angular distribution, from scattering by the atmospheric constituents.

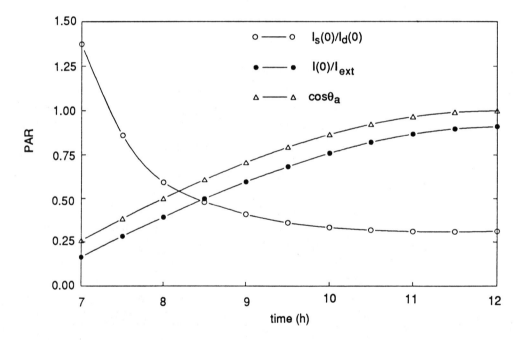

Figure 1.1. This figure shows an example of changes in the Photosynthetically Active Radiation (PAR) at the surface of the ocean, as a function of local apparent time. These are results of computations for a clear day at the equator, when the sun is directly overhead at noon. Computations were carried out using the spectral transmittance model of Bird (1984), and the results integrated to obtain PAR (400-700 nm). The ratio $(I(0)/I_{ext})$ is plotted, where $I_{ext}(= 8.66$ E m^{-2} h^{-1} in these computations), is the extra terrestrial PAR normal to the sun's rays. The ratio of diffuse to direct irradiance, on a horizontal plane facing upwards, at the sea surface $(I_s(0)/I_d(0))$ is also plotted. Notice that the increase in total irradiance $I(0)$ at the sea surface corresponds closely to changes in cos θ_a where θ_a is the zenith angle of the sun (in air).

Under clear-sky conditions, atmospheric transmission is determined primarily by the zenith angle of the sun, which is a function of latitude, date and local time (see Figure 1.1 for an example). Models of varying degrees of complexity are now available that compute spectral transmission through a cloud-free atmosphere for a given zenith angle, starting with the extra-terrrestrial solar irradiance (Brine & Iqbal, 1983; Bird, 1984). The computations for the visible part of the spectrum have to account for Rayleigh scattering by atmospheric gases, for aerosol extinction, and for absorption by ozone and water vapour.

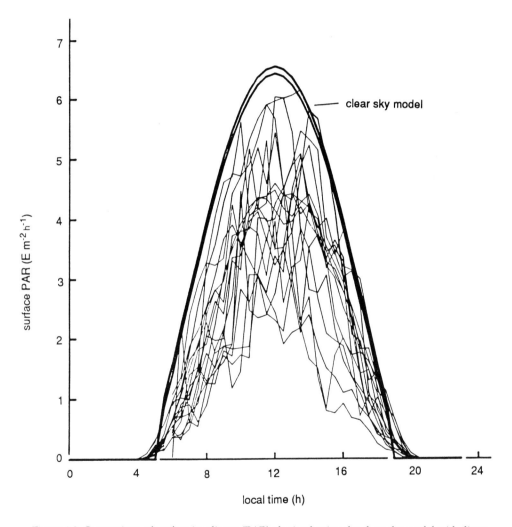

Figure 1.2. Comparison of surface irradiance (PAR) obtained using the clear-sky model with direct observations at the sea surface during a cruise in the Celtic Sea on board RRS 'Darwin' from 18 May to 7 July 1986. The two thick curves correspond to clear-sky computations for the first and last days of observation. The thin jagged lines represent all observations. The clear-sky computations envelope the upper limit of the observations. The discrepancies are mostly due to clouds, and the figure gives a good indication of the complex effect of clouds on the surface irradiance. We are indebted to Dr M. J. R. Fasham for use of the data.

In the visible, water vapour has an absorption band at 593 nm, but it is a weak one. The Rayleigh scattering component can be computed very accurately, given atmospheric pressure. Aerosol extinction depends on the concentration of aerosols, their size spectrum, refractive index and distribution with altitude. This component is therefore the most difficult one to model for routine computations treating clear skies. But some parameterisations are available (Paltridge & Platt, 1976; Brine & Iqbal, 1983; Bird, 1984) that allow reasonable approximations of their spectral extinction, given the aerosol turbidity at one wavelength (defined as the aerosol optical depth τ_a at that wavelength, and for the total vertical atmospheric column). These computations are usually easier for the sea than for the land, because of low concentrations of aerosols over the sea.

Modelling spectral transmission under cloudy skies has proved to be extremely difficult (see Figure 1.2 for an example). In the first place, the computations, which depend on the types and vertical height distributions of clouds, are very complex; in the second place, cloud data are rarely available in sufficient detail for such computations. By scattering more light in the direction of the sea, certain fields of small, loosely-distributed clouds can even cause an increase (5-10%) in radiation at sea level, compared to that under a clear sky. Extensive and thick clouds, on the other hand, can reduce the surface flux to about 20% of the clear-sky value (Kirk, 1983). Clouds also alter the partition of the total flux between direct and diffuse components. Paltridge & Platt (1976) have discussed parameterisation of the cloud effect. Most parameterisations now in use account for changes in the total flux and in the relative size of the direct and diffuse fractions, but do not offer any indication of the change in spectral shape.

THE AIR-SEA INTERFACE

The light reaching the sea surface is modified by refraction and reflection at the interface. Reflection (which can be computed according to Fresnel's Law) remains very low ($\approx 2\%$) for angles of incidence down to 45°, and then increases steadily to total reflection at grazing incidence. Note that, at Brewster's angle (53·1°), reflection drops to zero for light polarised parallel to the plane of incidence. The effect of refraction (computed according to Snell's Law as a function of incident angle), is to change the angular distribution of light under water. In the case of a flat sea surface, it limits the light field to a cone of semi-vertical angle of 48° (the critical angle). A rough sea surface diffuses the edge of this cone.

Computation of reflection and refraction requires a knowledge of the angular distribution of the light field reaching the sea surface. For diffuse sky light, it is common to assume either a uniform or a cardioidal distribution in air (see Jerlov, 1976). Reflection is also a function of polarisation, but computations are usually made assuming the light to be unpolarised (note however, that clouds have a depolarising effect; see Jerlov, 1976).

Another complication is that refraction and reflection are functions of sea surface roughness. It has been shown that the slope distribution at the sea surface approximates a Gaussian distribution when the sea surface is ruffled by winds, and that there is only a small difference between the along-wind and across-wind slope distributions (Preisendorfer, 1976*a*). Computations and observations have shown that wind-induced roughness has a noticeable effect on reflection only at solar elevations less than 30°, with reflection decreasing with increased wind speed (Preisendorfer & Mobley, 1986; Apel, 1987).

LIGHT TRANSMISSION UNDER WATER

The equation of radiative transfer

A complete description of the underwater light field would require that the flux be described for every angle, and for every wavelength. In other words, if one assumed the horizontal divergence of light flux to be zero, it would be necessary to specify the radiance L for every depth z, wavelength λ, zenith angle θ, and azimuth angle ϕ. Neglecting light sources within the water, the equation of radiation transfer, which describes the rate of change of radiance with distance r, can be written as (Jerlov, 1976):

$$\frac{dL\,(z\,,\lambda\,,\theta\,,\phi)}{dr} = -\,c\,(z\,,\lambda)\,L\,(z\,,\lambda\,,\theta\,,\phi) + L_*\,(z\,,\lambda\,,\theta\,,\phi) \qquad (1.1)$$

where c is the total volume attenuation coefficient and L_*, known as the path function, accounts for the light scattered into the direction (θ,ϕ) from all other directions (θ',ϕ'):

$$L_*\,(z\,,\lambda\,,\theta\,,\phi) = \int_{\phi'=0}^{2\,\pi} \int_{\theta'=0}^{\pi} \beta\,(z\,,\lambda\,,\theta\,,\phi\,;\,\theta'\,,\phi')\,L\,(z\,,\lambda\,,\theta'\,,\phi')\,\sin\theta'\,d\theta'\,d\phi' \qquad (1.2)$$

where $\beta(z,\lambda,\theta,\phi;\theta',\phi')$ is the volume scattering function at depth, z, for the wavelength, λ, and for the angle defined by the directions (θ,ϕ) and (θ',ϕ').

These equations imply that a complete description of the radiance field requires that we specify all the inherent optical properties of the medium: the volume scattering function β, whose integral over all angles yields the volume scattering coefficient b, and the volume absorption coefficient a (note that $c = a + b$). Integration of Equation 1.1 over long distances also has to take into account the effects of multiple scattering and the depth dependence of the inherent optical properties. Even if the inherent optical properties of the water column and the incident light field at the ocean surface are known with the necessary detail (which is often not the case), finding complete solutions to the radiative transfer equation represents a monumental task in computing. Fortunately, most applications do not require such a complete description of the light field.

For biological applications, such as the estimation of the light absorbed by phytoplankton, computations can be simplified considerably by dealing with the irradiance field rather than with the radiance field. The vector irradiance I, which defines flux per unit area, is given by:

$$I\,(z\,,\lambda) = \int_{\phi=0}^{2\,\pi} \int_{\theta=0}^{\pi} L\,(z\,,\lambda\,,\theta\,,\phi)\cos\theta\,\sin\theta\,d\theta\,d\phi \qquad (1.3)$$

In addition to the saving in computational time that results from dealing with the integrated flux, irradiance also offers the added advantage of relative ease of measurement at sea.

Parameterisation of the irradiance field

In computations, as well as in measurements, it is common to split the total irradiance into two parts: the downwelling irradiance, which represents the integration of radiance over the upper hemisphere, and the upwelling irradiance, which represents the integration over the lower hemisphere. Note that downwelling irradiance is that fraction of irradiance measured by a flat horizontal collector facing upwards, and that upwelling irradiance is measured by a similar collector facing downwards. The upwelling irradiance arises primarily from scattering, and also from biological activity, such as fluorescence and bioluminescence.

Scattering in sea water is of two kinds: molecular scattering by pure sea water, and particle scattering by suspended material in the sea water. An important property of molecular scattering is that forward scattering equals backward scattering: it is symmetrical about the plane of incidence. Particle scattering in sea water, which falls in the domain of Mie scattering, is characterised on the other hand by very pronounced forward scattering. It is strongly asymmetrical. In the oceans, scattering by particles always dominates molecular scattering. An important consequence is that the upwelling irradiance is usually only a small fraction of the downwelling irradiance, and can safely be neglected for many applications. We adopt this simplifying assumption here, and equate I with the downwelling irradiance.

The rate of decrease of the downwelling irradiance is determined by the attenuation coefficient for irradiance K, defined by:

$$K(z, \lambda) = \frac{-1}{I(z, \lambda)} \frac{dI(z)}{dz} \tag{1.4}$$

The parameter K is a function of the inherent optical properties of the sea water and of the angular distribution of the light field. Because it is dependent on the ambient light conditions, it is classed as one of the apparent optical properties. It is interesting to see how K can be expressed as a function of the inherent optical properties. For light incident normally, and for single scattering approximation,

$$K(z, \lambda) = a(z, \lambda) + b_b(z, \lambda) \tag{1.5}$$

where b_b is the backscattering coefficient, defined by

$$b_b(z, \lambda) = 2\pi \int_{\pi/2}^{\pi} \beta(z, \chi) \sin\chi \, d\chi \tag{1.6}$$

where χ is the scattering angle with respect to the initial direction of the light ray. Now, for light incident at zenith angle θ, $K(z, \lambda)$ becomes:

$$K(z,\lambda) = \frac{a(z,\lambda) + b_u(z,\lambda)}{\cos\theta} \qquad (1.7)$$

where b_u is the upward scattering. The $\cos\theta$ term accounts for the increased path length of light per unit vertical distance for oblique incidence. Sathyendranath & Platt (1988) have used b_b as an approximation of b_u. Because of the asymmetry in the volume scattering function of particles, b_u for particles is greater than b_b. But Sathyendranath & Platt (1989) have argued that it is a reasonable approximation for most open ocean waters, where a dominates over b_u, such that K is largely determined by a. For a diffuse light field, the $\cos\theta$ term has to be replaced by μ, the mean path length of the light rays per unit vertical distance. In this case, with the approximate expression for upward scattering, we have:

$$K(z,\lambda) = \frac{a(z,\lambda) + b_b(z,\lambda)}{\mu} \qquad (1.8)$$

To account for multiple scattering, we have to admit the possibility that some of the light scattered in the upward direction may be scattered downwards again. But here once again, due to the pronounced forward scattering by particles, the correction would be only minor. Another consequence of multiple scattering is that the angular distribution of light underwater, and therefore the average cosine μ, changes with depth. It has been established both experimentally and theoretically that, a few attenuation depths (1 attenuation depth = $1/K$) below the sea surface, the angular distribution approaches an asymptotic state, which depends only on the inherent optical properties of the medium (Preisendorfer, 1959). However, change in μ with depth is generally not more than ±10% for open ocean waters (Preisendorfer, 1976b; Prieur, 1976).

One of the primary sources of variability in $K(z,\lambda)$ is phytoplankton, the other sources being suspended sediments and dissolved organic matter. In fact, in open-ocean waters, phytoplankton can be considered to be the single independent variable controlling the magnitude of K. Parameterisations of the spectral values of K fall into two classes: (1) those that are based on statistical correlations between K and the phytoplankton concentration C (Smith & Baker, 1978a; Morel, 1988), and (2) those based on expression of a and b as functions of C, combined with parameterisation of μ (Dirks & Spitzer, 1987; Sathyendranath & Platt, 1988). Examples of results from the model of Sathyendranath & Platt (1988) are given in Figures 1.3 and 1.4.

The first class makes uses of the 'quasi-inherent' nature of K: that is, according to many observations in the open oceans, K is not strongly dependent on the underwater light distribution (partly explained by the fact that refraction at the sea surface limits maximum θ to 48° for a flat sea surface, and by the fact that light measurements at sea are usually made at high light-levels, that is, when solar elevation is high). The second approach offers some flexibility. For example, separation of absorption and scattering effects makes it relatively easy to compute the fraction that has been absorbed by phytoplankton, and is therefore available for photosynthesis. But on the other hand, it requires that μ be specified explicitly.

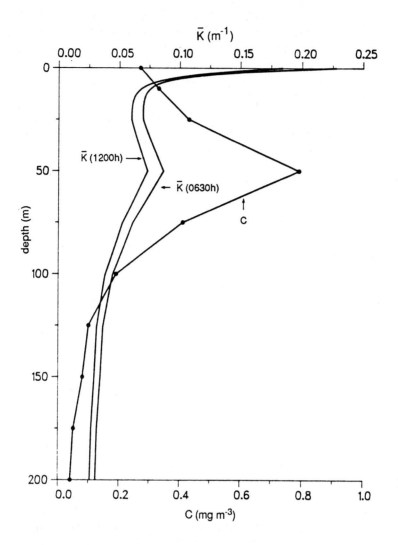

Figure 1.3. An example of computations of $\bar{K}$, the attenuation coefficient for total PAR, for a given vertical distribution of pigments C (from Sathyendranath & Platt, 1988; copyright by the American Geophysical Union). The computations of $\bar{K}$ were carried out for early morning (0630 h) and noon (1200 h) and the curves are labelled accordingly. Note that the changing angular distribution of incident light from morning to noon does introduce a significant effect on the attenuation coefficient. Also significant is the fact that in the surface layers, the attenuation coefficient does not covary with the concentration of pigments; instead, it is dominated by the high attenuation of the red wavelengths by pure sea water.

In the first approach, $K(z,\lambda)$ is expressed as:

$$K(z,\lambda) = K_w(\lambda) + C(z)\,k_c(\lambda) \tag{1.9}$$

where K_w is the attenuation of pure sea water, and k_c is the attenuation of phytoplankton and associated material, per unit concentration of phytoplankton. The concentration of phytoplankton, C, is commonly expressed as the sum of concentrations of the major pigment chlorophyll *a* and its degradation product, phaeophytin.

In the second approach, the absorption and backscattering coefficients are expressed as functions of C. Sathyendranath & Platt (1988) then use Equation 1.7 to compute attenuation of direct sunlight underwater (the sun zenith angle in water = θ), and Equation 1.8 to compute attenuation of diffuse sky light ($\mu = 0.83$, for a uniformly diffuse sky light above water, and a flat sea surface).

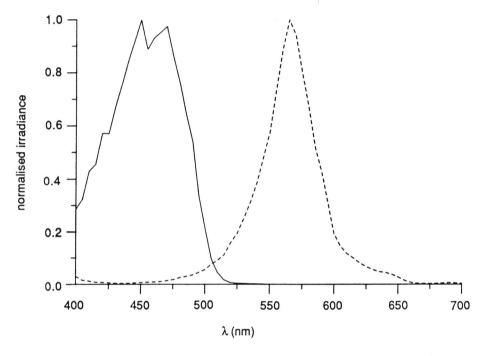

Figure 1.4. Theoretical computations on the effect of phytoplankton on the spectral distribution of irradiance at the base of the euphotic zone. Two cases are given: (A) Unbroken line - clear waters, with low pigment concentrations ($C = 0.01$ mg m^{-3}), and (B) Dashed line - phytoplankton-rich water ($C = 10$ mg m^{-3}). The residual light is predominantly blue in clear waters, and green in phytoplankton-rich waters (see Sathyendranath & Platt, 1988).

FLUORESCENCE BY PHYTOPLANKTON

The fluorescence emission of chlorophyll *in vivo* at 685 nm is well known (Goedheer, 1964). This property has been exploited in the *in vivo* fluorescence technique for measurement of chlorophyll concentration at sea (Lorenzen, 1966), and more recently, in the flow cytometer for sorting phytoplankton cells (Yentsch *et al.*, 1984; Li, 1986; see Figure 1.5). But only since 1977 has it been appreciated that fluorescence emission by chlorophyll *a* was responsible for a peak that was sometimes observed in the red part of the upwelling irradiance spectrum (Morel & Prieur, 1977; Neville & Gower, 1977;

Gordon, 1979). Earlier, this feature, centred at around 685 nm, had been attributed to anomalous dispersion in the vicinity of the chlorophyll absorption peak in the red (Mueller, 1973; Gordon, 1974). The fluorescence peak is now exploited in passive remote sensing, for monitoring chlorophyll concentration (Gower, 1980) and may be useful as an index of primary production (Topliss & Platt, 1986). Active remote sensing (LIDAR) uses a laser to stimulate the bulk phytoplankton fluorescence of the water column (Hoge & Swift, 1981; Yentsch & Yentsch, 1984).

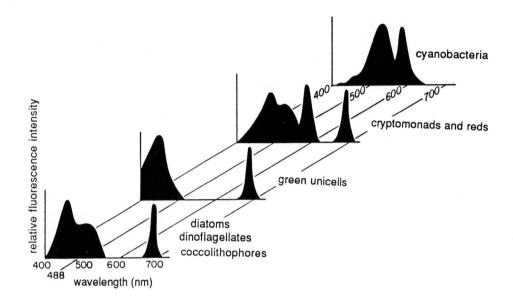

Figure 1.5. Characteristic fluorescence excitation and emission spectra of some phytoplankton groups (from Yentsch & Phinney, 1985, with permission of Oxford University Press). From front to back: (A) diatoms, dinoflagellates and cocolithophores, (B) green unicells, (C) cryptomonads and reds, and (D) cyanobacteria. (A,B,C) show chlorophyll *a* fluorescence emission in the red ($\approx$685 nm). (C) shows in addition an emission band due to phycoerythrin in the 570-590 nm region, whereas in (D), only the phycoerythrin emission is seen.

In flow cytometry (Yentsch *et al.*, 1984; Li, 1986), we can interrogate individual phytoplankton cells as to their fluorescence emission and light scattering at two angles (forward scattering, and scattering at 90°). Large numbers of cells can be processed in a short time. The possibility thus becomes real that the bulk or macroscopic optical properties of a given water mass can be connected with microscopic properties at the level of the individual particles contained therein, through some process of ensemble averaging. Furthermore, many biological properties (cell division, cell differentiation, gametogenesis, death by grazing) act at the level of the individual cell. By exploiting processes of light scattering and fluorescence in flow cytometry, we can begin to characterise the variance structure of these attributes in natural assemblages and interpret the population (ecological) response as the statistical limit of the responses of many individuals.

In chlorophyll-rich waters, the underwater light field in the red part of the spectrum cannot be described adequately without taking fluorescence emission into account. In such a case, Equation 1.1 would have to be modified, by addition of a fluorescence source on the right hand side (Gordon, 1979; Dirks & Spitzer, 1987). The fluorescence emission at any depth can be expressed as a function of light absorbed by phytoplankton at that depth and its fluorescence efficiency (Gordon, 1979). *In vivo* fluorescence efficiency is of the order of 10% of the *in vitro* efficiency, which in turn is estimated to be about 30% (see Kiefer, 1973; Gordon, 1979).

PHOTOSYNTHESIS-LIGHT RELATIONSHIP

Because of the exponential attenuation of light with depth, marine photosynthesis is confined to a surface layer (the euphotic zone), whose thickness, which varies regionally, is of order 100 m in the open ocean (see Tett, Chapter 4).

The photosynthetic rate is known to be a complex function of the intensity and spectral distribution of light. Sathyendranath & Platt (1989) have shown that marine photosynthesis also depends on the angular distribution of the light field underwater. For a given biomass B, the dependence of photosynthesis on light is described by the light saturation curve (see Figure 1.6 for some examples), which requires two parameters to describe it (three, if photo-inhibition is taken into account) (Platt *et al.*, 1977). The commonly-used parameters are α^B, the initial slope per unit biomass B, and P_m^B, the assimilation number, also normalised to biomass. The light dependence of photosynthesis can then be written as:

$$P^B(z) = \frac{\alpha^B I(z)}{\sqrt{1 + \left\{\alpha^B I(z) / P_m^B\right\}^2}} \tag{1.10}$$

where P^B is rate of primary production per unit biomass, and $I_k = P_m^B / \alpha^B$. Note that the biomass B here differs from the variable C. Biomass B is the photosynthetically-active part of the total phytoplankton content represented by C, which may also include detrital material.

In the above expression (Smith, 1936), $I(z)$ is the total flux at z in the photosynthetically-active range (400-700 nm), and any possible spectral effects are disregarded. However, the parameter α^B, characterising the photochemical reactions of photosynthesis, is known to have a strong wavelength dependence (the curve describing the spectral dependence of α is known as the action spectrum). The parameter P_m^B, on the other hand, characterises the enzyme-mediated, dark reactions of photosynthesis, and is therefore independent of wavelength, as far as is known. Techniques are now available for routine

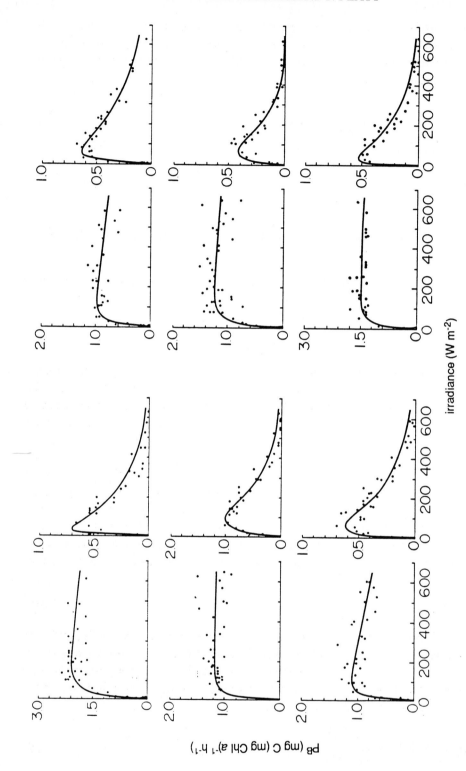

irradiance (W m⁻²)

PB (mg C (mg Chl *a*)⁻¹ h⁻¹)

Figure 1.6. Examples of curves fitted to experimental data on the photosynthesis-light response of natural sea water samples from the North-East Baffin Island, in summer. In each case, the left-hand graph is for surface samples, the right-hand graph for samples from the base of the euphotic zone. Deep samples are adapted to low light levels in the stable water column of the Arctic summer, and are therefore susceptible to photoinhibition (from Platt *et al.,* 1982).

measurements of the action spectrum of natural phytoplankton samples at sea (Lewis *et al.*, 1985). It has therefore become useful to refine Equation 1.9 to account for spectral dependence of α^B and of light penetration. According to Sathyendranath & Platt (1989), the spectral version of Equation 1.9 can be written as:

$$P^B(z) = \frac{\Pi(z)}{\sqrt{1 + \left\{\Pi(z)/P_m^B\right\}^2}} \tag{1.11}$$

where

$$\Pi(z) = sec\,\theta \int \alpha^B(z,\lambda) I_d(z,\lambda,\theta)\,d\lambda + (1/\mu) \int \alpha^B(z,\lambda) I_s(z,\lambda)\,d\lambda \tag{1.12}$$

In this equation, θ is the sun zenith angle in water, and $I_d(z,\lambda)$ and $I_s(z,\lambda)$ are the direct and diffuse components respectively of the irradiance at depth z and wavelength λ.

OCEAN COLOUR

The variances in the distributions of phytoplankton biomass and of primary production in the oceans cover a wide spectrum in both space and time scales. One of the major problems in biological oceanography has been the inability to make measurements of these variables at all relevant time and space scales. Remote sensing provides near-instantaneous large-scale coverage of the oceans, thereby yielding information at scales inaccessible to ship or buoy (see Robinson, Chapter 2).

The most common approach to remote sensing for biological oceanography is based on changes in ocean colour. The signal for remote sensing of ocean colour is the upwelling radiance at the sea surface. Even though it is only a small fraction of the incoming flux, it is of profound importance in this context. Theoretical relationships are available that relate the spectral structure of water-leaving radiance to ocean colour. In turn, ocean colour is determined by the spectral variation in reflectance R at depth zero, defined by:

$$R(\lambda) = I_u(0,\lambda) / I(0,\lambda) \tag{1.13}$$

where I_u is the upwelling irradiance, and I the downwelling irradiance (see Figure 1.7 for some examples of reflectance spectra). As in the case of K, it is possible to express R in terms of the absorption coefficient and the backscattering coefficient (see review by Sathyendranath & Morel, 1983). The simplest of these relationships is of the form (Prieur & Morel, 1975; Morel & Prieur, 1977):

$$R(\lambda) = 0.33 \frac{b_b(0,\lambda)}{a(0,\lambda)} \tag{1.14}$$

As we have already seen, values of a and b_b for the open ocean are determined primarily by the phytoplankton abundance, and it has been found that quantitative relationships can be developed linking changes in ocean colour with phytoplankton concentration (see review by Sathyendranath & Morel, 1983).

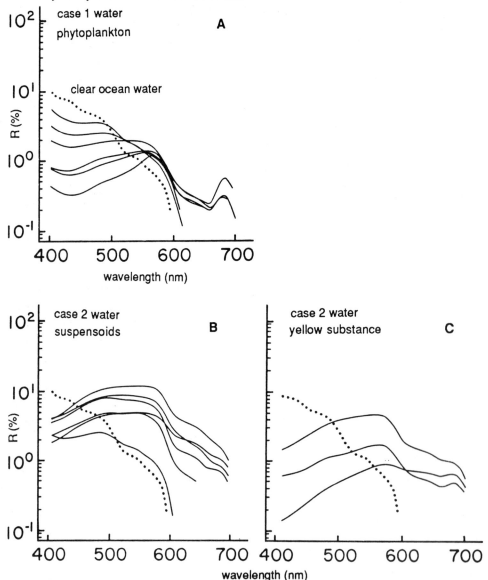

Figure 1.7. Examples of reflectance spectra from various marine regimes (adapted from Sathyendranath & Morel, 1983). (A) Case 1, represents phytoplankton-dominated waters. The peak in the reflectance spectrum shifts from blue to green with increasing pigment concentration. Also visible in phytoplankton-rich waters is the fluorescence emission peak of chlorophyll a in the red. Both of these characteristics are currently used in remote sensing for monitoring phytoplankton concentration. (B) & (C) Case 2, represent situations where the reflectance signal is determined mainly by the concentrations of suspended matter or dissolved organic substances (yellow substance). The presence of these substances complicate chlorophyll-extraction algorithms in coastal waters. Clear ocean-water reflectance is shown in all three panels by dotted lines.

Recent developments take remote sensing applications a step further, and use satellite-derived biomass to estimate primary production. Initial efforts in this direction were purely empirical, and depended on statistical relationships between biomass and primary production (Smith *et al.*, 1982; Platt & Herman, 1983; Eppley *et al.*, 1985). These were followed by a physiological approach (Platt & Sathyendranath, 1988), which is based on the spectral and angular structure of the submarine light field and the corresponding P^B-I relationship (Equation 1.11). Application of this approach at large horizontal scale for studies of the planetary carbon cycle has emphasised the need to establish an inventory of photosynthesis parameters and vertical distribution of pigments for a variety of oceanographic regimes.

CONCLUDING REMARKS

It is clear that marine organisms, pigmented or otherwise, exert a profound (usually dominating) influence on the light field in the ocean. Research on the processes responsible for these controls has suggested ways in which they may be exploited as tools for further study of the pelagic ecosystem. Other photophysiological processes, notably bioluminescence, await exploitation in this way. Application may be at the level of the individual cell, as in flow cytometry, or at the level of the entire water column, which may be probed actively, as in LIDAR, or passively as in remote sensing of ocean colour. Accumulated knowledge on the relationship between photosynthesis and light has been combined with what is known about the absorption and scattering of light by phytoplankton to yield algorithms for estimation of primary production from remote sensing. Against this background, it is easy to justify the view that study of the nature and control of light in the sea is among the safest of investments in modern biological oceanography.

This work was supported by a research grant to SS from the Office of Naval Research (USA) and the National Aeronautics and Space Administration (USA). Additional support from the National Science and Engineering Research Council (Canada) through operational grants to both TP and SS is also gratefully acknowledged.

APPENDIX 1.1. Glossary

Notation	Quantity	Typical units
α^B	Initial slope of P^B-I curve. Defined as $\partial P^B/\partial I\,\vert_{I\to 0}$	mg C (mg Chl a)$^{-1}$ h^{-1}(W m^{-2})$^{-1}$
$\beta(\chi)$	Radiant flux scattered through an angle, χ, from the direction of the incident beam, per unit incident flux, per unit solid angle and per unit thickness of the scattering element	m^{-1} sr^{-1}
θ	Zenith angle	Dimensionless
λ	Wavelength of light	nm
μ	Mean path length of light rays per unit vertical distance in water. For a ray with zenith angle θ, $\mu = 1/cos\,\theta$	Dimensionless
τ_a	Aerosol optical depth for the total vertical atmospheric column (air mass = 1)	Dimensionless
ϕ	Azimuth angle of light	Dimensionless
χ	Scattering angle. The angle between the initial direction and the scattered direction	Dimensionless
a	Total absorption coefficient	m^{-1}
b_b	Backscattering coefficient. $\qquad b_b = 2\,\pi\int_{\pi/2}^{\pi}\beta(\chi)\,sin\,\chi\,d\chi$	m^{-1}
b_u	Upward scattering coefficient	m^{-1}
B	Biomass. Usually, concentration of chlorophyll a	mg m^{-3}
c	Total volume attenuation coefficient. $c = a + b$	m^{-1}
C	Biomass and associated detritus. Usually, concentration of chlorophyll a plus phaeophytin	mg m^{-3}
$I(z)$	Downwelling irradiance at depth z. Irradiance has notation E in most current literature in optical oceanography. We are following the common notation used in biological oceanography literature	W m^{-2}
I_d	Direct component of irradiance	W m^{-2}
I_k	Adaptation parameter of the P^B-I curve. $I_k \equiv P_m^{\,B}/\alpha^B \equiv P_m/\alpha$	W m^{-2}
I_s	Skylight (diffuse) component of irradiance	W m^{-2}
I_u	Upwelling irradiance	W m^{-2}
k_c	Specific attenuation coefficient for phytoplankton	m^{-1}(mg m^{-3})$^{-1}$
K	Diffuse attenuation coefficient for downwelling irradiance $K = (-1/I)dI/dz$	m^{-1}
K_w	Diffuse attenuation coefficient for downwelling irradiance, for pure sea water	m^{-1}
L	Radiance. Flux from a given direction, per unit solid angle and per unit projected area	W m^{-2} sr^{-1}
L_*	Path function. Measure of increase in radiance from a given direction, due to scattering from all other directions	W m^{-2} sr^{-1}
P^B	Specific production. Primary production rate normalised to biomass	mg C (mg Chl a)$^{-1}$ h^{-1}
$P_m^{\,B}$	Assimilation number. Specific production at saturating light. In the absence of photoinhibition, $P_m^{\,B} = P^B\vert_{I\to\infty}$	mg C (mg Chl a)$^{-1}$ h^{-1}
R	Reflectance. $R = I_u/I$	Dimensionless
z	Depth	m

Chapter 2

REMOTE SENSING - INFORMATION FROM THE COLOUR OF THE SEAS

I.S. ROBINSON

Department of Oceanography, University of Southampton, University Road, Southampton, SO9 5NH

Remote sensing techniques from aircraft or satellite platforms are providing marine scientists with a new perspective of the oceans. This paper considers advances in methods which use visible wavelengths of light. The fundamentals of ocean colour-imaging techniques are reviewed, and their sampling capabilities are compared with conventional methods of marine observation.

Two approaches for extracting quantitative information from visible wavelength image datasets are discussed. The first uses ocean colour as a measure of the contents of the water. Algorithms are presented for measuring chlorophyll concentration, and illustrated with examples from the North Atlantic. Methods for the recovery of estuarine suspended particulate concentration from ocean colour are described, and further calibration possibilities are mentioned.

The second approach uses colour as a tracer of upper ocean dynamics. The patterns observed in colour images offer new insights into upper-ocean mesoscale processes. Information can be extracted by comparison with data from other types of sensor and with the output from numerical models. Techniques are presented for the measurement of velocities from sequential imagery. Conclusions are drawn about the potential value of ocean colour remote sensing, particularly if some of the outstanding problems can be addressed by future research.

INTRODUCTION

Remote sensing of ocean parameters from airborne and spaceborne platforms is now an accepted technique of observational marine science. However, the perspective and sampling capabilities provided by remote sensing methods differ significantly from conventional oceanographic techniques and progress to exploit their full potential has consequently been quite slow. This applies as much to visible wavelength remote sensing, the subject of this paper, as it does to infrared and microwave techniques. The objective of this review is therefore to explain the principles of ocean colour remote sensing and to present examples of applications, in order to demonstrate the potential of the techniques to a wider range of marine scientists than just remote sensing or ocean colour specialists.

The scope of the review is limited to passive measurement of the visible wavelength radiation emerging from the sea; mostly light from the sun which has been reflected

Herring, P.J., Campbell, A.K., Whitfield, M. & Maddock, L. (ed.). *Light and Life in the Sea.*
© 1990. Cambridge University Press.

within the sea although some of the observed light is emitted by fluorescence. Because it is the spectral composition of the light which conveys most information, attention will be focussed primarily on remote measurements of ocean colour, but monochrome and single broad-band image data will also be shown to have important oceanographic uses.

The basis for applying remote measurements of ocean colour in marine science is an understanding of the processes of light absorption and scattering which occur in the upper layers of the sea. The primary application of visible remote sensing to be considered is the calibration of ocean colour observations to retrieve measurements of other ocean properties such as the chlorophyll concentration. Another approach to the application of ocean colour data is the extraction of dynamical information from images. Examples range between qualitative insights from image patterns to quantitative measurements of flow from sequential images. A brief review can only outline the subject, point to some key landmarks of significant progress and aim to inspire a wider interest in the field of study. Maull (1985), Robinson (1985) and Stewart (1985) provide more detailed information.

OPTICAL PROCESSES IN OCEAN COLOUR REMOTE SENSING

Conventional marine optical theory, particularly as developed by those concerned to study the relationship between light and life in the sea, tends to concentrate on determining the amount of energy available for photosynthesis (Sathyendranath & Platt, Chapter 1; Tett, Chapter 4). The emphasis of optical measurements is either to measure the extinction rate with depth of a particular wavelength of light, to measure the beam attenuation coefficient from which it may be calculated, or to use the light emitted by fluorescence as a direct measure of chlorophyll concentration. Kirk (1983) has provided a comprehensive analysis of this approach.

In contrast, remote sensing is concerned with how much sunlight is back-scattered out of the sea, from what depth, and whether the magnitude and colour of the reflected light conveys information about the optically active constituents of sea water. The fundamental physical optical theory is of course the same in each approach, but the way in which the theoretical analysis is developed, and the optical parameters to be measured in experiments, are different. This is not an insignificant consideration, because it means that most marine optical measurements are of limited value to the improvement of calibration algorithms for remotely sensed ocean colour. Studies which have been able to relate remote sensing and conventional marine optics have been those on bio-optical classification of sea water (Smith & Baker, 1978*a*, *b*; Baker & Smith, 1982).

Figure 2.1 illustrates the various optical paths for light reaching a remote radiometer. A considerable proportion of the light, typically more than 80%, is atmospheric path radiance back-scattered by molecular and aerosol particle scattering in the atmosphere. In addition, not all the water-leaving radiance reaches the sensor since some is absorbed or scattered. A considerable body of work has been devoted to the problem of atmospheric correction of ocean colour data from satellites (see *e.g.* Gordon, 1978, 1981; Gordon & Clark, 1980; Sturm, 1981, 1983; reviewed by Robinson, 1983). Rayleigh scattering by atmospheric gas molecules is readily predictable and the most recent algorithms incorporate multiple scattering (Gordon & Castano, 1987; Gordon *et al.*, 1988). Aerosol scattering is harder to cope with since it is spatially and temporally variable, but the use of multispectral data can enable good correction estimates to be made and these should be improved by incorporating more suitable spectral channels in the design of future satellite sensors. For some aerosol removal algorithms, it is also necessary to predict the water-leaving radiance by initially assuming the water properties and then correcting iteratively. Provided sun-glitter is avoided, light reflected at the surface can be incorporated in the atmospheric corrections.

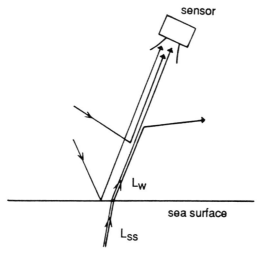

Figure 2.1. Light paths reaching a satellite sensor. L_w is the water-leaving radiance, of which some is absorbed or scattered by the atmosphere (L_{ss} is the subsurface radiance which forms L_w). The other rays represent sea-surface reflection and atmospheric scattering.

The light leaving the sea has two possible sources. It is normally back-scattered sunlight but it may have been emitted by fluorescence from chlorophyll pigments. This occurs within very narrow wavebands and is detectable above the background light if the sensor has a spectral sensitivity of 5 nm. For example Gower & Borstad (1981) and Grassl (1987) have used airborne spectroradiometers to measure chlorophyll by detecting the peak at 685 nm in the water-leaving spectrum due to the fluorescence emission of chlorophyll *a*. Another form of emission is that due to Raman scattering, in which some photons exchange energy and change their wavelength on interacting with a water molecule. The dominant peak of Raman scattering increases the wavelength by about 100 nm. This has been used in active remote sensing using lasers (*e.g.* Bristow *et al.*, 1981; Hoge & Swift, 1981), As more sensitive instruments are used in future, the Raman scattering of sunlight may also need to be taken into account in passive radiometry.

Across most of the spectrum, and in any case for broadband sensors which cannot detect fluorescence or Raman scattering, the water-leaving radiance can be considered in terms of back-scattering of solar radiation. Figure 2.2 illustrates the light field just below the sea surface. Photons directly from the sun or scattered in the atmosphere to become part of the skylight are refracted as they pass through the sea surface (2.2A) and then are scattered or absorbed as they interact with a water molecule or encounter dissolved or particulate material in the water. The behaviour of an individual photon is random, but the interactions of many photons with the medium are statistically governed by probabilities which can be expressed in terms of the inherent optical properties of the medium; the absorption coefficient, a, and the volume scattering function, $\beta(\theta)$, which describes the directional distribution as well as the magnitude of the scattered light.

In a general way, scattering acts to redistribute the direction in which the photons travel, whilst the absorption reduces their quantity. Just below the surface (Figure 2.2B), most of the photons are travelling downwards in the direction of refracted solar rays, giving an asymmetrical radiance distribution, whereas upward travelling photons have been scattered at least once and have a radiance distribution symmetrical about the vertical. The radiance is least in the vertically upwards direction and increases towards the horizontal. The remote sensing measurement L_w is that portion of the upward radiance field which, after refraction at the surface, is directed towards the sensor. Given a typically wavy surface, the remote sensor samples from a cone of the radiance field as shown in Figure 2.2B.

It is clear that the exact shape of the radiance distribution and hence the magnitude of L_w, is dependent on the value of a and the form of $\beta(\theta)$. Since these are inherent functions of the water and its content, information about the water can be extracted from L_w. $\beta(\theta)$ and a are also functions of wavelength, λ, and thus sampling L_w at several wavelengths (*i.e.* making colour measurements) in principle provides more ocean information.

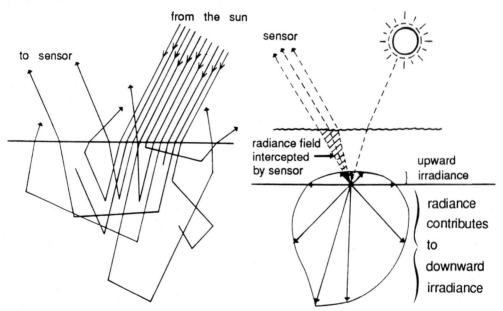

Figure 2.2. The underwater light field. (A) Photons from the sun are scattered and absorbed; a few leave the sea in the direction of the sensor. (B) The radiance distribution near the sea surface, indicating a typical locus of the radiance vectors.

It is also evident that L_w varies with the viewing direction and that the subsurface radiance distribution changes with the solar zenith angle and with the proportion of skylight which is present. However, the shape of the upward part of the radiance distribution is not strongly sensitive to the solar illumination conditions. Knowing the viewing geometry, and the solar illumination, it is therefore reasonable to derive from L_w the magnitude of R, the subsurface irradiance reflectance ratio. The downwelling irradiance, I_d, is the cosine weighted integral of the radiance field over the lower (downward directed) hemisphere, and the upwelling irradiance, I_u, is the integral over the upper hemisphere. $R = I_u/I_d$, and is a function of wavelength, i.e. $R(\lambda)$.

Because the radiance distribution varies not only with a and $\beta(\theta)$ but also with the illumination conditions, R, as defined and also as derived from measurements of $L_w(\lambda)$, is not an inherent property of the water. However, because similar open sky conditions are required for all satisfactory remote sensing measurements, and provided the sun is not too near the horizon, $R(\lambda)$ can be treated as if it depended only on the inherent properties. The validity of this assumption can be explored by Monte-Carlo modelling techniques (Gordon & Brown, 1973, 1975; Kirk, 1981) and is shown to be fairly robust. The effect of bottom reflection on the reflectance in shallow water can also be modelled (Gordon & Brown, 1974). Furthermore it is shown that R depends only on the ratio of b and a. Thus

$$R = f\,\{b/a\} \qquad\qquad (2.1)$$

where b is the back-scatter coefficient, i.e. b is the integral of $\beta(\theta)$ over the backward hemisphere. Thus if a and b vary because of changes in the water composition, so will R.

It is customary to treat the optical properties of a mixture of water constituents as the sum of their individual properties. Thus, for example, for shelf sea water:

$$R = f\{(b_w + b_p + b_s) / (a_w + a_p + a_y)\} \qquad\qquad (2.2)$$

where w represents the properties of the pure sea water, p those associated with a phytoplankton population, s suspended sediment load (assumed in this case not to absorb light), and y dissolved organic material (yellow substance) which does not scatter light. b_p, a_p, b_s and a_y and their spectral form vary with the concentration of the respective water constituents, so that $R(\lambda)$ also varies in a complex way with changes in water constituents.

The goal of ocean colour remote sensing is to measure R in sufficient spectral detail that it becomes possible to determine the concentrations of suspended sediment, dissolved organic substances and chlorophyll associated with the phytoplankton population (perhaps even differentiating between species if their pigment characteristics are sufficiently distinct). $a(\lambda)$ and $b(\lambda)$ are quite well known for several water constituents and it is possible (Figure 2.3) to predict from models how $R(\lambda)$ might vary with the constituent concentration (Sathyendranath & Morel, 1983). However, given the uncertainties and assumptions implicit in this approach it is not reasonable to suppose that an ocean colour calibration algorithm could be constructed on a theoretical basis alone. Instead algorithms must be derived empirically by relating measurements of water constituents to simultaneous remotely sensed observations of $R(\lambda)$. As an intermediate step, a relation can be derived between simultaneous *in situ* measurements of subsurface reflectance and water properties. Such measurements in support of remote sensing require a somewhat new approach to optical oceanography (Aiken & Bellan, 1986), and there is a need for appropriate instrumentation, *e.g.* towed near-surface reflectance meters (Booty, 1989).

OCEAN COLOUR IMAGING SENSORS

Having considered the theoretical basis for extracting useful ocean information from remotely sensed ocean colour, the practical constraints of the sensors must be noted before the achievements of the technique are reviewed. Digitally sampled visible wavelength imaging radiometers have been flown on aircraft and satellites for nearly two decades. Table 2.1 lists the important sampling capabilities of the sensors used most by marine scientists in the UK. Robinson (1985, 1988) provides references to more detailed information on these sensors and their use.

Early sensors such as the Landsat Multispectral Scanner had broad wavebands for overland applications. They have a high spatial resolution, but this is won at the cost of

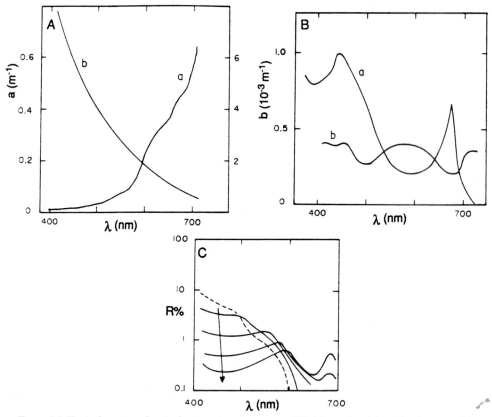

Figure 2.3. Typical spectra of optical properties of sea water. (A) Absorption (a) and back-scatter (b) coefficients of pure sea water. (B) Specific absorption and back-scatter coefficients for water containing chlorophyll (a is normalised to peak value, b has arbitrary units). (C) Typical reflectance of water containing phytoplankton and of pure sea water (dashed line). The different curves correspond to different concentrations of chlorophyll, increasing in the direction of the arrow.

infrequent sampling. The sensors produce images by scanning mechanically (electronically for more recent designs) across track while the satellite or aircraft moves along track. For high resolution sensors (MSS and TM on Landsat, HRV on SPOT) the swath width is limited to a few hundred kilometres. With between 14 and 15 satellite orbits per day, the earth coverage in one day is only a small fraction of the total surface, and 16-18 days elapse before every point is covered. Given the likelihood of cloud cover, there may be several months between cloud-free overpasses for a particular region. Landsat and SPOT image data thus have limited application in marine science. The broad bandwidth

Table 1. Specifications of some ocean colour sensors

Sensor		MSS	TM	HRV	CZCS	AVHRR	ATM	FLI	SeaWiFS	MERIS
Waveband	1		450-520	500-590	433-453	580-680	420-450	up to	402-422	400-
	2		520-600	600-680	510-530	725-1100	450-520	64 bands	433-453	1050 in
	3		630-690	790-900	540-560	+	520-600	of 5 nm	490-510	at least
	4	500-600	760-900		660-680		605-625	width	555-575	15 bands
	5	600-700	+		700-800		630-690	in a	655-675	with
	6	700-800			+		695-750	sparse	745-785	1.25 nm
	7	800-1100					760-900	spatial	843-887	resolution
	8						910-1050	mode	+	
	9						1550-1750			
Scanline field of view		7.5°	7.5°		39°	55.4°	36°		58°	41°
Satellite/ vehicle		Landsat 1-4	Landsat 4,5	SPOT	Nimbus-7	TIROS /NOAA	Aircraft	Aircraft	Satellite ?	Columbus
Orbit period, min		103	99	101	105	102	variable		100	100
Height, km		908	705	822	950	815-860	according		700	824
Pixel size, m		76	30	20 (10*)	825	1100	to the		1130	260-520
Swath width, km		185	185	60	1640	2580	operation		2800	1500
Typical revisit, d		18	16	26	2	1	of the		1	2
Period of operation		1972-83	1982-	1986-	1978-86	1978-	aircraft		1992?	1997?

+ The thermal infra-red bands are not included. * Band 1 (panchromatic) only.
MSS: Multi-Spectral Scanner. TM: Thematic Mapper. HRV: High-Resolution Visible sensor. CZCS: Coastal Zone Color Scanner. AVHRR: Advanced Very High Resolution Radiometer. ATM: Airborne Thematic Mapper. FLI: Fluorescence Line Imager. SeaWiFS: Sea-viewing Wide Field-of-view Sensor. MERIS: Medium Resolution Imaging Spectrometer

can detect only strong colour changes and they have been used only in coastal and estuarine regions where the high spatial resolution is necessary to detect small scale but strongly coloured features. Only a single snapshot of a phenomenon can be obtained. Its time evolution cannot be followed although fortuitous overlaps may provide a succession of two or three daily images.

The greatest achievement so far in satellite ocean colour remote sensing has undoubtedly been the Coastal Zone Colour Scanner on Nimbus 7 (Hovis *et al.*, 1980; Gordon *et al.*, 1985). It was designed with high radiometric sensitivity and narrow wavebands in order to measure the chlorophyll absorption peak at 443 nm and hence detect phytoplankton. Although the detectors may saturate over bright land or cloud, they can detect subtle changes in sea colour. The larger pixel size enables a much broader swath to be scanned, and earth coverage is obtained in two days. At mid and high latitudes daily sampling is possible. The visible waveband on the meteorological AVHRR instrument yields only a monochrome view of the sea, but the operational nature of the sensor, and its wide swath width, ensures the highest chance of cloud-free ocean viewing. CZCS and AVHRR data covering the seas off north-west Europe are archived at the NERC Satellite Receiving Station at the University of Dundee (Bayliss, 1981).

Airborne sensors are flown much lower and can achieve a spatial resolution down to 1 m. The swath widths are correspondingly narrow, but the flexibility of aircraft operation means that a higher sampling frequency can be obtained. The Daedalus ATM has relatively broad wavebands and relies for detecting marine phenomena on their having strong colour signatures. The Fluorescence Line Imager, as its name implies, is effectively a spectrophotometer which can resolve spectrally down to 5 nm and in principle can detect very subtle changes of sea colour, although in this mode its spatial sampling is poor. A major drawback of airborne remote sensing is the instability of the platform, leading to difficulties in reducing the data to regular geographic coordinates for comparison with other images or *in situ* data, and in applying atmospheric correction.

Figure 2.4 summarises the space/time sampling capabilities of the different types of ocean colour sensors. The lower boundaries on the diagram are the sampling frequencies. The upper boundaries correspond to the spatial coverage of a synoptic overview, and the time span over which consistent data coverage can be repeated. For satellites this is the lifetime of the sensor or series of sensors. A diagram like this can be used in comparison with the space/time scales of a particular phenomenon, to determine the

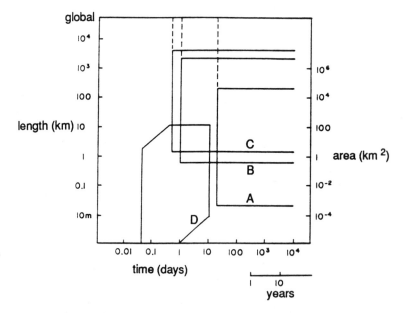

Figure 2.4. Space - time sampling characteristics of visible wavelength imaging sensors: (A) Landsat Thematic Mapper; (B) Nimbus-7 CZCS; (C) NOAA AVHRR; (D) Airborne thematic mapper. See Table 2.1 for further details.

sensor type most appropriate for monitoring it. It should be noted that *in situ* observing methods can sample much better in time, but generally cannot offer the same spatial sampling capabilities as remote sensing.

The final two columns in Table 2.1 describe satellite sensors which are not yet in operation. The Sea-viewing Wide Field-of-view Sensor (SeaWiFS) offers a somewhat improved version of CZCS which will contribute to the global phytoplankton studies initiated using the CZCS archive and provide continuity for ocean colour research activities, if it flies as planned in 1991 or 1992. The MERIS instrument (Cutter *et al.*, 1989) planned by ESA for launch in the mid-1990s on the Columbus or Freedom Polar Platforms, will be a new class of ocean colour sensor, capable of very high spectral resolution and promising exciting new views of the ocean.

CALIBRATION OF REMOTELY SENSED OCEAN COLOUR DATA

Case 1 and 2 waters

It should be evident that the potential for calibration of ocean colour data in terms of water content which was suggested in section 2 has not yet been realised because the spectral performance of the available sensors, as described above, is not adequate. With a limited number of narrow wavebands, it is not yet possible to distinguish between the subtle differences in spectral response due to dissolved organic compounds, live cells containing chlorophyll, and suspended particulates from river discharge, sludge dumping or seabed resuspension. However, some progress has been made in calibrating ocean colour data where only one optically active constituent is dominant in the water, particularly in the case of chlorophyll since the CZCS was designed to detect the absorption around 440 nm. Water is classified in relation to its optical constituents as either Case 1 or Case 2 (Morel & Prieur, 1977; Gordon & Morel, 1983). Case 1 water is that whose optical properties are controlled by phytoplankton and their derivative products. Thus the concentration of any dissolved organic or particulate material covaries with the chlorophyll concentration, and each of these variables can be shown to correlate with the green/blue colour ratio. Considerable success has been achieved in developing calibration algorithms for such waters, typically in the open ocean. Case 2 waters include all other situations, where there is optically active material from another source. Universal calibration algorithms are not valid for Case 2 waters, although where there is a strong correlation between water colour and a particular water constituent, a locally calibrated algorithm may be applicable.

Chlorophyll algorithms

Chlorophyll algorithms are based on the green/blue ratio between channels 3 and 1 of CZCS (550 and 443 nm). When the chlorophyll concentration is high the ratio between channels 3 and 2 (520 nm) is used because channel 1 yields such low water-leaving radiance. Based on a regression between the log of the chlorophyll concentration (C, mg m^{-3}) and the colour ratio, calibration equations have the form:

$$C = A \, (R_3 / R_1)^B \tag{2.3}$$

where R_x is the reflectance in waveband x.

The standard algorithms used for routine processing of CZCS data for Case 1 water are:-

$$C = 1.13 \, (L_{ss}(443) \, / \, L_{ss}(550))^{-1.71} \qquad\qquad [C < 1.5 \text{ mg m}^{-3}] \qquad (2.4)$$

$$C = 3.326 \, (L_{ss}(520) \, / \, L_{ss}(550))^{-2.439} \qquad\qquad [C > 1.5 \text{ mg m}^{-3}] \qquad (2.5)$$

Because spectral ratios are involved, it makes little difference that the algorithms are presented in terms of the subsurface radiance L_{ss} directed towards the sensor, rather than the reflectance ratio. These algorithms have been derived from *in situ* measurements from the Mid-Atlantic Bight (Gordon *et al.*, 1983). Figure 2.5 is an example of a North Atlantic image representing chlorophyll distribution which has been calibrated in this way. Work is in progress to generate basin and global scale maps of chlorophyll distribution based on the application of this calibration to the archive of CZCS data (Esaias *et al.*, 1986).

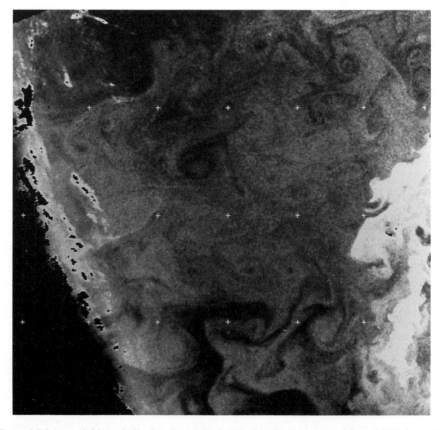

Figure 2.5. Image of chlorophyll *a* distribution in the North Atlantic recovered from CZCS data using the NASA algorithm. The image is approximately 500 km square, centred at 58°N 11°W, on 17 May 1980. The grey-scale varies logarithmically with chlorophyll, white being the maximum (0·063 mg m⁻³) and black the minimum (0·033 mg m⁻³). (Image processed by N. Ward, Southampton, University.)

Some caution is necessary when seeking to apply a universal algorithm in this way. A variety of different calibration equations have been derived by various workers using the green/blue ratio (Morel, 1980; Gordon & Clark, 1980; Smith & Wilson, 1981; Clark, 1981). Whilst they have similar exponents there is insufficient evidence that calibrations based on observations in lower latitude Case 1 waters are applicable in higher latitudes where very different species of phytoplankton may be encountered. It is already apparent (Holligan *et al.*, 1983) that the standard algorithms cannot be applied to coccolithophore blooms which dominate the chlorophyll concentration in some regions because their colour is strongly influenced by the highly reflective calcareous scales. It is therefore important that continuing effort is devoted to testing and improving calibration algorithms for Case 1 waters, by appropriate programmes of field measurements.

It is also possible to recover chlorophyll measurements from ocean colour in Case 2 waters, provided sufficient *in situ* measurements are made (Mitchelson *et al.*, 1986; Viollier & Sturm, 1984) but these calibrations are site specific.

A major problem with ocean colour remote sensing as a route to measuring primary production is that chlorophyll is detected only in the upper few metres where light levels remain high. Deeper phytoplankton fails to influence the apparent ocean colour viewed from above and if there is significant biomass stratification a deep chlorophyll layer may be missed altogether by a remote sensor. This factor has been addressed by Platt & Herman (1983) and Platt (1986) but remains a subject for further investigation if those studying primary productivity are to have confidence to use ocean colour image data. The other outstanding problem with the chlorophyll algorithms is their uncertain dependence on the phytoplankton species, the maturity of the population and the presence of other optical constituents, including dissolved organic compounds. Better spectral measurements and the detection of chlorophyll fluorescence may offer a solution.

Suspended sediment

Although suspended sediment plumes and patterns are revealed in visible wavelength images (Figure 2.6) it has nevertheless been much harder to find a reliable way of measuring concentrations of suspended particulates from remotely sensed data. This may be because when the quantity of suspended material in the water is increased both absorption (a) and back-scattering (b) increase so that R hardly changes, unless other factors affecting the optical properties also change, such as the size distribution, composition and component pigments of the suspended material.

The most successful calibration of sediment loads has been where the sediment itself has strong coloration, so that spectral ratios can be used to quantify sediment concentration. A good example is the Bay of Fundy (Alfoldi & Munday, 1978) where red sediments are added by local rivers. Chromaticity-transformed data from bands 4 (green) and 5 (red) of Landsat MSS plot to form a locus on which sediment concentration can be differentiated (Figure 2.7). A method like this can avoid the need for atmospheric corrections, but relies entirely on simultaneous *in situ* water sampling.

Figure 2.6. Atmospherically corrected CZCS channel 3 image of the Irish Sea, 17 May 1980, showing streaks and plumes of high reflectance due primarily to suspended sediment. A false colour composite based on channels 1, 2 and 3 would show more detail. (Image processed by N. Ward, Southampton University.)

Limited success has been achieved in calibrating suspended sediment loads in UK estuaries (MacFarlane & Robinson, 1984). Lees (1989), using an extensive dataset of near-surface suspended sediment in the Bristol Channel (Kirby, 1986) matched *in situ* and Landsat data, not in true time but according to tidal state. Figure 2.8 is an example of the result from one of the Landsat images. Analysis of a tidal sequence of such maps added complementary information to Kirby's description of sediment dynamics. It is to be expected that the improved spectral, radiometric and spatial sensitivity of the thematic mapper will improve this approach.

Recovery of other parameters from ocean colour

There are some other ways in which ocean colour data can be interpreted. An alternative approach to calibration is that which relates reflectance colour ratio to the diffuse attenuation coefficient, $K(\lambda)$, at a specific wavelength λ. Jerlov (1974) and Højerslev (1980) have demonstrated a strong relation between water colour and the light penetration depth, and it is not surprising therefore that Austin & Petzold (1981) were able to develop the following algorithms for CZCS data:-

$$K(490) - 0.022 = 0.088 \ (L_w(443)/L_w(550))^{-1.491} \qquad (2.6)$$

$$K(520) - 0.044 = 0.066 \ (L_w(443/L_w(550))^{-1.398} \qquad (2.7)$$

By working with K it should be possible to obtain a calibration which is less sensitive to the water type, more nearly universal, and yet readily interpreted by marine scientists familiar with measuring K and relating it to the particular phytoplankton species or other water constituent being studied. Another use for K measured in this way is to relate it to the light penetration depth which is directly relevant, both to productivity studies (Platt, 1986; Eppley *et al.*, 1985) and to modelling of the thermal structure of the mixed layer and the thermocline (Martin, 1985).

Finally, another parameter which can be directly recovered from suitable colour images is the bathymetric depth. If the water is largely clear of scattering or absorbing material, then penetration of the blue light can reach 30 m or more. If the water depth is less than this, and if the sea-bed has a high reflectance, the bed influences the surface reflectance. The reflectance is enhanced more in the blue, since more light penetrates to the sea-bed to be reflected than in the green or red. In these circumstances the colour provides a reliable indicator of the sea depth down to 30 m (Bullard, 1983). Since it is shallow depths which are of most concern to shipping, this application has considerable potential for charting safe shipping routes in remote and unsurveyed waters.

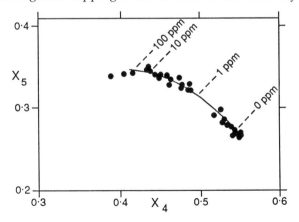

Figure 2.7. Radiance values from Landsat MSS channels 4 and 5 viewing Minas Basin in the Bay of Fundy, Nova Scotia, plotted on a chromaticity transform plane. $X_4 = L_4 \ / \ (L_4 + L_5 + L_6)$ where L_n is the radiance measured in band n of the MSS. There is a correlation between the *in situ* suspended particulate concentration and the position along the chromaticity locus. (Adapted from Alfoldi & Munday, 1978.)

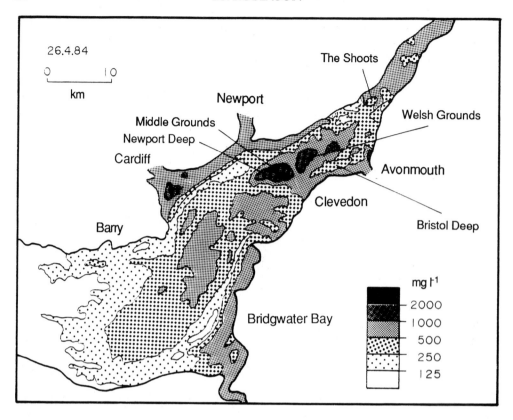

Figure 2.8. Near-surface suspended sediment distribution in the Bristol Channel and Severn Estuary, 6 hours before high water Avonmouth for a neap tide, derived from a calibrated Landsat MSS image of 26 April 1984. (From Lees, 1989.)

DYNAMICAL CONTENT OF OCEAN COLOUR IMAGES

So far in this review, the information content of remotely sensed ocean colour has been considered only in terms of what can be determined about the ocean separately for each individual observation by the sensor (*i.e.* at each pixel). In addition, imaging sensors provide a novel synoptic perspective which allows spatial patterns to be detected in the ocean. This is valuable information in its own right. If the cause of the patterns can be quantified in terms of water content or bathymetry as discussed above, the information content is maximal, but even if calibration of the tracer is not possible there remains usable information relating to the dynamical or other processes which determine the shapes which appear on the image. Because the analysis of spatial pattern has not played a large part in conventional marine science, it can easily be dismissed as merely qualitative interpretation, but this is to miss its true potential.

Turbulence structure and patchiness

Figure 2.9 shows atmospherically corrected channel 3 data from the CZCS over the North Atlantic, south-west of Iceland. Cloud-free images of this region are rare, but this image spans several hundred kilometres of open sea. It reveals a wealth of structure which can be interpreted in terms of two-dimensional turbulence in the upper layers of the ocean. Counter-rotating vortex pairs and characteristic hammer-head shapes are evident. So too is the stretching and elongation of some of the tracer patterns. Gower *et al.* (1980) have analysed similar patterns (from higher resolution Landsat data) to determine the spectrum of the length scales of the turbulence. Figure 2.10 is another example of turbulent structure, being revealed this time by reflectance of coccolithophores in a broadband AVHRR channel 1 image.

For patterns to be evident at all on images (*e.g.* Abbott & Zion, 1987; Barale *et al.*, 1986), there must be gradients of colour present. In Figures 2.9 and 2.10 they are presumably due to phytoplankton production which is patchy. The mesoscale velocity field associated with dynamical eddy-like instabilities then advects these patches, the shear in the

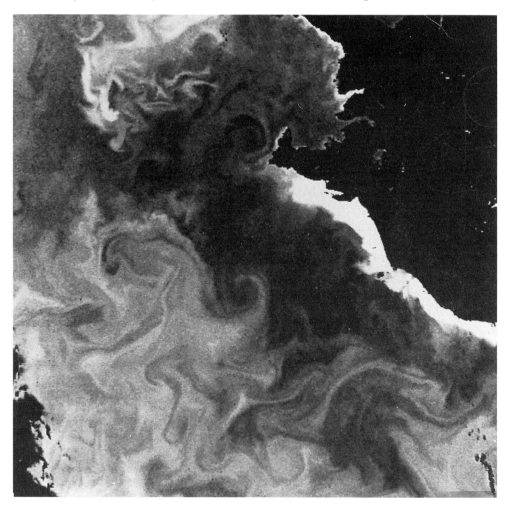

Figure 2.9. Oceanic eddy structure south-west of Iceland, revealed on a CZCS atmospherically corrected channel 3 image, 21 August 1980. The corresponding composite of channels 1, 2 and 3, which cannot be reproduced in black and white, shows further detail in the colour patterns.

currents drawing them out into long streaks. Turbulent processes at a much shorter length scale provide the mechanism for diffusing the colour tracer and smoothing the sharp gradients produced by advection. From a dynamical point of view, the colour is acting as a passive tracer, since the presence of phytoplankton does not significantly influence the density of the water. However, it is not necessarily a conservative tracer, nor is the pattern of colour entirely due to the fluid dynamics, since the patchiness, required in the first place for there to be a pattern at all, is controlled by the factors affecting biological population growth and decay.

In order fully to exploit the information content of images such as these it is suggested that comparison be made with models of the dynamical processes of upper ocean turbulence. For example the model output in Figure 2.11 (Barkmann & Richards, 1989) shows some patterns similar to those present on the images. A realistic goal for such work would be to use image data to improve the parameterisation of factors in the model which maintain the balance between advection and diffusion, and thus control the streakiness of passive tracers.

Ocean colour patterns compared with other tracers

Figure 2.12 shows (a) a visible wavelength and (b) a thermal infra-red image from the CZCS channels 3 and 6 respectively of the Iceland-Faroes front. Here the colour is not only revealing eddy-like turbulence structures associated with the front, but comparison with the temperature patterns on the thermal image demonstrates that it is offering complementary information. The thermal image clearly delineates the sharp frontal temperature step of 4°C between the cold water to the north and the warmer to the south. It also reveals the meanders of the front, some of which have broken off to form eddies. However, the thermal image gives no direct indication of the flow along the front. The pattern on the colour image is dominated by the high reflectance of water emerging from the narrow continental shelf margin off the south-east coast of Iceland. Without needing to know whether this is the signature of phytoplankton or suspended sediment, it acts as a tracer which clearly marks the eastward jet flowing along the frontal boundary and entraining the reflective water. It clearly reveals how the flow meanders along the front, and characterises the detached warm core ring to the north of the front.

It is important to note that the combination of colour and thermal data offers a richer source of dynamical information than could have been deduced from the sum of each image type interpreted individually. Ocean colour images have been used in conjunction with other data to study, for example, UK continental shelf fronts (Holligan, 1981), the Californian Coastal current (Simpson et al., 1986), the western Mediterranean (Arnone & LaViollette, 1986) and island mixing processes (Simpson et al., 1982).

Feature tracking

When a time-series of image data is available, there are two benefits. Firstly, contamination of ocean data due to atmospheric effects or other artefacts of the remote sensing technique become apparent from comparisons between images of the same region

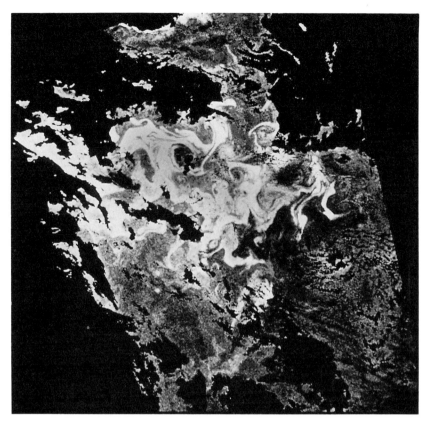

Figure 2.10. Oceanic eddy structure revealed south of Iceland in the visible band (channel 1) of an AVHRR image, 15 June 1985. (Image processed by S. Groom, NERC Image Analysis Unit, Plymouth.)

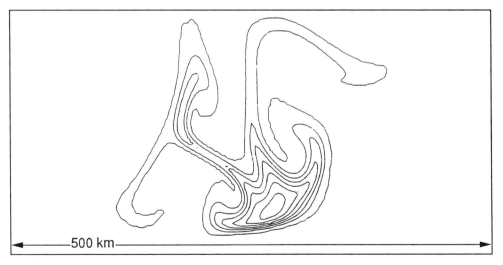

Figure 2.11. The distribution of a passive tracer in the surface layer of a quasi-geostrophic model of upper ocean dynamics. The horizontal length scale of the box is 500 km, and the tracer was introduced as a Gaussian blob at the centre, 30 days earlier in the model run. (Plot produced by W. Barkmann, Southampton University.)

closely spaced in time. Ephemeral features are not given undue attention as they might on a single image, whilst more permanent features require an oceanographic explanation. Secondly, if features are observed to change position between images, this can be used to measure their speed of movement. Features which appear to move too fast to be in the water may be of atmospheric origin. In satellite images a day apart there is little ambiguity between the movement of a colour pattern in the water and the much greater movement of cloud or haze in the atmosphere, but airborne images separated by minutes or hours require more care. Garcia (1989) has analysed a tidal sequence of airborne images of Southampton Water, calibrated in terms of *Mesodinium* cell count. These reveal the detailed complex structure which is characteristic of *Mesodinium* distribution,

Figure 2.12. (A) Channel 3 CZCS image on 1 May 1980 of the North Atlantic to the east of Iceland, geometrically rectified to a Mercator projection and atmospherically corrected. The image is about 500 km wide. (B) Sea surface temperature image derived from the thermal channel (6) of the CZCS at the same time as (A). The darker shading corresponds to warmer water. (Images processed by N. Ward, Southampton University.)

including linear streaks related to the wakes of ships which have traversed the region some time earlier. Although the *Mesodinium* in the upper metre of the water column is not a conservative biological tracer because of the rapid vertical migration which occurs, the movement of the patterns can still be traced from one image to another. The value of this technique is that it provides a direct, Lagrangian, measurement of complex water movements in estuaries and is capable of tracing the dispersion of polluting discharges.

Maximum cross-correlation technique

The feature-tracking method discussed above has also been applied to satellite imagery, for which the length scales of the tracked features must be increased to match the image resolution. Medium resolution sensors must be used to obtain images one or two days apart, given cloud-free conditions. However, feature tracking is subjective, relies on the presence of clearly defined patterns, and is not very accurate because distortion of the tracked features makes it difficult to match up the corresponding points precisely. In order to extract the maximum information from colour images having smoother, less distinct patterns, Garcia & Robinson (1989) have adapted an automatic statistical matching technique previously applied to cloud motion and images of sea-surface temperature gradients (Emery *et al.*, 1986).

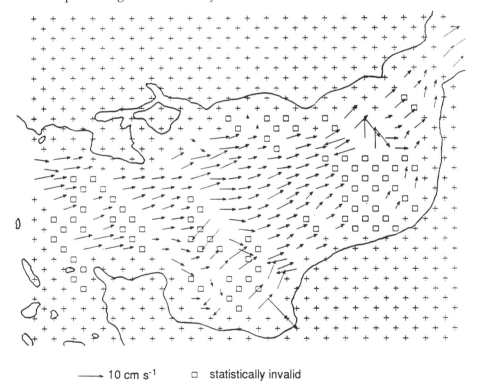

→ 10 cm s⁻¹   □ statistically invalid

Figure 2.13. Residual sea surface velocity vectors derived from CZCS images of 10 and 11 May 1980 by the maximum cross-correlation method. A square indicates a rejected result which failed a significance test. (From Garcia & Robinson, 1989.)

For example, Figure 2.13 shows the motion derived from two CZCS band 3 images of the English Channel, obtained 24·5 hours apart, which look very similar. The movements of patterns within each subarea are represented as vectors. With almost exactly two tidal cycles between the images, the vectors correspond to the residual water movement advecting the optical water constituent, in this case suspended sediment. The resulting field of vectors is quite coherent and represents a measurement which would be very difficult to make by other means. Because of the stringent requirement for accurate overlaying of one image on another, the method is practicable only for seas surrounded by coastal landmarks.

CONCLUSION

The diversity of analysis methods available indicates the potential for ocean colour remote sensing. Attention has been focussed on examples of UK work. ESA (1987) has published a survey of the international importance of ocean colour remote sensing, whilst NASA's proposals for future colour scanning sensors are driven by the demands of the global earth observing programme (NASA, 1987).

Large international experiments, such as the Joint Global Ocean Flux Study, are dependent upon the spatial coverage provided by medium resolution satellite sensors measuring near-surface chlorophyll if the detailed process studies of primary productivity are to be extrapolated to basin scales. Regional and local studies of water quality, pollution and productivity will equally benefit from regular ocean colour image data.

No successor to the CZCS is expected until 1991 at the earliest, but an active research programme is needed now to ensure the maximum benefits to marine science when ocean colour monitoring from space becomes operationally routine. The research programme should include:- (i) Continuing analysis of the large archive of CZCS image data, yielding valuable information about the spatial and temporal distribution of ocean productivity and regional water quality. (ii) The development of techniques for the dynamical interpretation of the spatial information in colour images, (a) synergistically in relation to thermal, radar and other satellite data and (b) in conjunction with models of upper ocean dynamics, including new mathematical tools for comparing the spatial statistics of numerical model output and satellite images. (iii) Improvement of the existing calibration algorithms for satellite colour data by measurement programmes of *in situ* optical parameters and water constituents, and further modelling of optical processes. The results can be applied straight away to archived satellite data and will contribute to the specification of new satellite sensors. (iv) Development of new *in situ* optical instruments (exploiting recent optical fibre technology) for measurements at high spectral resolution, in support of a new generation of imaging spectrometers which are being built for airborne and eventual spaceborne deployment. New calibration methods are required to derive estimates of *in vivo* fluorescence, dissolved organic compounds, suspended particulates and polluting substances, using multiple narrow-band radiance measurements.

Given a modest investment of research effort, the remote sensing of ocean colour will continue to make significant contributions to the study of the distributions of light and life in the sea.

Chapter 3

OPTICAL OCEANOGRAPHY: AN ASSESSMENT OF A TOWED METHOD

J. AIKEN AND I. BELLAN

Plymouth Marine Laboratory, Prospect Place, West Hoe, Plymouth, PL1 3DH

Measurements of the optical properties of the oceans are assessed in the context of the requirements for the interpretation of satellite remotely sensed images of ocean colour in terms of the biogenic constituents of the water, particularly dissolved compounds and phytoplankton pigments. Conventionally, optical measurements are obtained by profiling instruments vertically from a stationary research vessel, limiting the number of data acquired to a few per day, spread thinly over a small geographic area. In this chapter, measurements of the optical properties of several large oceanic areas with diverse phytoplankton populations are presented, acquired by means of a suite of light sensors attached to the Undulating Oceanographic Recorder (UOR), towed at speeds of 4-6 m s^{-1} (typically 20 km h^{-1}, covering 300-400 km d^{-1}). The data acquired are compared with measurements by conventional instruments at stations and analysed for artefacts or bias in the towed measurements.

INTRODUCTION

Satellite remote sensing of the oceans is a core element of current international research programmes for physical oceanography, the World Ocean Circulation Experiment (WOCE) and biological oceanography, the Joint Global Ocean Flux Study (JGOFS), which require basin scale, even global scale sampling strategies to meet their prime objectives. For biological oceanography, satellite sensors can detect changes of ocean colour which arise directly from spectrally selective absorption and scattering by the major biogenic constituents of the water column; these are, dissolved organic compounds, phytoplankton pigments, phytoplankton cells and suspended organic detrital material (see Robinson, Chapter 2). In simple terms, the colour of the ocean is blue in winter or in oligotrophic waters, because water absorbs blue light least and back-scatters blue light preferentially and it is green in the spring and summer in the highly productive, temperate oceans and shelf seas, because chlorophyll, the major and ubiquitous pigment in all plants (phytoplankton) absorbs blue light strongly and phytoplankton cells back-scatter the residual green light preferentially.

Conventionally, bio-optical measurements are obtained by profiling instruments vertically from a stationary research vessel, limiting the number of data sets acquired to

Herring, P.J., Campbell, A.K., Whitfield, M. & Maddock, L. (ed.). *Light and Life in the Sea.*
© 1990. Cambridge University Press.

a few per day, spread thinly over a relatively small geographical area. Consequently the data base of bio-optical properties is sparse, with little information on seasonal or inter-annual variability, particularly for the temperate and sub-polar oceans, which have diverse phytoplankton assemblages, often with accessory pigment complexes besides chlorophyll. These oceanic areas are highly productive, and with a large exchange of CO_2 between atmosphere and ocean they are the focus of JGOFS and the UK Biogeochemical Ocean Flux Study (BOFS).

Addressing the problem of satellite biological oceanography, the National Academy of Sciences (1982) recommended: "As a complement to the development of a satellite system, a towed sensor is needed that can remotely sense in the upper layer of the ocean, where biological activity is most concentrated (*i.e.* top 100 metres approximately). This sensor should have the capability to measure a major feature of the physical structure (temperature), obtain a measure of phytoplankton (fluorescence), ...".

The Undulating Oceanographic Recorder (UOR, Aiken, 1981*a*, 1985; Aiken & Bellan, 1986*a*) meets most of these requirements. It is an automatic, depth-profiling sampler with sensors for physical properties (temperature and conductivity) and bio-optical properties (chlorophyll fluorescence, particulate scattering and water transmission) and uniquely it has a suite of light sensors to measure downwelling, $E_d(\lambda)$, and upwelling, $E_u(\lambda)$, irradiance in several wavebands (λ) and provide the means to interpret satellite, remotely-sensed images of ocean colour. The reflectance $R(\lambda) = E_u(\lambda)/E_d(\lambda)$ is proportional to the radiance leaving the water as detected by satellite sensors.

It is the objective of this chapter to assess the relative advantages or disadvantages of the undulating towed method compared to conventional stationary profiling methods. The apparent optical properties, the diffuse attenuation coefficient $K(\lambda) = d/dz(\ln E_d(\lambda,z))$, and the reflectance $R(\lambda)$ derived from UOR measurements, are compared with measurements taken by conventional state-of-the-art radiometers, profiled vertically on station. Each is examined for artefacts or bias caused by the vessel, *e.g.* ship-shadow when stationary or the ship's wake when under way. Along track the UOR data are examined for discrepancies with respect to the station measurements in regions of laterally homogeneous physical and biological structure. Data from a variety of oceanographic sites which are influenced by different physical and biological factors are analysed for consistency spectrally and compared with published data. Likewise, the inherent optical properties, the absorption coefficient, $a(\lambda)$, and the scattering coefficient, $b(\lambda)$, are calculated to test the validity of the UOR data.

INSTRUMENTATION AND METHODS

The UOR is a multi-sensor oceanographic sampler (Figure 3.1, which profiles automatically from near surface to *ca* 70 m (optimum range) when towed on unfaired steel cable at speeds in the range 4-6 m s^{-1} (8-12 knots); the maximum depth is restricted to *ca* 40 m at the higher speeds of 9-10 m s^{-1}. The undulation pitch length is typically 1·6 km (adjustable from 0·8-4·0 km). Hemispherical 2π scalar irradiance sensors measuring the downwelling hemispherical scalar irradiance, E_{od}, and the upwelling hemispherical scalar irradiance, E_{ou}, (Aiken & Bellan, 1986b) are used to ensure that the direct and diffuse components of the light field are measured with minimum error, independent of the pitch-angle or pitch fluctuations of the UOR body. For comparative purposes, conventional irradiance sensors, with cosine collectors which measure E_d and E_u have been used; these are subject to 'noise' due to variations of the pitch-angle of the UOR from horizontal ($\pm15°$). Scalar irradiance reflectance, $R_o(\lambda)$, and conventional irradiance reflectance, $R(\lambda)$ are not equal, though they are related through the inherent optical properties, and the average cosine of the light field μ (defined by Kirk, 1981, 1986). Values of the inherent optical properties $a(\lambda)$ and $b(\lambda)$ derived independently from $R_o(\lambda)$ and $R(\lambda)$ should be equal.

Data are logged *in situ* using two 8-channel miniature digital tape recorders (25 h duration, 96k data capacity) or, from 1987, a microprocessor controlled solid-state data logger (SSDL) at rates of 5·4, 10·8, 18, 54k data h^{-1} (10, 5, 3 or 1 s scan rates) from up to 16 sensors. The SSDL has capacity for 64,000 measurements giving the system a duration of 11 h for 16 sensors at 10-s scan rate; the data capacity can be increased four-fold, with an X4 memory board.

Upon recovery of the UOR, the data are transferred from the logger to an IBM PC for data processing and analysis. At sea, data plots, vertical profiles and vertical sections (distance *vs* depth) can be produced for any measured parameter, within a few minutes

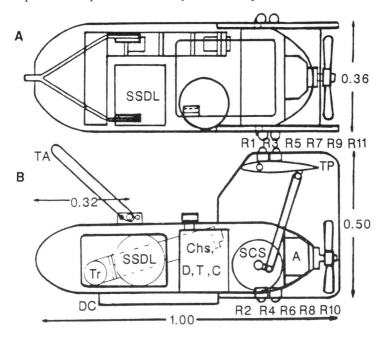

Figure 3.1. Schematic diagram, (A) plan, (B) elevation, of the UOR Mark 2 (dimensions in m): A = alternator; SCS = servocontrol system; TA = towing arm; DC = diving chute; TP = tail plane. The UOR has 16 sensors: for depth, D; temperature, T; conductivity, C; chlorophyll fluorescence, Chs (Aiken, 1981b); particulate scattering, Sct; water transmission, Tr (Sea Tech, transmissometer, 0·25 m, 660 nm); and 8-12 light sensors, R1-R11 (downwelling and upwelling pairs at 412, 445, 490, 520, 560 nm and PAR); the 412 nm band has been added in 1988 along with new sensors for pH and DOC fluorescence, as yet unevaluated.

of dumping the data. Analysis programmes can determine the average properties of the surface mixed layer for each undulation, for both measured (*e.g.* temperature, chlorophyll) and computed parameters (*e.g.* $K(\lambda)$, $R(\lambda)$) and inter-relate these for statistical assessment or algorithm development.

On research cruises, the UOR is towed between stations to provide a description of the horizontal variability and vertical structure of the euphotic zone along the course of the tow. Approaching or leaving a station, the UOR is towed at slow speed (2-4 knots, 1-2 m s^{-1}) for a double oblique profile to 100 m or more, for comparison with the station measurements of temperature, chlorophyll concentration and optical properties. At this speed, the UOR maintains a horizontal attitude, yet it is well astern of the vessel and avoids ship-shadowing effects. Conventional measurements, even by state-of-the-art radiometers such as the Marine Environmental Radiometer (MER, Smith *et al.*, 1984) from a stationary research vessel, may be corrupted by ship-shadow effects, though these are minimised if the instruments are deployed from the stern of the vessel. Set against this, towing the UOR at speeds of 4 m s^{-1} or more might be expected to perturb the water it samples optically, due to scattering from air-bubbles in the ship's wake, possibly increasing systematically the apparent value of the diffuse attenuation coefficient, $K(\lambda)$, and the reflectance, $R(\lambda)$, particularly near the sea surface.

There are two other instrumental factors which may have an influence on UOR optical measurements. The upwelling light sensors are fitted to the underside of the UOR, at the rear of the tail fins, to obtain an unobstructed view of the upwelling light field while remaining screened from contamination by the downwelling light field. Although the UOR is small in area (0·32 m^2), it may still shadow the upwelling light field partially. Pilgrim (1988) has analysed the problem, using a simple model, showing that the effect is small except when the diffuse attenuation coefficient is high, in high turbidity water or for red light. Secondly, the upwelling light sensors are situated 0·5 m deeper in the water column and measure a lesser upwelling light value than for the corresponding depth of the downwelling sensors. Again this error is small, except for high values of K, though as K is known, measured or can be estimated, a correction can be applied: the correction for E_u equivalent to a depth 0·5 m shallower is $E_u(+0·5) = E_u(1+0·5K)$. Both these instrumental effects are additive and would lead to an underestimation of $R(\lambda)$.

RESULTS

Table 3.1 lists the bio-optical data acquired by the UOR since 1984, ranging from the oligotrophic, tropical, Indian Ocean, and the productive temperate areas (western English Channel, North Sea and eastern North Atlantic Ocean) to the Polar Seas (Norwegian, Greenland and Barents Seas) and the Antarctic Ocean.

In this paper, the UOR measurements are analysed in absolute terms (*e.g.* base-line measurements in the Indian Ocean), in comparative terms (*e.g.* against state-of-the-art radiometers such as the MER) and in relative terms, both spectrally and spatially, against other bio-optical measurements and satellite imagery.

Table 3.1. *Data base for the ocean colour studies*

date	vessel	area	sensor suite	miles towed	data total (K)	no.of profiles (undulations)
May 1984	RV 'Squilla'	WEC	A	50	19	55
June 1984	RV 'Venturous'	BC	A	67	30	91
June 1984	RV 'Squilla'	WEC	A	310	122	350
July 1984	RV 'Squilla'	WEC	A	190	78	240
April-May 1985	RV 'Squilla'	WEC	A	150	89	170
July 1985	RRS 'C Darwin'	CS	A	155	81	190
Sept 1985	RV 'Squilla'	WEC	A	40	27	45
				962	446	
May-June 1986	G A 'Reay'	N Sea	A	1036	650	600 FD
July 1986	RV 'Squilla'	WEC	A	120	75	132
Aug 1986	USNS 'Lynch'	Arctic	B	1411	735	1472
Sept 1986	RRS 'C Darwin'	Indian Ocean	A	1795	880	1912
				4366	2340	
June 1987	RRS 'Challenger'	NEA	C	2063	1100	2100
July 1987	RRS 'Challenger'	N Sea	D	1921	1000	1900 + FD
Aug 1987	USNS 'Lynch'	Arctic	C	1364	1700	1407
May-Sept 1987	OWS 'Cumulus'	NEA	D	~1000	200	FD
				6348	3900	
Jan-March 1988	RRS 'John Biscoe'	SO	A	2421	2200	2400
			Total	~14121	8900	

WEC = western English Channel, CS = Celtic Sea, BC = Bristol Channel, NEA = north-east Atlantic, FD = fixed depth, SO = Southern Ocean

A = T, Chs, E_d, E_u (445, 520, 550, 670) B = T, Chs, E_d, E_u (450, 490, 490E, 550, 550E) C = T, Chs, E_d, E_u (450, 490, 490E, 520, 550, 670) D = T, Chs, E_d, E_u (450, 550). All light sensors hemispherical, 2π collectors, except E= Irradiance (cosine collectors).

Corrupted measurements

Perturbed optical measurements are observed when the UOR is towed shallow, close astern, in the wake of the towing vessel. Measurements from the UOR towed by RV 'Squilla' at 5 m s^{-1} (10 knots) are shown in Figure 3.2. The reflectance data are extremely noisy and anomalously high, except for the periodic depth profiles to 30-40 m, when on the longer cable length the UOR is further astern of the vessel and out of the turbulent part of the wake. The vertical profiles show that the water is well mixed, though particulate scattering increases with depth (probably due to increasing suspended sediment concentrations) so that the decreasing reflectance with depth is probably real. For a medium-sized vessel as 'Squilla' (21 m in length) the noisy effect practically disappears when the UOR is about 15-20 m astern or >5 m deep. Analysis of the residual noise indicates that any apparent increase in the reflectance should not exceed 2-5% of

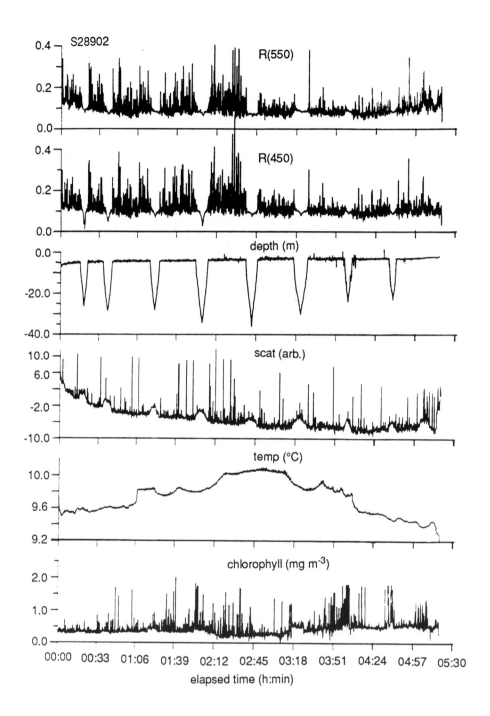

Figure 3.2. Timeplot (data value *vs* time) of *R(550)*, *R(450)*, depth, scattering, temperature and chlorophyll concentration from the UOR towed on a short cable, near surface, in the vessel's wake, with periodic profiles, from RV 'Squilla' in mixed water in the western English Channel, February 1989.

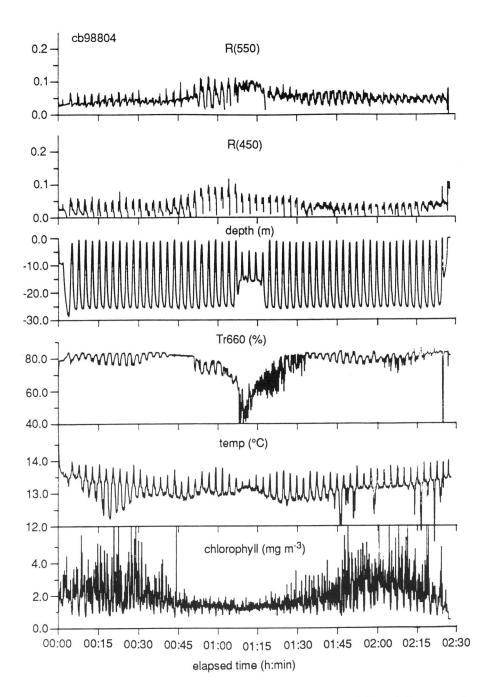

Figure 3.3. Timeplot of R*(550)*, R*(450)*, depth, transmission, temperature and chlorophyll concentration from the UOR towed at 5 m s⁻¹ with 150 m cable deployed from RRS 'Challenger,' offshore to inshore to offshore in the North Sea, August 1988.

the measured value. Normally the UOR is well astern of the towing vessel with 200-500 m of cable deployed and the optical data are free of noise, indicating that there is no observable perturbation from the vessel's wake, so that any increases in K or R are probably insignificant. Figure 3.3 shows measurements from the UOR towed by RRS 'Challenger 'at 5 m s^{-1} (10 knots) with 150 m cable deployed, in the North Sea off Whitby in August 1988, between stratified water offshore with high chlorophyll and high transmission, and nearly mixed inshore water with low chlorophyll and high turbidity.

Base-line optical measurements

The measurements in the Indian Ocean in September 1986 (RRS 'Charles Darwin'), represent 'blue-water' oceanography, with very low chlorophyll concentrations (surface values of ca 0·05-0·1 mg m^{-3} and peak values of ca 0·2-0·5 mg m^{-3} in the thermocline at depths of 50-80 m), providing base-line optical conditions. In the clearest ocean water, just north of the Equator and ca 350 miles west of the Maldives (7°56·4'N, 67°2·5'E), blue light penetrated to greater depth than green light; the chlorophyll concentrations were 0·05 mg m^{-3} surface to 50 m, and the diffuse attenuation coefficients were $K(445) = 0.046$, $K(490) = 0.032$ and $K(550) = 0.069$ m^{-1} These values are only slightly greater than the values for the clearest ocean water especially for green light ($K_w(550) = 0.065$ m^{-1}, Smith & Baker, 1981) suggesting that there are no green-absorbing pigments present. The values for blue light are slightly greater than would be expected for pure water plus the 0·05 mg m^{-3} of chlorophyll pigment, suggesting that there may be additional absorbing material at these wavelengths, e.g. dissolved organic matter.

The relatively high blue reflectance ($R(445)$ ~5-6%) compared to the uniformly low green reflectances ($R(550)$ ~1%) are almost exactly equivalent to the water-leaving radiances, $L_w(\lambda)$, measured by Gordon & Clark (1981) for very low chlorophyll (<0·1 mg m^{-3}) oligotrophic, blue waters. Using the conversion given by Viollier et al. (1980), $R(\lambda) = L_w(\lambda) (\pi / 185 \times 0.52)$, where the solar irradiance at the top of the atmosphere is 185 mW cm^{-2} μm Sr, their measurements of $L_w(450) = 1.75$ and $L_w(550) = 0.3$ mW cm^{-2} μm Sr give reflectance values of 5·7% and 1·0% respectively.

Diffuse attenuation coefficients K_{od} and K_d

The measurements of the diffuse attenuation coefficients at all sites showed that the values determined by the hemispherical, scalar irradiance sensors, $K_o(\lambda)$ and the irradiance sensors, $K(\lambda)$ were equal, within the accuracy of the measurements, in agreement with the predictions of Højerslev (1975) and Kirk (1983) and the observations of Aiken & Bellan (1986b), Weidermann et al. (1985) and Weidermann & Bannister (1986). From the many examples, Table 3.2 summarises the regressions of $K_o(490)$ vs $K(490)$ for the measurements from six successive UOR tows in the north-east Atlantic Ocean in June 1987 from RRS 'Challenger' (see cruise track, Figure 3.4). Tows C68725 to C68729 were through waters dominated by moderate to high concentrations of coccolithophores and free coccoliths and tow C68730 across the shelf break north-west of Ireland was dominated by phytoplankton chlorophyll at the shelf edge and on-shelf. Note in Table

3.2, the near zero intercepts, the near unity coefficients and the high significance (R^2 value) of the relationships.

Table 3.2. *Statistical relationships between the diffuse attenuation coefficients $K_o(490)$ and $K(490)$ measured for six UOR tows, in the north-east Atlantic Ocean, June 1986*

tow no.		$R^2\%$
C68725	$K_o(490) = 0.0035 + 0.934\ K(490)$	99.3
C68726	$K_o(490) = 0.0043 + 0.975\ K(490)$	97.0
C68727	$K_o(490) = 0.0013 + 0.961\ K(490)$	99.1
C68728	$K_o(490) = 0.0014 + 0.972\ K(490)$	98.9
C68729	$K_o(490) = 0.0027 + 0.961\ K(490)$	98.9
C68730	$K_o(490) = 0.0072 + 0.930\ K(490)$	96.0

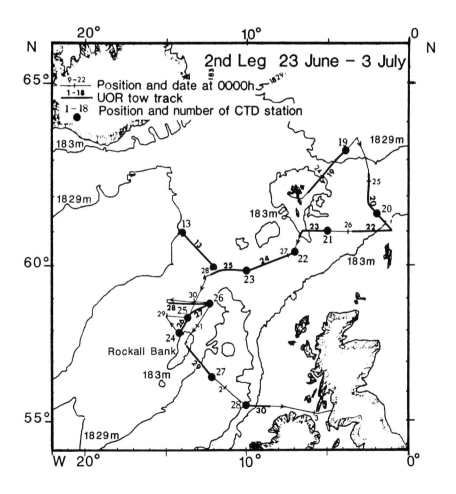

Figure 3.4. Chart of the cruise track RRS 'Challenger' in the north-east Atlantic Ocean, leg 2, 23 June - 3 July and tow 13, 18 June 1987; the 183 m (100 fathom) and 1829 m (1000 fathom) isobaths are marked.

Diffuse attenuation coefficients, UOR vs MER

The measurements of $K(\lambda)$ by the UOR (wavelengths 445, 488, 520 and 550 nm) were compared with measurements by the MER (wavelengths 438, 461, 485, 536 and 558 nm) and the Visibility Laboratory Irradiance Meter (VLIM at 493 nm), profiled vertically at stations throughout the Arctic Seas, during August 1986 (L886) and July/August 1987 (L787), from the research vessel USNS 'Lynch'; both the MER and the VLIM were operated by scientists from the Visibility Laboratory of Scripps Institution of Oceanography. Figure 3.5 shows the vertical profiles of downwelling irradiance (E_d) and the diffuse attenuation coefficient (K) derived from the depth differential of E_d ($K = d/dz \ln E_d$) from separate deployment of the MER (485 nm) and the VLIM (493 nm) at Station 11

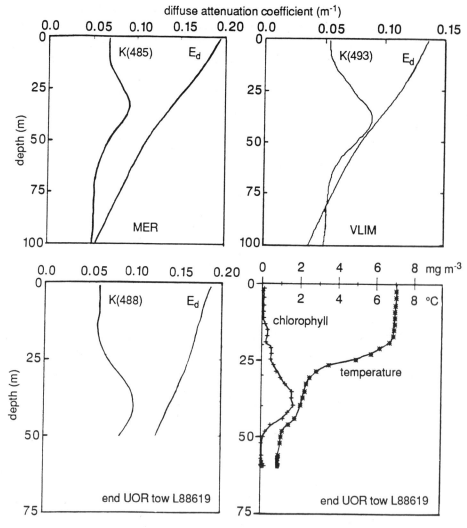

Figure 3.5. Vertical profiles of E_d and K_d (= d/dz ln E_d) for MER (485 nm), VLIM (493 nm) and UOR (488 nm) at Station 11, in the Barents Sea (74°42·5'N, 44°32'E), 12 August 1986; the corresponding profiles of temperature and chlorophyll from the UOR measurements are shown.

in the Barents Sea (74°42·5'N, 44°32'E, 12 August 1986) and the vertical profile of $K(488)$ from the final up-cast of the UOR at the end of tow L88619 adjacent to Station 11. The vertical profiles of temperature and chlorophyll concentration from the UOR up-cast show that the maximum in $K(\lambda)$ in each case corresponds to the chlorophyll concentration maximum in the thermocline at 35-40 m. The data for all three measurements are comparable, though the value of $K(485)$ from the MER is significantly greater than the other two measurements in the surface mixed layer; this is perhaps an indication of contamination by ship-shadow which would tend to increase the value of K at the surface. The regression analyses of these data (Table 3.3a) show that there is a significant relationship between all three measurements with significance at <5% level, even in the worst case (UOR *vs* MER). The divergences arise almost entirely from the discrepancy between the data for the surface layer discussed above. The difference between the data from MER and the VLIM emphasises the variability which can arise from measurements made on separate deployments, due to changing conditions such as changing water masses and, by the same token, the highly significant agreement between the UOR and the VLIM can be considered as somewhat fortuitous.

The summary regressions, of the surface layer $K(\lambda)$ data for all 20 stations in cruise L886 are listed in Table 3.3b. The variability explained by the regressions (R^2) between the UOR and the MER data (85%-91%) is comparable to the variability between the MER and the VLIM (92%) taken on separate casts, though considerably less than the variability between the simultaneous measurements at different wavelengths by the MER (98-99%), or the simultaneous measurements by the UOR (96-99% *cf.* Table 3.2).

Table 3.3. *Regression analyses of measurements by the UOR(U), the MER(M), and the VLIM(V) for USNS 'Lynch'*

(a) K(λ,z) at Station 11 (74°42·5'N, 44°32'E) in the Barents Sea, 12 August 1986

	R^2	N	TR	SL
K(488)U = -0.0174 + 1.225K(485)M	38.4%	11	2.42	<0.05
K(488)U = -0.0086 + 1.198K(493)V	93.1%	11	10.99	<0.001
K(485)M = -0.0253 + 0.678K(493)V	55.8%	13	4.05	<0.002

TR = the T-ratio and SL = the significance level.

(b) K(λ) for all 20 stations of cruise L886 in the Arctic Seas, August 1986.

	R^2 %
K(445)U = 0.0052 + 0.964 K(438)M	90.9
K(445)U = 0.0124 + 1.030 K(461)M	88.1
K(550)U = -0.0033 + 1.029 K(536)M	88.9
K(550)U = -0.0508 + 1.362 K(558)M	85.2
K(485)M = 0.0010 + 1.044 K(493)V	91.8
K(438)M = 0.0063 + 1.079 K(461)M	99.0
K(536)M = -0.0477 + 1.337 K(558)M	97.9

Diffuse attenuation coefficient UOR towed measurements

Having established the agreement between the measurements of $K(\lambda)$ by the UOR and the other irradiance sensors at stations, it is still necessary to confirm the veracity of the UOR measurements when towed at operational speeds (4 m s^{-1} or faster). There are no comparable towed optical instruments, for direct inter-comparison, so it is possible to assess the UOR data only in areas of extreme homogeneity of physical, biological and optical properties, where station measurements are representative of the surrounding area. Figure 3.6, shows the diffuse attenuation coefficients, $K(\lambda)$, of the surface mixed

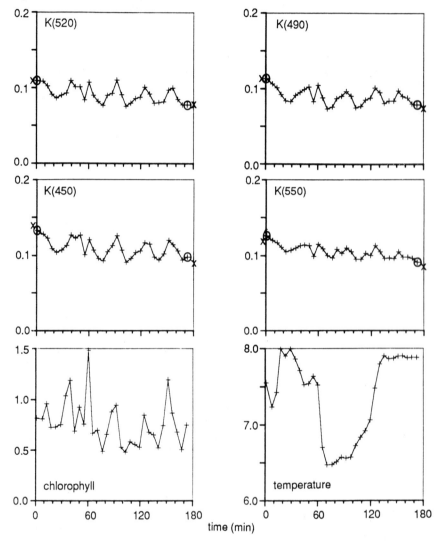

Figure 3.6. Surface layer (0-20 m) diffuse attenuation coefficients $K(450)$, $K(550)$, $K(520)$, $K(490)$, average chlorophyll concentration and temperature for tow L78720 in the Greenland Sea ca 71°30′N, 3°W, 7 August 1987; each cross (+) represents the value derived for each undulation, the circled cross $\oplus$ the value derived from the down-cast or up-cast adjacent to each station and x-symbols (x), the measurements from the MER profiled at the corresponding stations.

layer (0-20 m) for tow L78720 in the Greenland Sea, north of Jan Mayen Island in August 1987. This had moderately homogeneous conditions at the start and end, corresponding to the station measurements (station numbers 13 and 14 respectively); the UOR measurements from the down-cast or up-cast adjacent to each station position (circled) are in close agreement with the MER measurements at each station (bold cross), exactly equal in some cases.

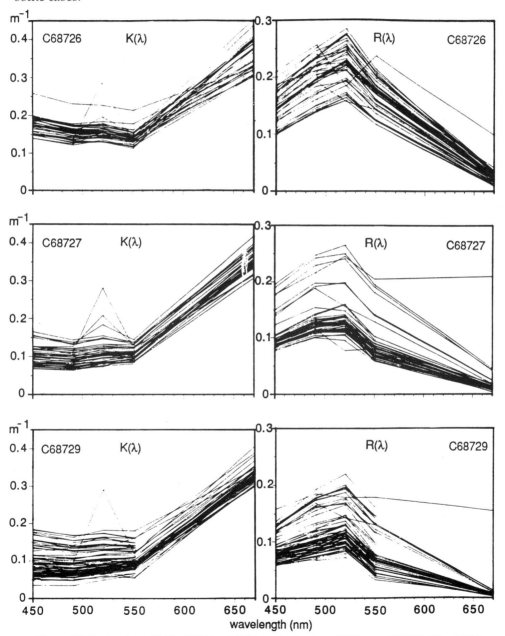

Figure 3.7. Surface layer (0-20 m) $K(\lambda)$ and $R(\lambda)$ spectra for three UOR tows (C68726, 27 and 29) in the north-east Atlantic Ocean, June 1986.

Spectral measurements of K(λ) *and* R(λ)

The spectra of $K(\lambda)$ for 3 tows in the north-east Atlantic in June 1986 (cruise track Figure 3.4) are shown in Figure 3.7 (one record for each undulation). Generally the spectra are smooth, though anomalous data can be easily identified and eliminated if necessary. The $K(\lambda)$ spectra have a flat minimum, generally around 490 nm though sometimes higher, and a maximum at 670 nm (*cf.* Morel, 1988) and the $R(\lambda)$ spectra have a maximum at 520 nm and a minimum at 670 nm (*cf.* Morel & Prieur, 1977).

The inherent optical properties a(λ), b(λ)

Kirk (1981, 1986) has developed empirical relationships, using Monte Carlo model-ling, linking R and R_o with b/a and μ. Kirk's curves, fitted to polynomials give:

$$b/a = 0{\cdot}0367 + 58{\cdot}0255R_o - 86{\cdot}56101R_o^2 + 571{\cdot}52388R_o^3$$
$$\mu = 0{\cdot}839 - 3{\cdot}3705R_o + 10{\cdot}49125R_o^2 - 14{\cdot}6574R_o^3$$
$$b/a = 0{\cdot}1486 + 109{\cdot}86786R - 185{\cdot}43508R^2 + 1152{\cdot}293R^3$$
$$\mu = 0{\cdot}8374 - 5{\cdot}7556R + 29{\cdot}73864R^2 - 60{\cdot}99536R^3$$

Since $a = \mu K_d$ and $K_d \approx K_{od} \approx K_e \approx K_u$; $a(\lambda)$ and $b(\lambda)$ can be derived independently from measurements of R_o and R.

Analysis of the measurements of $R(490)$ and $R_o(490)$ for selected tows from the north-east Atlantic in June 1986 (cruise track, Figure 3.4), shows that the values of $a(490)$ and $b(490)$ derived thus, are comparable within the accuracy of the measurements. For example for tow C68726:

$$a_o(490) = 0{\cdot}0039 + 0{\cdot}958\, a(490);\ R^2 = 97{\cdot}1\%;\ n = 41$$
$$b_o(490) = -0{\cdot}0045 + 0{\cdot}985\, b(490);\ R^2 = 99{\cdot}1\%;\ n = 41.$$

Using these methods for calculating $a(\lambda)$ and $b(\lambda)$, spectra can be derived as for $K(\lambda)$ and $R(\lambda)$.

Both the spectral measurements $K(\lambda)$ and $R(\lambda)$ and the derivation of the inherent optical properties $a(490)$ and $b(490)$ do not prove positively that the UOR measurements are free of any bias, but taken with the earlier comparisons with conventional radiome-ters (*e.g.* the MER and VLIM), they show that the UOR measurements are spectrally and numerically consistent with other observations.

Comparison of K(λ) *and UOR, with beam transmissometer measurements*

The beam transmissometer, from which the beam attenuation coefficient ($c = a + b$) can be calculated ($c = 1/L \ln Tr$, where Tr is the water transmission (%) and L is the path length in m) is the only optical instrument which can be operated continuously in parallel with the UOR, either towed, attached to the UOR, or measuring pumped surface water in flow-through mode on board the towing vessel. The Sea Tech 0·25m 660 nm transmis-someter, was used to measure the transmission of pumped surface water in cruises in the north-east Atlantic in June 1987 (cruise C687) and in the Arctic Seas in July and August 1987 (cruise L787). Overall the measurements have been shown to be in qualitative agreement: increases of the diffuse attenuation coefficients at blue wavelengths, $K(445)$ and $K(490)$, which co-vary with increases in chlorophyll concentration, are inversely related to the transmission at 660 nm (*i.e.* c increasing due to increasing a); likewise increases in $K(\lambda)$ and $R(\lambda)$ where coccolithophores and detached coccoliths are abundant are inversely related to transmission (*i.e.* c increasing due to increasing b).

Pilgrim (1988) has determined the relationship between $K(670)$ from measurements by the UOR and $c(660)$ from the Sea Tech Transmissometer, for the surface layer (0-20m) for a UOR tow in the Greenland Sea across the Arctic front east of Jan Mayen Island (Tow L78715, overnight 5-6 August 1987, 69°49′N, 4°38′W to 71°27′N, 4°54′W) giving:

$K(670) = 0{\cdot}203 + 0{\cdot}230\ c(660)$; $R^2 = 77{\cdot}3$; n = 40.

The average value for the ratio $c(660)/K(670)$ is *ca* 2·23 which is comparable to the average value of 2·7 reported by Tyler (1968) for oceanic conditions, though smaller values are possible (*e.g.* in low scattering waters).

The Sea Tech 0·25m transmissometer has been fitted to the UOR for measurements in the North Sea in August 1988 (cruise C888). The timeplot of transmission (660 nm) for tow C88804, offshore (stratified) to inshore (mixed and turbid) to offshore is shown in an earlier Figure (Figure 3.3). Note, the transmissometer measurements are generally free from noise, have maximum values (83%) in the thermocline, reduced values (77%) in the chlorophyll rich (2-3 mg m^{-3}) surface layer but are attenuated to less than 50% in the most turbid water inshore, presumably due to high concentrations of suspended sedimentary material. From the data, the average transmission, $Tr(660)$ and the attenuation coefficient $K(670)$ for the surface layer are derived (see Figure 3.8). Note the inversely related

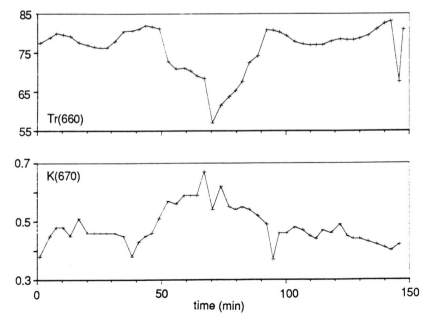

Figure 3.8. Surface layer (0-10 m) values of $Tr(660)$ and $K(670)$ from UOR tow C88804 in the North Sea, August 1988; each cross represents the value derived for each undulation

covariance between Tr(660) and K(670) when chlorophyll concentrations are high (offshore) or when turbidity is high (inshore).

Comparison of R(UOR) with satellite imagery

Measurements of reflectance by the UOR (R445, R520, R550) have been compared with contemporary CZCS imagery for the western English Channel (June 1984) and the North Sea (June 1986). In both cases the along-track reflectances were in good agreement qualitatively with the satellite imagery even though the measurements were separated by a few days. Since June 1986, only the AVHRR sensors (NOAA-n satellites) have a visible band (580-680 nm) which can be compared with in-water optical measurements,

Figure 3.9. (A) Atmospherically corrected AVHRR, channel 1, visible band image of the north-east Atlantic Ocean (5°W-25°W, 55°N-64·5°N) for 18 June 1987; UOR tow C68713 for 18 June is marked.

and during the RRS 'Challenger' cruise in the north-east Atlantic Ocean (June 1987) contemporary images were acquired for 18, 23, 24, 27 and 28 June, coinciding with UOR tows through two massive blooms of coccolithophores north of Rockall Bank and south of Iceland (see cruise track, Figure 3.4). Figure 3.9A shows the atmospherically corrected AVHRR channel 1, visible band image for 18 June, coincident with UOR tow 13 (tow track overlaid on image) and Figure 3.9B shows the corrected channel 1 image for 23 June, overlaid with the tracks of UOR tows 26, 27, and 29 a few days later. Figure 3.10 shows the UOR reflectances *R(550)* and the along-track satellite radiances (in digital numbers) for tows 13, 26 and 29 with excellent qualitative agreement in all cases: the track of tow 13 shows moderate reflectance, low reflectance, high reflectance; the track of tow 26 shows high, low, high reflectance; the track of tow 29 shows high reflectance at the start

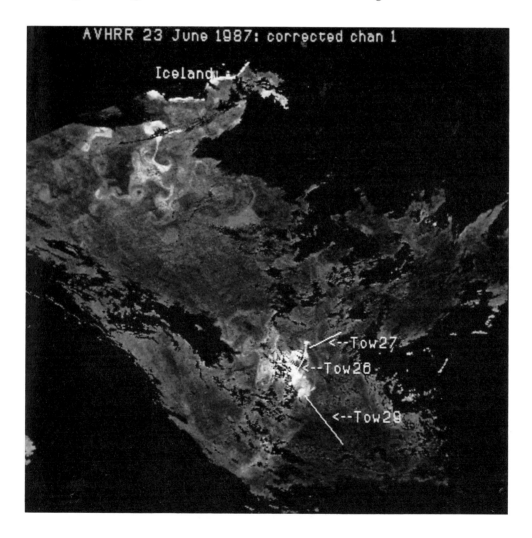

Figure 3.9. (B) Atmospherically corrected AVHRR, channel 1, visible band image of the north-east Atlantic Ocean (5°W-25°W, 55°N-64·5°N) for 23 June 1987; UOR tows C68726, 27 & 29 are marked.

and low reflectance thereafter. Using the conversion $R(550) = 6.66\,R(670)$, derived by Groom & Holligan (1987) from CZCS and AVHRR imagery and verified by the present UOR data sets, the maximum reflectance in the image, corresponding to a position close to the end of tow 26, was calculated to be equivalent to 23% at 550 nm, approximately equal to the maximum values recorded by the UOR on tow 26. Only tow 13 and the image for 18 June were coincident and statistical analysis showed the variability explained by the regression (R^2) at 50%; a possible error of 10 km in geo-referencing the image and the tow position, could cause a mis-match in the measurements to give this low value.

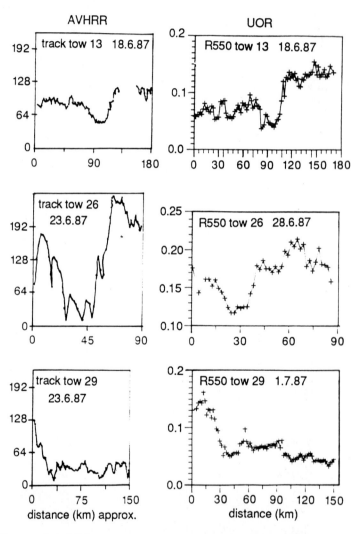

Figure 3.10. Along-track AVHRR radiances (arbitrary scale) and UOR reflectance $R(550)$ for tows 13, 26 and 29 in north-east Atlantic, June 1987.

CONCLUSIONS

It has been shown that the optical measurements with the UOR are corrupted when the instrument system is towed shallow, close-astern, in the turbulent wake of the towing vessel, though normally the UOR is 200-500 m astern of the towing vessel and the data are unperturbed. The full suite of optical measurements with the UOR has been shown to be consistent, spectrally and numerically, for a range of oceanographic sites, from base-line, blue water oligotrophic Indian Ocean, to coccolithophore blooms in the northeast Atlantic Ocean and frontal areas in the North Sea and the Arctic Ocean. Comparisons of UOR measurements with measurements by conventional radiometers at stations are significant and UOR towed-measurements are equal to station measurements in ocean waters that are homogeneous horizontally.

There are no exactly comparable optical instruments which can be towed or which provide along-track measurements for comparison with UOR towed-measurements. The Sea Tech 0·25m, 660 nm beam transmissometer, measuring $c(660)$ provides along-track measurements which can be compared with the UOR measurement of $K(670)$, though the sharp change in the water spectra in this region makes it difficult to relate these measurements directly. Likewise, coincident measurements of UOR reflectance and satellite imagery show similar along-track optical signatures, though inaccurate geo-referencing can decrease the perceived comparability in statistical tests.

This work forms part of the programme of the Plymouth Marine Laboratory (Natural Environment Research Council, UK) and has been supported, in part, by the Admiralty Research Establishment, Portland (Ministry of Defence, UK) and the Visibility Laboratory, Scripps Institution of Oceanography (SIO), La Jolla, Ca. USA. We would like to thank the officers and crews of the many vessels which have been used in this work: RV 'Squilla' (UK), USNS 'Lynch' (USA), RRS 'Charles Darwin' (UK) and RRS 'Challenger' (UK). Ron Davies, Ewan McLeod, Rob Walker and Eleanor Russel have assisted with computing and data processing. Satellite imagery was received from the station at the University of Dundee and processed at the NERC Image Analysis Unit at Polytechnic South West with the assistance of Stephen Groom. We acknowledge valuable discussions on many aspects of the programme with Patrick Holligan, Stephen Groom, Chuck Trees (SIO) and numerous colleagues.

Chapter 4

THE PHOTIC ZONE

PAUL TETT

School of Ocean Sciences, University College of North Wales, Menai Bridge, Gwynedd, LL59 5EY

The photic zone is the illuminated surface layer of the oceans wherein the absorption of heat energy by water and light energy by phytoplankton helps to regulate many aspects of the global environment. This review particularly considers the way in which optical conditions and vertical turbulent mixing determine the maximum depth to which micro-algal populations can grow. The historical development of compensation and critical depth theory is reviewed and the theory re-examined. Factors affecting the compensation illumination for a range of algae and water types are discussed, and the effect of mi-croheterotroph respiration is exemplified. The euphotic zone is defined as including all mixed layers in which photoautotrophic production exceeds heterotrophic consumption on the time-scale under investigation. The lower limit to the zone is set sometimes by the compensation depth and sometimes by the critical depth.

INTRODUCTION

Each second the sun emits about 2×10^{45} photons. The Earth, occupying only a tiny part (0.46×10^{-9}) of the sky as seen from the sun, intercepts relatively few of these quanta of radiant energy, and one in three are immediately reflected back into space by clouds, ice and atmospheric scattering. About a fifth of the intercepted radiation is absorbed by the atmosphere and a sixth by the land. The remaining third falls on the oceans, so that about 2×10^{35} photons enter the sea each second. They are absorbed in the layer, typically a few tens or hundreds of metres deep, which is the photic zone and the subject of this review.

The illuminated region is only a small part of the 3·7 km mean depth of the oceans, yet it houses several of the great engines of planetary control. The absorption of heat energy by water and light energy by phytoplankton has the profoundest consequences for life on Earth, helping to determine global climate and the chemical composition of oceans and atmosphere, as well as the nature of marine food chains and the abundance of marine organisms.

These aspects of the planetary ecosystem may be linked in ways as yet imperfectly comprehended (Lovelock, 1987). It is certainly demonstrable that physical, chemical and biological processes are intimately related within the photic zone. For example, solar warming of the superficial ocean creates buoyancy which dampens wind-generated turbulence and hence may assist phytoplankton to remain within the zone of strong

Herring, P.J., Campbell, A.K., Whitfield, M. & Maddock, L. (ed.). *Light and Life in the Sea.*
© 1990. Cambridge University Press.

illumination. Conversely, the resulting stratification may separate planktonic microalgae from nutrient-rich deep water, so impairing their ability to harvest light photosynthetically.

This review is concerned with planktonic photosynthesis in the photic zone, and, in particular, the way in which optical conditions and vertical turbulent mixing regulate the maximum depth at which microalgal populations can grow.

LIGHT IN THE SEA

Photosynthetically active radiation (PAR) comprises photons with wavelengths between 400 and 700 nm (350-700 nm, according to UNESCO, 1966) or energies of between 170 and 300 kJ E^{-1}. An Einstein (E) is a mole (6·022 x 10^{23}) of photons. On the basis of the figures given in the Introduction, the maximum rate at which photons arrive at the sea surface from an overhead sun in a sky of average cloudiness is about 4000 $\mu E\ m^{-2}\ s^{-1}$. Less than a third of these photons are however PAR.

Indeed, given the interest in its heating effect, it is usual to report solar irradiance in terms of energy flux (or power) density and hence as Watts (or Joules per second) per square metre (W m^{-2}). At sea level, PAR is 0·42 to 0·50 of the solar energy flux, depending on cloud cover, atmospheric state and sun angle (Baker & Frouin, 1987; Jitts *et al.*, 1976). Morel & Smith (1974) give an energy-to-photons conversion factor corresponding to 4·15 (±0·42) $\mu E\ J^{-1}$ for underwater PAR in the sea. Thus if I_o refers to PAR photon flux density ($\mu E\ m^{-2}\ s^{-1}$) immediately beneath the sea surface, and E_o to power density (all wavelengths) just above the sea surface, the conversion

$$I_o = 1·91 \ . \ m_1 \ . \ E_o \tag{4.1}$$

is good to within 15%, plus any errors in the factor m_1 allowing for losses due to sea-surface reflection. For most sun angles this loss is 4-6%, so a typical value of m_1 is 0·95.

Submarine daylight is diffuse and a full description of its properties would be complex (Sathyendranath & Platt, Chapter 1). The coefficient of diffuse attenuation, λ, for downwards irradiance may be defined as the relative rate of change of photon flux density with distance along the downwards component of the paths traversed by the photons under consideration:

$$\lambda = (dI/dz) \ . \ (1/I) \tag{4.2a}$$

Downwards irradiance, I, is the flux density on the upper surface of a plane parallel to the sea surface. If λ is assumed constant with depth, equation (4.2a) may be integrated between the sea surface and depth, z, to give photon flux density as a function of depth:

$$I = I_o \ . \ e^{-\lambda.z} \tag{4.2b}$$

That is, irradiance declines exponentially with depth (Figure 4.1). The dimensionless term $\lambda.z$ is called the 'optical depth', and enables the comparison of photosynthetic potential in waters of different transparency.

The assumption of λ unaffected by depth is correct only for monochromatic radiation of constant angular distribution in a uniform water column. As Figure 4.1 shows, equation (4.2) is thus an inexact description of the decay of illumination. A convenient modification is:

$$I = m_2 \cdot I_0 \cdot e^{-\lambda(min).z} \qquad (4.3)$$

where $\lambda(min)$ is a minimum value of the diffuse attenuation coefficient and m_2 corrects for more rapid attenuation of polychromatic light near the sea surface. The value of m_2 depends on the method used to estimate $\lambda(min)$. Thus, when the attenuation coefficient was that for the wavelength band of greatest submarine transparency, and the decay curves for monochromatic light and PAR were adjusted to coincide at 1% of I_0, m_2 was found to range from 0·34 to 0·39 for turbid coastal, fjordic and oceanic waters (Tett, unpublished).

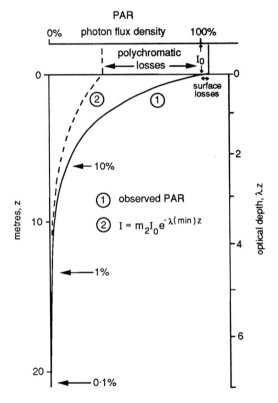

Figure 4.1. Penetration of light into the sea. Curve (1) is a profile of relative photon flux density synthesized from measurements of diffuse attenuation for 5 wavelengths of near-monochromatic light in Loch Striven, 2-4 June 1988. The attenuation coefficients (and the proportion of sea-surface PAR photons contained in each waveband) were 670 nm: 0·62 m⁻¹ (0·19); 595 nm: 0·33 m⁻¹ (0·17); 552 nm: 0·27 m⁻¹ (0·10); 521 nm: 0·31 m⁻¹ (0·23); 444 nm: 0·42 m⁻¹ (0·31). The attenuations were estimated from downwards irradiance measured by a cosine-response instrument described by Mitchelson *et al.* (1986). Curve (2) was back-calculated from the depth of 1% I_0 using equation (4.3) with $\lambda(min) = 0·27$ m⁻¹, enabling the correction factor m_2 to be estimated as 0·34.

Finally, submarine irradiance, I, is the flux density of downwelling photons only. Planktonic algae also harvest sideways- or upwards- scattered photons. In many coastal waters the ratio of scalar irradiance (that arriving at a point from all directions) to downwards planar irradiance (that measured by a flat detector whose response diminishes as the cosine of the angle between the incident photon and the vertical) is about 1·3 (Højerslev, 1981). This ratio, termed m_3, will be used to enhance predicted photosynthetic efficiency when this is related to planar irradiance data.

THE COMPENSATION DEPTH

Photosynthesis is evidenced by the evolution of oxygen or the assimilation of inorganic carbon. 'Biomass-related' rates, symbolized p^B, might be cited in ml O_2 (million cells)$^{-1}$ day^{-1}, or mg C (mg Chl)$^{-1}$ h^{-1}. Units of nanomoles O_2 released, or CO_2 fixed, (mg Chl)$^{-1}$ s^{-1} are sometimes convenient. Algal respiration, r^B, can be expressed in the same units. The rate of photosynthesis depends on the illumination to which photosynthesizing algae are exposed. Figure 4.2A illustrates a typical relationship. A graph can be drawn for gross photosynthesis (Figure 4.2B), on the assumption that respiration continues at the same rate in light as in dark bottles.

Figure 4.2A shows that there is a minimum illumination for net photosynthesis; since algal biomass is constructed from the products of photosynthesis, there must therefore also be a minimum illumination for growth. Accurate measurements of the depth to

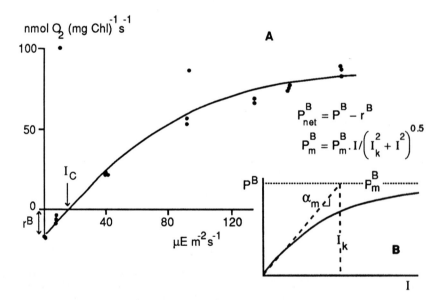

Figure 4.2. Photosynthesis-irradiance curve. A plot of biomass-related photosynthesis, p^B, against photon flux density, I, for a water sample from Loch Creran, 21 June 1978, containing planktonic microheterotrophs as well as algae. Data, for oxygen changes in light and dark bottles exposed in a light gradient incubator, from Tyler (1983). (A) Net photosynthesis. (B) Gross photosynthesis. The curves were drawn after fitting equation (4.13) nonlinearly to gross photosynthetic data, giving estimates of α_m of 1·14 nmol O_2 (mg Chl)$^{-1}$ (μE m^{-2})$^{-1}$, and p^B_m of 114 nmol O_2 (mg Chl)$^{-1}$ s^{-1}. The other parameters were: r^B: 17 nmol O_2 (mg Chl)$^{-1}$ s^{-1}; I_c: 15 μE m^{-2} s^{-1}; I_k: 100 μE m^{-2} s^{-1}.

which such illumination penetrates became possible with use by Gaardner & Gran (1927) and Marshall & Orr (1928) of the Winkler titration for dissolved oxygen. Marshall & Orr (1928) suspended clear and opaque glass bottles containing laboratory-grown cultures of the diatom, *Coscinosira polychorda*, at various depths in the Scottish sea-loch Striven. Their '*in situ* light and dark bottle' incubations were each about 24 hours long, the duration of the natural day-night cycle. It can thus be argued that oxygen change in light bottles truly indicated microalgal growth potential at each depth in Striven. Above the 'compensation point', algae would end each 24 hours with more organic material than they started, enabling cell division and population growth. Below this depth "photosynthesis may still go on, but there is a continuous loss of oxygen and life becomes impossible".

These results can be summarized by plotting daily rates of gross photosynthesis and respiration against depth (Figure 4.3). The 'compensation depth' is that at which the photosynthetic and respiratory graphs intersect. As might be expected, this depth

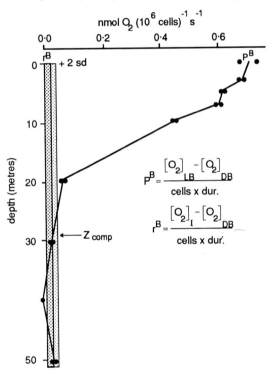

Figure 4.3. The compensation depth. Profiles of gross photosynthesis p^B and respiration r^B of algal cultures incubated in Loch Striven, 13-14 June 1927. Data from Marshall & Orr (1928); the culture contained 3700 cells of *Coscinosira polychorda* cm^{-3} and the incubation lasted 23·8 hours from 1100 h during a period of more than 16 hours bright sunshine. Two light and one dark bottle were incubated at each depth; respiration rates have been averaged as there was no evidence of systematic variation with depth. The intersection of the photosynthesis and respiration graphs indicates the compensation depth, at 29 m, or a minimum of 25 m if the respiration rate is increased by two standard deviations. Marshall & Orr were not able to estimate irradiance at the compensation depth. $\lambda(min)$ of 0·20 m^{-1} is, however, compatible with their values for the decrease in gross photosynthesis between 10 and 20 m. Given I'_o of 600 µE m^{-2} s^{-1} averaged over 24 hours of a sunny June day, and m_2 of 0·34, the irradiance at the compensation depth in June 1927 can be estimated as about 1 µE m^{-2} s^{-1}.

depends on time of year. Marshall & Orr (1928) found it to range from 5-15 m in Loch Striven in March 1927, increasing to 29 m on 13-14 June 1927, a 24-hour period with more than 16 hours sunshine.

Why did Marshall & Orr use algal cultures for their investigations? Winkler's titration has limited sensitivity but cell densities sufficient to give significant diurnal oxygen changes could be achieved when cultures were incubated in the sea. With hindsight their approach has further benefits. Much recent debate on the measurement of phytoplankton photosynthesis (*e.g.* Colijn *et al.*, 1983) concerns the 'bottle effects' that result from separating a small volume of the planktonic ecosystem from the natural regime of turbulence and nutrient recycling. Since Marshall & Orr's *Coscinosira* were adapted to culture conditions it is unlikely that they suffered any extra ill-effects when incubated in bottles in Loch Striven. A second advantage concerns the relative purity of their cultured material: in samples of untreated sea water the respiration of algae is often much augmented by that of microheterotrophs.

There are, however, potential differences between the properties of *Coscinosira* cultures and the natural phytoplankton. Thus, the spring bloom in Loch Striven is dominated by the smaller diatom, *Skeletonema costatum*, with *Coscinosira* (now *Thalassiosira*) *polychorda* playing only a small part (Marshall & Orr, 1927, 1930; Tett *et al.*, 1986). Furthermore, the nutrient status and light-exposure history of laboratory algal cultures might well differ from those of phytoplankton.

The introduction of the carbon-14 method (Steemann Nielsen, 1952) allowed the measurement of photosynthetic carbon uptake by phytoplankton under almost all conditions. Later improvements in the precision of oxygen measurement (Bryan *et al.*, 1976) enabled the light- and dark-bottle oxygen technique to be used with natural phytoplankton in coastal waters, and sometimes with that of the oligotrophic oceanic gyres. In the 'simulated *in situ*' method, bottles are exposed behind optical screens (corresponding in attenuation to a range of optical depths) in a temperature-controlled bath on-board ship or in a shore laboratory.

These procedures are widely used to estimate primary productivity but less often to determine the compensation depth. An indirect approach is to derive from first principles an expression for the compensation depth and to evaluate this from optical data and generalizations about algal photosynthesis and respiration. The expression is

$$z_{comp} = \ln (m_2 . I'_o / I_c) . (1 / \lambda(min)) \tag{4.4}$$

The main steps in its derivation are the calculation of 24-hour mean PAR at the compensation depth:

$$I_c = m_2 . I'_o . e^{-\lambda(min) . z_{comp}} \tag{4.5}$$

and the equation of 24-hour mean photosynthesis with respiration at this compensation illumination:

$$\alpha . I_c = r^B \tag{4.6}$$

α is photosynthetic efficiency in nmol C (mg Chl)$^{-1}$ (μE m^{-2})$^{-1}$. It is assumed that neither this efficiency, nor respiration, vary diurnally. Equation (4.5) derives from equation (4.3) and thus allows through m_2 for additional attenuation of polychromatic light near the sea surface. I'_o is 24-hour mean PAR quantum flux density immediately beneath the sea surface.

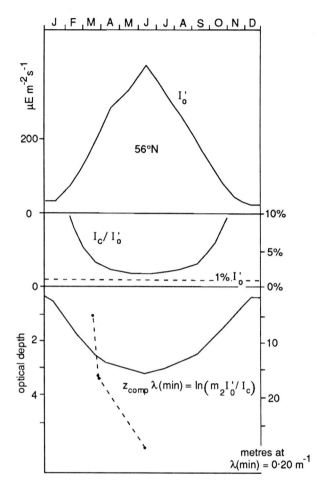

Figure 4.4. Predicted compensation depths in Loch Striven. Given against scales of optical depth and actual depth, the latter calculated with $\lambda(min)$ of 0·20 m^{-1} as deduced for June 1927. Compensation illumination I_c was taken as 6·3 µE m^{-2} s^{-1} after Jenkin (1937) and Sverdrup (1953). Marshall & Orr's (1928) estimates for compensation depth in 1927 are shown by the dashed line; their June value implies I_c of about 1 µE m^{-2} s^{-1} (see legend to Figure 4.3). Values of I'_o, 24-hour mean subsurface irradiance, are monthly means for Dunstaffnage, about 50 km NW of Striven, for 1970-75 (Anon., 1980).

Figure 4.4 illustrates the use of equation (4.4). The product $\lambda(min).z_{comp}$ is the 'optical compensation depth', and the proportion of $m_2.I'_o$ remaining at this depth is given by:

$$e^{-\lambda(min) \cdot z_{comp}} = (1 \ / \ m_2 \cdot I'_o) \cdot (r^B \ / \ \alpha) = I_c \ / \ m_2 \cdot I'_o \qquad (4.7)$$

The Figure includes seasonal variation in relative compensation illumination, I_c/I'_o, at 56°N, roughly the latitude of Loch Striven. Marshall & Orr were unable to measure PAR adequately, and so the value of 6·3 µE m^{-2} s^{-1} used for I_c is derived from oxygen incubations carried out in the English Channel by Jenkin (1937). This value of I_c was later used by Sverdrup (1953) in his study of the critical depth, and leads to predictions (such as those in Figure 4.4) which may be the source of the widely held view that the compensation depth occurs at roughly 1% of surface PAR, at least during the season of phytoplankton growth. The discrepancy between the compensation depth of about 12 m shown in Figure 4.4 for Loch Striven in midsummer, and the depth of 29 m obtained with Marshall & Orr's data in Figure 4.3, suggests however that I_c may in some cases be substantially less than Jenkin's value (see legend to Figure 4.3).

THE CRITICAL DEPTH

The seasonal and diel cycles of mixing and stratification in the ocean result mainly from the interaction between heat input, convection, and wind-driven turbulence (Woods & Barkman, 1986). In shelf seas tidal turbulence, usually generated near the sea-bed, augments wind and convective mixing (Pingree et al., 1975; Simpson & Bowers, 1984). So far as phytoplankton are concerned the result is to structure water-columns into layers where the mixing timescale is a few hours (the surface wind-mixed, and tide-mixed, layers), a few days (the diurnal thermocline) or a substantial fraction of a year (the seasonal thermocline) (Figure 4.5).

Turbulent eddies are weak and of small extent in the seasonal thermocline (and beneath it in oceanic water columns), and are unlikely to transport algal cells more than 1-2 m vertically in 24 hours. These algae thus experience changes in illumination mainly as a result of the diurnal variation of attenuated sea-surface PAR. In contrast, mixed-layer phytoplankton suffer additional fluctuation in light as a result of transport in eddies with vertical velocities of metres or tens of metres per hour.

In a mixed layer which extends below the compensation depth, turbulence may transport an algal cell to a region where even in daytime the illumination is less than that at the compensation point. In order to decide whether or not the cell will be capable of growth, it is, in principle, necessary to monitor (or model) its cumulative photosynthesis and respiration during a 24-hour random walk through the mixed layer (e.g. Woods & Onken, 1982; Wolf & Woods, 1988).

The classical approach (Gran & Braarud, 1935; Sverdrup, 1953) is instead to consider the fate of a homogeneous phytoplankton population evenly spread throughout a mixed layer of thickness, h. Given constant photosynthetic efficiency, layer photosynthesis is $\alpha.I'$, and the inequality to be satisfied for growth to be possible is:

$$\alpha . I' / r^B > 1 \tag{4.8}$$

Layer illumination, I', is averaged over depth as well as 24 hours. It can be estimated from:

$$I' = I'_0 . e^{-\lambda(1) . z(1)} . (1 - e^{-\lambda(2) . h}) / (\lambda(2) . h) \tag{4.9}$$

The PAR attenuation coefficients $\lambda(1)$ and $\lambda(2)$ refer to, respectively, the overlying water and the layer in question; $z(1)$ is the thickness of overlying water. In the case of an optically deep layer extending to the sea surface, equation (4.9) can be approximated by:

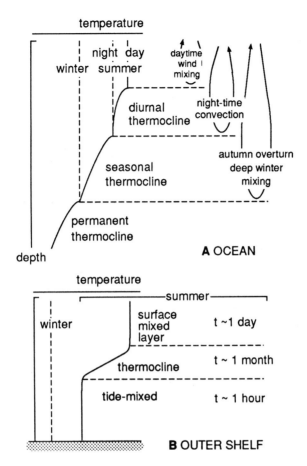

Figure 4.5. Mixing and stratification in the sea. The mixing structure of the sea is usually deduced from density or temperature layering, as indicated schematically here. Temperature profiles, however, reflect past as well as present heating and stirring, although once a thermocline is formed its associated density gradient resists penetration by turbulent eddies and so helps to maintain stratification. In (A) the 'mixed layer', in the terminology of Woods & Barkman (1986) extends only to the top of the diurnal thermocline. In more general usage the 'surface mixed layer' extends to the greatest depth exchanging with the surface every 24 hours. In (B) for deep water on a continental shelf where tidal mixing creates an isothermal near-bottom layer, t refers to the approximate timescale for an algal cell to be mixed across a layer (Pingree *et al.*, 1975). Such time-scales are not easily calculable for winter conditions, but it seems likely that temperate shelf seas are thoroughly convectively stirred at least once every 24 hours in winter, and perhaps more frequently when tidal mixing reaches to the surface.

$$I' = m_2 \cdot I'_o / (\lambda(min) \cdot h) \qquad\qquad (4.10)$$

Sverdrup (1953) defined the critical depth as that of a surface-mixed layer in which depth-integrated photosynthesis exactly equalled depth-integrated respiration (Figure 4.6). Transforming (4.8) and combining with (4.10), leads to an equation for the critical depth:

$$z_{crit} = (\alpha / r^B) . (m_2 . I'_o / \lambda(min))= (m_2 . I'_o / I_c) . (1 / \lambda(min)) \qquad (4.11)$$

In his paper, Sverdrup compared values of the critical depth with that of the mixed layer at 66°N in the Norwegian Sea, arguing that the spring phytoplankton increase began when the critical depth first exceeded the mixed-layer depth. Figure 4.7 illustrates this for the seasonal cycle of PAR and mixed-layer depth at Scottish oceanic and shelf stations at 59°N.

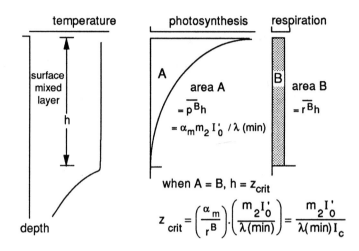

Figure 4.6. The critical depth. When the depth of the surface mixed layer is such that area A (photosynthesis) equals area B (respiration), the mixed layer depth is the critical depth. In this explanation, photosynthesis is supposed not to saturate near the sea surface.

On the shelf, phytoplankton growth should commence, assuming no other than respiratory losses, in April, when the critical depth first exceeds the depth of the sea bed, and before the commencement of stratification later in the month. This prediction is supported by data in Steele & Henderson (1977), which show chlorophyll concentration on the Fladden Ground first exceeding 1 mg m^{-3} on about 10 April.

At the oceanic station, a limit to vertical mixing is set by the permanent thermocline at about 500 m (Levitus, 1982; Robinson *et al.*, 1979 give 900 m). Clearly, phytoplankton could not grow here without seasonal stratification, which is predicted to commence towards the end of April. The prediction that the spring bloom should commence at this time, is supported by the observations of Williams & Robinson (1973) at Ocean Weather Station India, where chlorophyll concentrations first exceeded 1 mg m^{-3} on about 3 May.

Sverdrup (1953) listed several requirements for the satisfactory estimation of critical depth. They can be restated as follows:

(1) *Homogeneity and light limitation*: phytoplankton must be evenly distributed throughout the layer in question, and their production must not be limited by lack of plant nutrients such as nitrate.

(2) *Linearity*: the production of organic matter by photosynthesis must be linearly proportional to the photon flux density in all parts of a layer. Furthermore, optical attenuation must be constant throughout the layer.

(3) *Consumption*: Sverdrup required only that the compensation illumination be known. A more rigorous assumption is that algal respiration be the only process destroying the organic products of photosynthesis, so that compensation illumination estimated from experiments with algal cultures truly indicates the balance between algal gains and losses.

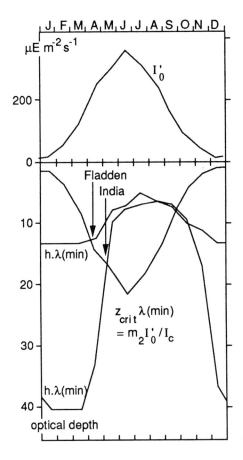

Figure 4.7. Predicting the start of the spring phytoplankton increase. The optical critical depth predicted for 59°N is compared with the depth of the surface mixed layer on the Fladden Ground in the North Sea and at Weather Station India in the north-eastern Atlantic Ocean. Surface irradiance data from mean solar radiation for Aberdeen (weighted 1/3) and Lerwick (weighted 2/3) (Anon., 1980). Mixed layer depth was calculated with the model of Clarke (1986). A limit of 500 m was set to deep winter-mixing at India (Levitus, 1982). Mixed layer depths were converted to optical depths using minimum diffuse attenuation coefficients made up of a constant (0·08 m^{-1} for India, 0·10 m^{-1} for Fladden) plus a contribution from phytoplankton pigments. The latter were estimated from an unpublished model of E. Woods & P. Tett. Jenkin's (1937) value of 6·3 µE m^{-2} s^{-1} was used for compensation illumination. The vertical arrows give the observed date when chlorophyll concentrations first exceeded 1 mg m^{-3} at India in 1971 (Williams & Robinson, 1973) and on the Fladden Ground 1961-1970 (Steele & Henderson, 1977).

PHOTOSYNTHESIS-IRRADIANCE CURVES

The relationship between biomass-related photosynthesis, p^B, and irradiance, I, shown in Figure 4.2 is characteristic of results obtained by incubating cultured microalgae or natural phytoplankton at a range of constant illuminations. At low photon flux densities, photosynthesis increases in proportion to light, but saturates at moderate illuminations. Not shown in Figure 4.2 is the inhibition of photosynthesis evident on further increases in illumination. Whereas Richardson *et al.* (1983) suggested that the onset of photoinhibition occurs at 200 µE m^{-2} s^{-1} or less, the evidence presented by Harris (1978), Platt *et al.* (1980) and Ryther (1956) shows that substantial photoinhibition seldom takes place below about 600 µE m^{-2} s^{-1}. Phytoplankton in a surface mixed layer are likely to be only intermittently exposed to such high irradiances, and photoinhibition will be neglected here.

The two most important properties of photosynthesis-irradiance, or 'P/I', curves are thus (1) the efficiency of biomass-related photosynthesis under low illuminations, which determines the initial slope of the curve, and (2) the rate of light-saturated photosynthesis. Following Jassby & Platt (1976) the symbols α for efficiency and p^B_m for saturation photosynthesis have become widely used. For reasons which will become apparent, it is convenient to use the symbol α_m, formally defined by

$$\alpha_m = dp^B \, / \, dI \; : \; (I \rightarrow 0) \qquad\qquad (4.12)$$

for maximum photosynthetic efficiency, represented graphically by the initial slope of a P/I curve.

Jassby & Platt (1976) concluded that it was possible to select a best equation containing p^B_m and α_m for natural phytoplankton photosynthesis as a function of illumination. Lederman & Tett (1981) argued that it was not in fact possible to distinguish the goodness-of-fit of most two-parameter P/I models, and the widely-used equation of Smith (1936) and Talling (1957a) could thus remain in employment. The equation is most often written as:

$$p^B = p^B_m . \, I \, / \, (I_k^2 + I^2)^{0.5} \qquad\qquad (4.13)$$

and is illustrated in Figure 4.2B. I_k is the irradiance above which photosynthesis is noticeably saturated. It is not an independent third parameter of P/I curves, being defined by:

$$I_k = p^B_m \, / \, \alpha_m \qquad\qquad (4.14)$$

and equation (4.13) can in fact be written using any two of these three parameters. According to Harris (1978) most published values of I_k are between 50 and 120 µE m^{-2} s^{-1}.

The variable of most relevance to the estimation of compensation and critical depths is not the rate of photosynthesis but the efficiency with which PAR is converted to assimilated carbon at a range of photon flux densities. Defining this variable efficiency as

$$\alpha = p^B / I \qquad\qquad (4.15)$$

equation (4.13) can be rewritten:

$$\alpha = \alpha_m / (1 + I^2 / I_k^2)^{0.5} \tag{4.16}$$

Figure 4.8 shows the graph of this equation.

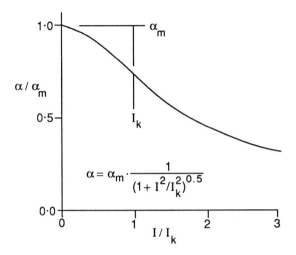

Figure 4.8. Photosynthetic efficiency and illumination. Relative photosynthetic efficiency α/α_m plotted against I/I_k, PAR photon flux density standardized to saturation illumination.

Finally, photosynthetic efficiency can be related to the main photosynthetic processes through:

$$\alpha = m_3 . k . \varepsilon . \phi / q \tag{4.17}$$

where k is a constant whose value depends only on the units employed for the other terms. It is convenient to make k unity, and thus eliminate it, by expressing α in nmol C (mg Chl)$^{-1}$ (μE m^{-2})$^{-1}$, and the other terms in corresponding units. The factor m_3, value 1·3, corrects for the upwelling and sideways-scattered light absorbed by algae but not normally included in measurements of I (see p. 62).

Absorption cross-section, ε, also known as the *in vivo* specific extinction coefficient for chlorophyll a, measures the ability of the photosynthetic pigment complex to absorb PAR photons and transfer their energy to active centres. Its units are m^2 (mg Chl)$^{-1}$.

Quantum yield, ϕ, defines the efficiency with which the energy of absorbed photons is used in the 'light reactions' of photosynthesis to split water molecules, releasing oxygen and providing reducing power. The units used here are nanomoles oxygen released per microEinstein absorbed. The maximum value allowed by photosynthetic theory is 125 nmol O$_2$ μE^{-1} (Radmer & Kok, 1977). This is denoted here by ϕ_m. Bannister (1974) suggested a normal maximum corresponding to about 80 nmol O$_2$ μE^{-1}.

Photosynthetic quotient, q, relates carbon fixation by the photosynthetic dark reactions to the supply of reducing power. Its units are nanomoles oxygen released per nanomole carbon fixed, and its value is normally greater than 1·0, due to the diversion of reducing power to nitrate assimilation. The maximum quantum yield for carbon is thus likely to be less than about 100 nmol C μE^{-1}; Bannister (1974) argued for about 60 nmol C μE^{-1}.

LINEARITY

The equations for critical depth require each photon arriving in the neighbourhood of an algal cell to have exactly the same chance of being absorbed by the pigments of that cell, and to make the same contribution to photosynthesis after absorption, as any other photon. This assumption is put in doubt by Figure 4.8, which shows photosynthetic efficiency decreasing with increasing photon flux density, especially when ambient illumination exceeds I_k.

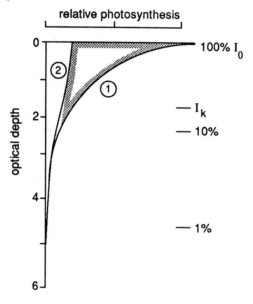

Figure 4.9. The problem of nonlinearity. Relative photosynthesis plotted against optical depth, with I_k/I'_o for May at 56°N. Curve (1) assuming a constant photosynthetic efficiency, curve (2) assuming saturation as in Figure 4.8. In both cases the decay of photon flux density with depth is that for polychromatic irradiance in Loch Striven in June 1988 (Figure 4.1).

Figure 4.9 illustrates the problem in regard to critical depth. Curve (1) assumes constant photosynthetic efficiency. Curve (2) shows saturation of photosynthesis near the sea surface, especially at illuminations above I_k. The numerical value of a photosynthesis-depth integral is proportional to the area 'under' each curve: that is, between it and the vertical axis in Figure 4.9. The extent of inaccuracy in Sverdrup's derivation of the critical depth is measured by the area of the shaded region as a proportion of the total area under curve (1). If this proportion is $(1-f)$, where f is the ratio of the area under curve (2) to that under curve (1), the equation for critical depth can be modified:

$$z_{crit} = (\alpha_m / r^B) . (f . m_2 . I'_o / \lambda(min)) \tag{4.18a}$$

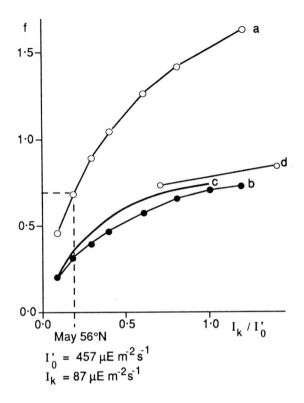

Figure 4.10. Corrections for nonlinearity. Correction factor, f, plotted against the ratio, I_k/I'_o, of algal saturation irradiance to 24-hour mean subsurface PAR. The surface irradiance data were for those recorded for a typical day in May 1984 at 56°N by the solarimeter at Dunstaffnage (as given in Tett et al., 1988), and converted as equation (4.1). I'_o was 457 μE m^{-2} s^{-1}. Submarine optical data were those for Loch Striven, June 1988, as used in Figure 4.1. Details of curves as follows: (a) Numerical integration over depth and time was used to compute a mean photosynthetic rate p$'^B$ from (i) depth and time dependent photon flux density synthesized for polychromatic light with wavelength dependent attenuation and (ii) equation (4.16) for light-dependent photosynthetic efficiency and an arbitrary value of α_m. f was then calculated from p$'^B$ divided by $\alpha_m.m_2.I'_o/\lambda(min)$. $\lambda(min)$ was 0.27 m^{-1}, m_2 was 0.34. (b) The same as (a) except that submarine photon flux density was calculated using a mean attenuation λ' of 0.34 m^{-1} and f was p$'^B$ divided by $\alpha_m.I'_o/\lambda'$. (c) As predicted by equation (4.19), derived from Talling (1957b), and using daytime irradiance $I^* = 687$ μE m^{-2} s^{-1}. (d) From Platt (1986), Figure 4.2, for relative error e, with maximum surface PAR of 1288 μE m^{-2} s^{-1}. f was the ratio $1/(1 + e)$.

Figure 4.10 shows values for f as a function of I_k/I'_o for a typical day in May at 56°N. Curves a, b and d result from the use of numerical integration to take into account diurnal variation in I_o as well as depth variation in photosynthesis predicted by equation (4.13). Curve d is based on Platt (1986). Curve c used the following equation, derived from Talling's (1957b) semi-empirical solution to the photosynthesis-depth integral:

$$f = (I_k / I'_o) . (0.9 / 1.33) . \ln(2 . I^*_o / I_k) \qquad (4.19)$$

In this equation, which is accurate only for $I^*_o > I_k$, I^*_o is the mean PAR just below the sea surface during the hours of daylight.

Curve (a) is distinguished from the others by substantially higher values of f. The explanation is that (a) allows for the effect of the additional (polychromatic, rapidly-attenuated) light that is available near the sea surface, and hence provides the appropriate value for use in equation (4.18a), where I'_o is already attenuated by m_2. Curves (b), (c) and (d) were based on simple exponential decay of PAR at a depth-invariant rate λ', and would be appropriate in an alternative form of equation (4.18):

$$z_{crit} = (\alpha_m / r^B) . (f . I'_o / \lambda') \qquad\qquad (4.18b)$$

An apparently less tractable problem concerns the extent to which algal adaptation to changing illumination invalidates calculations that assume constant photosynthetic parameters. Values of the parameters are usually thought to depend on previous illumination history (*e.g.* Steemann Nielsen & Jorgensen, 1968), but the extent of their variation must, however, depend on the ratio of the algal photoadaptive and layer-mixing time scales (Falkowski, 1983; Lewis *et al.*, 1982, 1984). Planktonic microalgae typically need several days to adapt to substantial changes in illumination (*e.g.* Steemann Nielsen & Park, 1964; Falkowski, 1980; Harding, 1988). The timescale of irradiance fluctuations experienced by an algal cell in a well-mixed layer is likely to be a few minutes to a few hours. Photoadaptation should thus not be a major source of error in computing photosynthesis-depth integrals for well-mixed layers.

Lewis *et al.* (1984) report a direct test of this hypothesis. Values of P^B measured near the top and near the bottom of a mixed layer differed significantly only when the layer-mixing timescale exceeded about 10 hours (Figure 4.11). Gallegos & Platt (1982) compared the effect of constant and varying light in a shipboard incubator and reported that simulated "vertical mixing has little quantitative effect on water column primary production". Mixing time was about an hour. In the absence of photoadaptation,

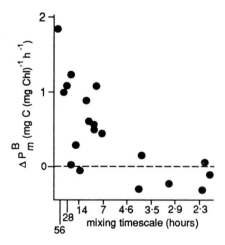

Figure 4.11. Photoadaptation and mixing timescale. Adapted from Lewis *et al.* (1984). $\Delta p^B{}_m$ is the difference between contemporaneous measurements of maximum photosynthetic rate at the top and bottom of a mixed layer during spring in Bedford Basin, Nova Scotia. Mixing timescale estimated from observations of turbulent energy dissipation rate.

photosynthesis-depth integrals calculated from incubations at fixed depths should not differ from those calculated using data from bottles moved vertically during incubation. Marra (1978) carried out an *in situ* experiment, with a period of several hours. He found that in "most experiments, the vertically cycled bottles have rates of integral photosynthesis 19 to 87% higher than estimates from the [fixed] series", implying some photoadaptation in the fixed bottles. More investigation is needed.

EFFICIENCY

The parameters of P/I equations such as (4.13) may be treated simply as descriptions of graphs of photosynthesis against irradiance. As equation (4.17) implies, however, it is also possible to interpret them, and especially α_m, in terms of the processes comprising photosynthesis. For example Platt & Jassby (1976) estimated their mean α_m of 0·21 mg C (mg Chl)$^{-1}$ h^{-1} (W m^{-2})$^{-1}$ to be 44% of the theoretical maximum, given an upper limit to quantum yield of 0·10 mol C E^{-1} and a photosynthetic pigment absorption cross-section of 0·016 m^2 (mg Chl)$^{-1}$. Finally, photosynthetic parameters in algal growth models have sometimes been synthesized from descriptions of basic properties such as quantum yield (*e.g.* Bannister, 1974).

Such shifts of perspective may lead to erroneous conclusions unless the nature and extent of errors in parameter estimates are properly appreciated (Lederman & Tett, 1981). Nevertheless, analysis and synthesis are unavoidable, since it is seldom possible to estimate quantum yield directly for natural phytoplankton. Two papers, one dealing with *in situ* incubation of oceanic phytoplankton, the other with pure algal culture, will be considered in order to illustrate methods and problems of relating photosynthetic efficiency, absorption cross-section and quantum yield.

Tyler (1975) used the results of ^{14}C *in situ* incubations by the SCOR 'Discoverer' expedition in May 1970 to estimate photosynthetic quantum yield as a function of depth in the eastern equatorial Pacific. Neglecting minor constants, and the effect of the photosynthetic quotient, quantum yield is the ratio of carbon fixed to photons absorbed by chlorophyll and accessory pigments:

$$\phi = p/I_a \tag{4.20}$$

where p is in nmol C m^{-3} s^{-1}. The problem is to estimate absorbed photon flux concentration, I_a µE m^{-3} s^{-1}. Tyler's solution was to partition the diffuse attenuation coefficient into components due to water and pigments, and hence to estimate the share of chlorophyll in removing downwelling photons. He regressed attenuation coefficients on corresponding chlorophyll concentrations, using data from a number of stations and depth increments to give a slope of 0.042 m^{-2} (mg Chl)$^{-1}$. This value was taken as an estimate of the absorption cross-section, ε, of the mixed phytoplankton populations encountered during the 'Discoverer' expedition. Tyler was then able to compute absorbed photon flux concentration from:

$$I_a = (\varepsilon \cdot X / \lambda) \cdot (\Delta I / \Delta z) \tag{4.21}$$

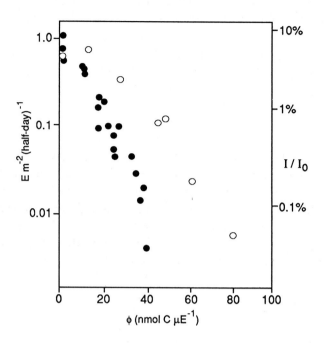

Figure 4.12. Quantum yield in the sea. Photosynthetic quantum yield, ϕ, as a function of incident quanta received during an incubation from dawn to midday, and depth shown as a percentage of surface PAR. Adapted from Tyler (1975).

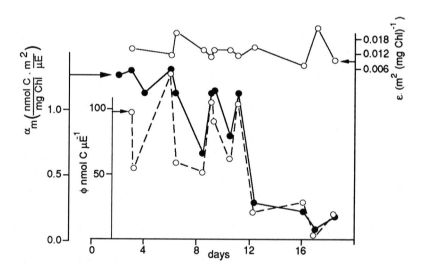

Figure 4.13. Efficiency changes during a culture. Changes in maximum photosynthetic efficiency, α_m, quantum yield, ϕ, and absorption cross-section, ε, during a batch culture of the diatom *Thalassiosira pseudonana* grown by Welschmeyer & Lorenzen (1981). The culture seems likely to have become nutrient limited by about day 12.

where ΔI was the reduction in flux density over depth interval Δz. The chlorophyll concentration in this interval was X mg m^{-3} and the overall diffuse attenuation, to which phytoplankton chlorophyll contributed $\varepsilon.X$, was λ m^{-1}.

Tyler's results (Figure 4.12) show quantum yield increasing with depth to a maximum of 40-80 nmol C μE^{-1} at ambient photon flux densities less than about 1 μE m^{-2} s^{-1}. Other authors who have concluded that deep-water photosynthetic quantum yield approaches the theoretical maximum include Morel (1978), whose greatest value was 125 nmol C μE^{-1}, and Tilzer (1984), who calculated a maximum yield of 92 nmol C μE^{-1} for freshwater phytoplankton. Bannister & Weidemann (1984) argued however that values of in situ quantum yield greater than 100 nmol C μE^{-1} "are almost certainly due to imprecision or systematic error" and that true values are in the range 30-70 nmol C μE^{-1}.

Welschmeyer & Lorenzen (1981) report experiments with bacteria-free cultures of several marine algae, one of which was the small diatom *Thalassiosira pseudonana*. Their time series for a batch culture is shown in Figure 4.13. Absorption cross-section remained effectively constant at about 0·013 m^{-2} (mg Chl)$^{-1}$. Photosynthetic efficiency commenced high, as did quantum yield at a mean value of 84 nmol C μE^{-1} prior to day 12, after which it is likely that the diatoms became nutrient depleted. After day 12 photosynthetic efficiency and quantum yield were much less, ϕ having a mean of only 19 nmol C μE^{-1}.

The reduction in quantum yield at increasing illuminations in Tyler's results probably resulted from increasing light saturation of photosynthesis. The decrease in yield and efficiency with increasing age of Welschmeyer & Lorenzen's culture cannot be thus explained, however, since increasing self-shading with increasing biomass should reduce available light. The suggestion of Welschmeyer & Lorenzen (1981) that nutrient depletion may cause reduced quantum yield is supported by the results of experiments carried out by Droop *et al.* (1982) with the prymnesiophyte flagellate *Pavlova lutheri*.

Values of absorption cross-section for a number of algal cultures grown under white light are listed in Table 4.1; they range from 0·004 to 0·017 m^2 (mg Chl)$^{-1}$. This range does not, however, include Tyler's value of 0·042 m^2 (mg Chl)$^{-1}$. It is not unusually high for estimates obtained by regressing attenuation coefficient on chlorophyll concentration (*e.g.* Smith & Baker, 1978a). Morel (1988) provides an explanation. First, attenuation measurements include photons lost by back-scattering from the downwards beam, as well as by absorption. The ratio of back-scattering to absorption is, however, small for most algae (Bricaud *et al.*, 1983). Second, the concentration of detrital material, which scatters and absorbs light, is often significantly correlated with the concentration of living phytoplankton. The attenuation ascribed to chlorophyll by the regression method is thus often substantially greater than absorption by active photosynthetic pigments. The enhancement may be most marked under oligotrophic conditions, when the ratio of detritus to chlorophyll tends to be high.

Figure 4.14 is a mean absorption spectrum for 14 species of eukaryote microalgae (Morel, 1988). It suggests that the absorption cross-section must depend on the wavelength distribution of the light to which an alga is exposed. In considering the efficiency of photosynthesis in relation to compensation and critical depths, it thus seems appropriate to choose values of absorption cross-section higher than many of those given in Table 4.1 or by Atlas & Bannister (1980). The most penetrating photons in oceanic waters are those with wavelengths between 450 and 500 nm. In clear coastal waters peak transmission is found between 500 and 525 nm (Jerlov, 1976). For these water types, deep-water absorption cross-sections of, respectively, 0·030 and 0·025 m² (mg Chl)⁻¹ seem appropriate. The shift of the wavelength of peak transmission towards 600 nm and a worse match with the absorption spectra of most algae, suggests cross-sections perhaps as low as 0·010 m² (mg Chl)⁻¹ in turbid coastal water. An appropriate value for algae in Loch Striven, which had peak transmission at 552 nm in June 1988, would seem to be 0·015 m² (mg Chl)⁻¹.

Finally, nutrient-replete marine algae are likely to be so because they are supplied with adequate nitrate-nitrogen by vertical mixing, rather than with ammonium-nitrogen by zooplankton excretion. Conventional stoichiometric calculations assuming a molar C:N ratio of 6:1 indicate a photosynthetic quotient of about 1·4 under these conditions. This value will be used henceforth, notwithstanding the case for higher values presented by Williams *et al.* (1979).

The evidence from studies in culture that nutrient-replete, light-limited marine algae photosynthesize with a quantum yield approaching the theoretical maximum seems convincing. Accepting the maximum as 100 nmol O_2 µE⁻¹ indicates a yield in terms of carbon of 70 nmol C µE⁻¹, close to the value of 60 nmol C µE⁻¹ adopted by Bannister (1974) and the maximum of about 80 nmol C µE⁻¹ observed by Tyler (1975) and Welschmeyer & Lorenzen (1981), and at the upper limit of Bannister & Weidemann's (1984) range of 30-70 nmol C µE⁻¹. An upper limit to photosynthetic efficiency α_m is thus 3·9 nmol O_2 (mg Chl)⁻¹ (µE m⁻²)⁻¹, or 2·7 nmol C (mg Chl)⁻¹ (µE m⁻²)⁻¹. In coastal waters maximum photosynthetic efficiency at the compensation depth seems likely to range from 0·9-2·3

Table 4.1. *Absorption cross-sections for algal cultures*

species	cell volume µm³	absorption cross-section m² (mg Chl)⁻¹	reference
BACILLARIOPHYCEAE			
Ditylum brightwelli	39000	0.0059	Welschmeyer & Lorenzen, 1981
Phaeodactylum tricornutum	150	0.0041	Geider et al., 1986
Skeletonema costatum	170	0.0098	Welschmeyer & Lorenzen, 1981
Thalassiosira fluviatilis	1400	0.0171	Laws & Bannister, 1980
Thalassiosira pseudonana	25	0.0129	Welschmeyer & Lorenzen, 1981
mean = 0.010 (s.d. 0.005, n=5)			
DINOPHYCEAE AND PRYMNESIOPHYCEAE			
Amphidinium sp.	1400	0.0069	Welschmeyer & Lorenzen, 1981
Gonyaulax polyedra	14000	0.0043	Welschmeyer & Lorenzen, 1981
Isochrysis sp.	25	0.0097	Welschmeyer & Lorenzen, 1981
Pavlova lutheri	60	0.0120	Droop et al., 1982
mean = 0.008 (s.d. 0.003, n=4)			
CHLOROPHYCEAE			
Nannochloris atomus	90	0.0045	Geider & Osborne, 1986
Dunaliella tertiolecta	64	0.0052	Welschmeyer & Lorenzen, 1981
mean = 0.005 (s.d. 0.001, n=2)			

nmol C (mg Chl)$^{-1}$ (μE m^{-2})$^{-1}$, the lowest values relating to the most turbid waters. An appropriate value for Loch Striven would be 1·9 nmol O$_2$ μE^{-1}. For comparison, the mean value obtained by Platt & Jassby (1976) in St Margaret's Bay, Nova Scotia, corresponds to 1·2 nmol C (mg Chl)$^{-1}$ (μE m^{-2})$^{-1}$, with a maximum of 3·5 nmol C (mg Chl)$^{-1}$ (μE m^{-2})$^{-1}$.

It may be possible further to refine these values of α_m. Thus, the algal culture data in Table 4.1 suggest that, although there is considerable variation within taxonomic groups, light absorption by brown algae is better than by green algae (*cf.* Bricaud *et al.*, 1988; Falkowski & Owens, 1980).

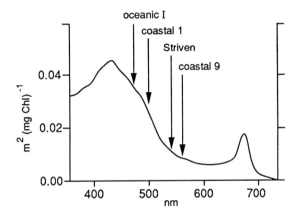

Figure 4.14. Absorption cross-section and water type. Mean *in vivo* absorption cross-section spectrum for 14 species of eukaryote algae in culture, after Morel (1988).The arrows indicate the wavelengths of maximum transparency of various types of oceanic and coastal sea water, according to Jerlov (1976) except for Loch Striven, which is based on the measurements described in Figure 4.1.

CONSUMPTION

Compensation and critical depths depend on the reciprocal of the compensation illumination, I_c, or the ratio α_m / r^B. Although algal respiration rate is usually held to depend on temperature (*e.g.* Ryther & Guillard, 1962) and inversely on cell size (Banse 1976), the most convincing relationship is that between respiration and growth rate (*e.g.* Droop *et al.*, 1982; Laws & Bannister, 1980). A general argument may be made: irrespective of exact taxonomic composition and water temperature, nutrient-replete phytoplankton near the compensation point will, as a result of genotypic selection or physiological adaptation, respire at the lowest possible rate. The data in Table 4.2 suggest that this basal rate is about 1·3 nmol C (mg Chl) s^{-1} for diatoms and 3-12 nmol C (mg Chl) s^{-1} for flagellated algae. The cause of this difference is unclear (Raven & Beardall, 1981).

A minimum value of compensation illumination may thus be computed for diatoms in clear ocean water. It is

$$I_c = r^B / \alpha_m = 1\cdot3 / 2\cdot7 = 0\cdot5 \ \mu\text{E m}^{-2} \text{ s}^{-1} \tag{4.22}$$

when both photosynthesis and respiration measurements relate to carbon flux. Because the respiratory quotient is generally about 1·0 mol O$_2$ (mol C)$^{-1}$, the minimum compensation illumination is a little less, about $0.3 \mu E m^{-2} s^{-1}$, for oxygen-based measurements. In coastal waters such as those of Loch Striven the minimum compensation illumination for diatoms should be about $0.7 \mu E m^{-2} s^{-1}$ (oxygen data), close to the approximate value of $1 \mu E m^{-2} s^{-1}$ calculated from Marshall & Orr (1928). For flagellated algae the minimum compensation depth would range from about $2 \mu E m^{-2} s^{-1}$ in clear coastal seas to about $6 \mu E m^{-2} s^{-1}$ in turbid waters.

These conclusions predict compensation illuminations for oceanic phytoplankton closer to 0.1% than 1% of typical growth season 24-hour mean subsurface PAR of $400 \mu E m^{-2} s^{-1}$, and thus help to explain the occurrence of midwater chlorophyll maxima extending below 1% of surface illuminations (Venrick *et al.*, 1973; Fasham *et al.*, 1985; Geider *et al.*, 1986; Figure 4.15).

On the other hand, Jenkin's (1937) compensation illumination of $6.3 \mu E m^{-2} s^{-1}$ is substantially greater than the theoretical, oxygen-based, value of $0.3 \mu E m^{-2} s^{-1}$ for diatoms in clear coastal water, and corresponds to 1.6% of growth-season subsurface

Table 4.2. *Basal respiration rates for light-limited cultures*

name	pg C cell^{-1}	temp., °C	$\mu E m^{-2} s^{-1}$ (1)	μ, d^{-1} (2)	r^B ($\mu=0$), nmol (mg Chl)$^{-1}$ s^{-1} (2)	(3)	(4)	reference
DIATOMS								
Leptocylindrus danicus	87	5-20	8	0.2	1.7	O	a	Verity, 1981, 1982*a, b*
				0.0*	1.3*			
Phaeodactylum tricornutum	17	23	0.8	0.00	1.4	O	a	Geider *et al.*, 1986
Skeletonema costatum	20	15	0.7	0.19	1.8	O	a	Falkowski & Owens,
				0.0*	1.6*			1980
Thalassiosira fluviatilis	88	20	4.4	0.054	1.1	C	b	Laws & Bannister,
				0.0†	0.8†			1980
mean r^B ($\mu = 0$) = 1.3 (s.d. 0.3, n=4)								
FLAGELLATES								
Dunaliella tertiolecta	39	15	8	0.00	3.0	O	a	Falkowski & Owens,
								1980
Gonyaulax polyedra	4500	21	18	0.05	15.2	O	d	Prezelin & Sweeney,
			11*	0.0*	11.8*			1978
Pavlova lutheri	11	20	9			O	c	Droop *et al.*, 1982
				0.0†	3.7†	O	c	
mean r^B ($\mu = 0$) = 6.2 (s.d. 4.9, n=3)								
COCCOID GREEN								
Nannochloris atomus	16	23	1	0.00	1.4	O	a	Geider & Osborne, 1986

Notes: (1) Minimum illumination at which observations made, not corrected for self-shading. (2) Where no additional comment, respiration rates measured by authors at irradiance for zero growth in batch culture. † extrapolated to zero growth by authors; * my extrapolation to zero growth. (3) Respiration measured by reduction in either dissolved oxygen concentration (O) or ^{14}C label of algal particulates. (4) Light-limitation of potentially nitrogen-limited batch (a), or continuous (b), cultures confirmed by published algal N:C equalling or exceeding Redfield ratio. Light-limitation of potentially B12-limited continuous cultures (c) affirmed by Droop *et al.* (1982) on the basis of known *Pavlova* B12 kinetics. Light-limitation of low-light batch culture (d) weakly affirmed on the basis of nutrient sufficiency in medium.

PAR. Nevertheless, it seems to be a correct value for I_c in critical depth equations predicting the start of the spring phytoplankton increase.

It is thus necessary to consider losses of organic production which might augment algal respiration and so increase the effective compensation illumination. The effect of these additional losses was examined by Tett *et al.* (1988) in a study of microplankton dynamics in a quarry flooded by the sea on the Scottish island of Easdale.

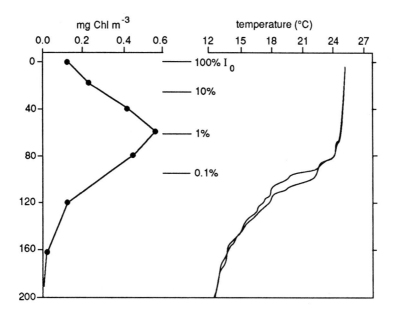

Figure 4.15. A deep chlorophyll maximum. Vertical distribution of chlorophyll concentration in relation to temperature layering and relative submarine illumination in the western Indian Ocean, 9°S 46°E, 2-3 July 1987. Relative photon flux densities based on measurements of downwards irradiance at five wavelengths (see legend to Figure 4.1). Temperature and chlorophyll profile made about 3 hours after sunset: note the midwater chlorophyll maximum at and below 1% of I_o and located in the diurnal thermocline.

The method recommended by Lederman & Tett (1981) was used to fit equation (4.13) to ^{14}C assimilation rates measured in a light-gradient incubator and assumed to estimate gross photosynthesis. The mean value of I_k was 87 (s.d. 33, n=6) μE m^{-2} s^{-1}, coincidentally similar to the value of Jenkin (1937) and in the middle of Harris's (1978) normal range of 50-120 μE m^{-2} s^{-1}. The mean value of α_m was 0·84 (s.d. 17, n=4) nmol C (mg Chl)$^{-1}$ (μE m^{-2})$^{-1}$, a little less than the value of 1·1 nmol C (mg Chl)$^{-1}$ (μE m^{-2})$^{-1}$ calculable from the theory of the previous section assuming that the spectral attenuation properties of water in the Quarry resembled those in Loch Striven.

The rate of utilization of oxygen was measured in dark bottles incubated *in situ* for 24 hours. Normalized to bottle chlorophyll and converted to carbon at a respiratory quotient of 1·0 nmol C (nmol O$_2$)$^{-1}$, the respiration rate was 12 (s.d. 3, n=2) nmol C (mg

Chl)$^{-1}$ s^{-1}. I estimate the algal respiration rate as 4·5 nmol C (mg Chl)$^{-1}$ s^{-1} for a phytoplankton of diatoms, dinoflagellates and flagellates, growing at about 0·3 d^{-1}; the rate, equivalent to 9·7 nmol C (mg Chl)$^{-1}$ s^{-1}, assumed by Tett et al. (1988) seems in retrospect too high. Although the incubated water was screened to exclude zooplankton above 125 μm, it contained substantial numbers of heterotrophic microorganisms as well as photoautotrophs, and these must have been responsible for the rest of the observed respiration. The vertical distributions of photosynthesis and microplankton respiration are shown in Figure 4.16B, for 21 May (a sunny day) and 22 May (a cloudy day). Gross photosynthesis p was estimated by multiplying the chlorophyll concentration at each depth by chlorophyll-related photosynthetic rate p^B, derived from the irradiance profile and equation (4.13). A photosynthetic quotient of 1·33 (growth on mixed nitrate and ammonium) was assumed for the calculation of oxygen production, and a respiratory quotient of 1·0 for oxygen consumption. The latter was estimated by multiplying observed ATP concentrations (used as a measure of microplankton biomass) by a mean ATP-normalised respiration rate.

The physical structure of the Quarry water column, as deduced from temperature profiles (Figure 4.16A), was unusual for Scottish coastal waters at this time, in that stirring was weak at depths greater than 8 m. Physical turnover times for layers of thickness 4 m (or about 1 optical depth given mean PAR attenuation of 0·26 m^{-1}) were estimated as a few days at 8-12 m, increasing to several weeks at 22-26 m. It was thus possible to analyse the growth potential of the phytoplankton or microplankton communities using compensation, rather than critical, depth equations. Several compensation depths are shown in Figure 4.16B. Where the profiles of gross photosynthesis, p, for 21 and 22 May cross that for phytoplankton respiration $r(Chl)$ are the algal compensation depths, with a mean at 18 m, or about 1·5% of subsurface PAR. Where the profiles for photosynthesis and microplankton respiration $r(ATP)$ cross are the community compensation depths, with mean at 12 m, or about 6% of subsurface PAR.

The intervening layer, occupying about a third of the zone above the algal compensation depth, is of considerable interest. Bottle measurements suggested that it should not sustain a viable phytoplankton. Free-water observations allow this conclusion to be qualified. Stronger stirring by spring tides extended to about 18 m every fortnight, and this can be taken as the depth of the surface mixed layer when the timescale is that of the summer period of four months (see Figure 4.16A). The appropriate critical depth equation is:

$$z_{crit} = (f \, / \, I_c) \, . \, (I_o \, / \, \lambda') \qquad\qquad (4.23)$$

since irradiance was measured with a broad-band PAR sensor. Community compensation illumination is given by:

$$I_c = (r^B + l^B) \, / \, \alpha_m \qquad\qquad (4.24)$$

where l^B refers to microheterotroph respiration normalized to autotroph chlorophyll. In fact, r^B and l^B were not measured separately, but the joint loss was known from free-water changes in oxygen concentration as well as bottle measurements. It averaged 14·7 nmol C (mg Chl)$^{-1}$ s^{-1} between May and August.

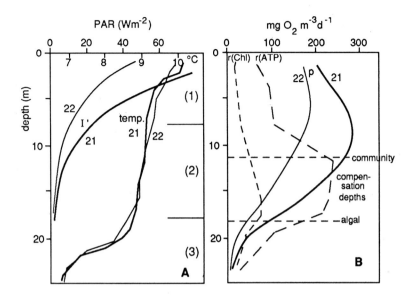

Figure 4.16. Community and algal compensation depths. Profiles in Easdale Quarry, 21-22 May 1984 (Tett *et al.*, 1988). 21 May was a day of bright sunshine; 22 May was overcast; these differences in I'_o are evident in the profiles of 24-hour mean irradiance, I', and of photosynthesis, p. (A) Irradiance and temperature profiles. Layer (1) was that thoroughly mixed by wind, night-time convection, and neap tidal stirring during each 24 hours. Layer (2) was part of the thermocline on 21-22 May, but was subsequently incorporated into the mixed layer by spring tidal stirring. Layer (3), the seasonal thermocline, became isolated in early May and thereafter exchanged only slowly with the overlying water. (B) Respiratory and photosynthetic profiles. Respiration was calculated either from chlorophyll concentration, r(Chl)) assuming 4·5 nmol O_2 (mg Chl)$^{-1}$ s^{-1} or from ATP concentration, r(ATP), assuming 50 nmol O_2 (mg ATP)$^{-1}$ s^{-1}. The profiles thus indicate the relative vertical distribution of phytoplankton autotrophic and microplankton heterotrophic biomasses. Biomass-related gross photosynthesis was computed for each hour of the day for observed diel changes in surface irradiance, summed over all hours, and multiplied by chlorophyll concentration to obtain the photosynthesis profile. The algal compensation depth is where the algal respiration profile intercepts the average photosynthesis profile for 21 and 22 May; the community compensation depth is where the microplankton respiration profile intercepts the average photosynthesis profile.

Mean attenuation during the summer was 0·25 m^{-1} and mean subsurface PAR was 364 µE m^{-2} s^{-1}. From equation (4.19), with daytime PAR of 546 µE m^{-2} s^{-1} and I_k of 87 µE m^{-2} s^{-1}, f was 0·41. For α_m of 0·84 nmol C (mg Chl)$^{-1}$ (µE m^{-2})$^{-1}$, the compensation illumination was 18 µE m^{-2} s^{-1}. The critical depth was thus 34 m, greater than the observed mixed-layer depth of 18 m. Thus theory and observation agree; the surface-mixed layer was able to support an active microplankton during summer despite losses additional to algal and microheterotroph respiration. The main additional loss was sedimentation, estimated to remove about a quarter of gross primary production each day. This and minor losses were not taken into account in the calculations above. Since the mean chlorophyll concentration of the Quarry mixed layer did not change significantly during the summer, including all losses in I^B should predict a critical depth equal to the mixed-layer depth. In fact the prediction was 24 m, compared to an observed mixed-layer depth of 18 m.

THE EUPHOTIC ZONE

Most texts define the euphotic zone as the region in which there is sufficient light for photosynthesis; in particular, the zone above the compensation depth at 1% of subsurface PAR. A better definition would be *all layers in which photoautrophic production exceeds heterotrophic consumption on the time-scale under investigation*. The condition satisfied, in the euphotic zone thus defined, is that

$$f . I' > I_c \qquad (4.25)$$

where compensation illumination is given by equation (4.24). In the inequality (4.25), layer mean photon flux density is calculable from equation (4.9). For an optically deep surface mixed layer, equation (4.9) can be replaced by equation (4.10). Such simplification gives an equation for the critical depth:

$$z_{crit} = f . m_2 . I'_o / (I_c . \lambda(min)) \qquad (4.26)$$

Alternatively, for an optically thin subsurface layer, equation (4.9) simplifies to:

$$I' = m_2 . I'_o . e^{-\lambda(min).z} \qquad (4.27)$$

as in the equation for compensation depth:

$$z_{comp} = \ln(m_2 . I'_o / I_c) . (1 / \lambda(min)) \qquad (4.28)$$

The 'timescale under investigation' is an important part of the definition. When the time of averaging was a day, the water column in Easdale Quarry during neap tides in May was only weakly mixed below 8 m, and the lower limit to the validity of inequality (4.25) was set by the community compensation depth at 10% of subsurface PAR on a cloudy day and 4% of subsurface PAR on a sunny day. When the time of averaging was the period from May through August, the euphotic zone was co-extensive with the mixed layer extending to 1·5% of subsurface PAR.

The loss terms implicit in (4.25) are those due to the respiration of autotrophs and heterotrophs. They do not include export (or import) of organic material by sinking, horizontal exchange, or food-web transfer. If all primary production were eventually consumed at the depth at which it was formed, as in the hypothetical case of a bottle incubation lasting one year, the long-term euphotic zone depth would always be zero, since inequality (4.25) could never be satisfied. The import of deep-water nitrate into the euphotic zone, and subsequent export of 'new production' (Dugdale & Goering, 1967; Eppley & Petersen, 1979) favours autotrophic at the expense of heterotrophic processes in the superficial waters of the ocean, however, and it is thus meaningful to consider the depth of the euphotic zone on an annual or semiannual timescale. Since much of the ocean exhibits persistent seasonal stratification, and midwater chlorophyll maxima such as that shown in Figure 4.15, it is appropriate to use compensation depth arguments in such considerations. A value of 400 μE m^{-2} s^{-1} may be adopted for 24-hour mean sea surface PAR under these conditions, and a value of 0·05 m^{-1} for $\lambda(min)$ in 'average' ocean water (intermediate between types IB and II of Jerlov, 1976). The compensation depth is then given by:

$$z_{comp} = 100 - 20 . \ln(I_c) \quad \text{metres} \qquad (4.29)$$

when I_c is in μE m^{-2} s^{-1}. According to the theory developed in previous sections, the lowest possible value of the compensation illumination is 0·5 μE m^{-2} s^{-1} for diatoms, corresponding to a maximum compensation depth of 114 m or about 0·13% of subsurface PAR. Assuming a phytoplankton without diatoms and with microheterotroph respiration equal to that of the photoautrophs suggests a compensation illumination of about 4 μE m^{-2} s^{-1} and a compensation depth of 72 m or about 1% of subsurface PAR. Such depths are compatible with many descriptions of midwater chlorophyll maxima, including the profile in Figure 4.15.

A little speculation is now in order. It is based on current theories of the part played by photoautotrophs in the origin of the present terrestrial atmosphere with its thermodynamically unlikely oxygen content (Berkner & Marshall, 1965; Ruttner, 1971; Lovelock, 1987; see also Canuto et al., 1982) and recent speculation about marine biological controls on atmospheric carbon dioxide content and hence climate (McElroy, 1983; Peterson & Melillo, 1985).

The early Earth lacked oxygen in its atmosphere, and hence substantial fluxes of 240-280 nm photons must have reached the sea surface and penetrated up to 10 m with potentially lethal consequences for near-surface organisms. It may thus have been essential for the earliest phytoplankton to photosynthesize efficiently, so that they could grow in the thermocline without the risk of exposure to damaging ultra-violet radiation resulting from turbulent transport towards the surface.

The subsequent development of an oxygen-containing atmosphere, as a result of the activities of these earliest plants, led to a reduction in sea-surface ultra-violet photon flux densities. Phytoplankton could thus colonize the mixed layer. Under oxygenic conditions, however, combined nitrogen is in relatively short supply in the superficial oceans. High photosynthetic efficiency thus remains of use to phytoplankton in stratified oceans, enabling them to grow in the upper thermocline in proximity to the nitrocline.

Consider also the likely history of oceanic plants on a world with stronger winds than present-day Earth. Under steady-state conditions, and neglecting convective mixing, the depth of the wind-mixed layer is given approximately by:

$$h = k_1 . w^3 / E'_o \qquad\qquad (4.30)$$

where w is mean wind speed, E'_o is sea-surface radiant heat flux density, and k_1 is a constant of proportionality. If mean winds were twice those of recent Earth, but heat flux was the same, the mixed layer would be eight times as deep, extending perhaps to depths of several hundred metres even at low latitudes. If exposure to solar ultra-violet remained a danger in the wind-mixed layer, the shallowest zone not mixed to the surface would be deeper than the compensation depth unless the protophytoplankton on this hypothetical world acquired a photosynthetic mechanism twice as efficient as that of our own plants. This may not be possible. Thus, unless these photoautotrophs could resist the destructive effects of ultra-violet light, these hypothetical oceans would remain devoid of organic photosynthesis and hence the biosphere would continue anoxic.

Related arguments may be applied to the dynamics of the spring phytoplankton bloom in temperate latitudes on our own world. According to critical depth theory, the

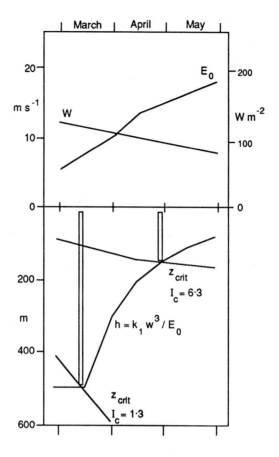

Figure 4.17. Nutrient capture at the start of the spring increase. Predicted mixed layer depth at 59°N 11°W is superimposed on critical depth predicted from equation (4.26) with $\lambda(min)$ of $0{\cdot}10$ m^{-1}, m_2 of $0{\cdot}34$, I'_o derived from E_o using equation (4.1), and f from Figure 4.10 curve (a) given I_k of 87 µE m^{-2} s^{-1}. With a compensation illumination of $6{\cdot}3$ µE m^{-2} s^{-1}, the predicted spring increase begins in late April and traps nutrients from a water column about 150 m deep, whereas a compensation illumination of $1{\cdot}3$ µE m^{-2} s^{-1} predicts algal growth starting in March and potentially trapping nutrients from a 500 m water column. The equation used to predict mixed layer depth h was derived from the Monin-Obukov length $-u_*^3/k.B$, where u_* is a friction velocity, B a buoyancy flux and k Von Karman's constant (Turner, 1973). In the present case u_* was taken to be related to (cube root mean) wind speed, w (from Clarke, 1986, for 59°N 11°W) and B to solar energy flux density, E_o (for 59°N, source as for Figure 4.7). Mixed layer depth was predicted taking k_1 as 25 kg s m^{-2} and a limit of 500 m (Levitus, 1982).

bloom commences when layer mean illumination first exceeds the compensation illumination for the wind-mixed layer. That is, when the following inequality is first satisfied in a particular year:

$$(f.\alpha_m / r^B) . (k_2 . I'^2_o / \lambda . w^3) > 1 \tag{4.31}$$

where k_1/E'_o is replaced by $1/(k_2.I'_o)$, and algal respiration is assumed to be the only loss term. Notice the strongly non-linear term I'^2_o/w^3. Its magnitude must increase rapidly as spring progresses, notwithstanding slower changes in sea-surface photon flux or wind speed. This ensures a speedy transition from a water column unable to support phyto-plankton growth to one in which conditions are increasingly favourable. The inequality thus explains the often abrupt start of the spring bloom.

The greater the ratio α_m/r^B (the less the compensation illumination), the earlier in the year will the bloom start and the deeper will be the mixed layer at this time. This has geochemical implications. Planktonic algae are capable of luxury uptake (Droop, 1983) of nutrients. Hence an early bloom will tend to accumulate more combined nitrogen than one starting late, simply because the early bloom can exploit a deeper layer of nitrate at winter concentrations (Figure 4.17). A rapidly-growing bloom is more likely to outpace potential grazers, with the consequence that much of the bloom production will subse-quently sink into deep water. In essence, the earlier a bloom starts, the more nitrogen it will acquire and the greater the amount of new production, and hence carbon, exported from the photic zone by sinking. Previous sections have suggested that diatoms have a lower compensation illumination that other algal groups. They certainly contribute substantially to the organic material reaching the bed of the North Atlantic a few weeks after the spring increase (Lampitt, 1985). They may thus be responsible for more profound reductions of photic zone concentrations of dissolved inorganic carbon than other types of algae. Diatoms are abundant only in the last 100 million years of the fossil record (Lipps, 1970; Loeblich, 1974; Parsons, 1979). Is it plausible that their success contributed to a reduction in atmospheric carbon dioxide concentrations and hence to the global cooling of the Tertiary?

CONCLUSION

Much of the previous section is speculation. The intent was to show that the 2%, at most, of sea-surface photons absorbed by phytoplankton (Morel, 1988) may well be as important an input in relation to life on Earth as the 98% that directly controls ocean heating and circulation. The depth of the euphotic zone may be one of the most important determinants of the way in which phytoplankton carry out this planetary role, and it would seem more than a coincidence that the length-scales set by the compensation and critical depths are often comparable with the depth of the mixed layer. One possible conclusion is that the phytoplankton, photic zone and mixed layer of the Earth's oceans have evolved together.

I am grateful to Professor Peter J. leB. Williams for a public reading of the first draft of this review.

Chapter 5

LIGHT HARVESTING AND PIGMENT COMPOSITION IN MARINE PHYTOPLANKTON AND MACROALGAE

M.J. DRING

Department of Biology, Queen's University, Belfast, BT7 1NN

Several new pigments have been identified in the chloroplasts of marine plants during the last decade, including chlorophyll c_3 and two derivatives of fucoxanthin. Most of these pigments, together with more familiar ones (*e.g.* violaxanthin, siphonaxanthin, prasinoxanthin, antheraxanthin) have also been shown to transfer absorbed energy to chlorophyll *a*, so that they can be added to the list of accessory pigments for photosynthesis. The discovery of these pigments has meant that some details of the diagnostic pigment complements for certain algal groups have had to be revised. However, the new techniques which revealed these pigments have also been applied to natural water samples, and have enabled the taxonomic composition of phytoplankton populations to be estimated without direct microscopic examination. The characterisation of light-harvesting pigment-protein complexes from many groups of marine algae has produced additional data of chemosystematic value, and the protein components of these complexes may reveal more about the phylogeny of the algae than the pigments themselves. The effects of growth irradiance and light quality on the light-harvesting apparatus of marine algae is reviewed, and the role of these two factors in their ecology is discussed. The recent claims that chromatic adaptation is important in marine phytoplankton are poorly supported by the ecological evidence, and it is concluded that irradiance is of greater significance than light quality in controlling the depth distribution of both phytoplankton and benthic algae in the sea.

INTRODUCTION

Our understanding of light-harvesting in the sea has made remarkable progress in the 1980s. For workers at sea, this has been the decade in which picoplankton were discovered and their importance was established, and in which the potential of pigments as markers for specific algal groups in the plankton began to be appreciated. In the laboratory, it has been the decade in which pigment-protein complexes have become more relevant to marine botanists, as biochemists have steadily worked through the algae, group by group. The more powerful analytical techniques of HPLC and HPTLC have also been applied to the pigments of algae, either in culture or in natural communities, and have revealed new pigments (albeit, mostly minor variants of the old, familiar ones) which are widespread among marine phytoplankton and which have forced a revision of some chemosystematic ideas. Consequently, although several major reviews

Herring, P.J., Campbell, A.K., Whitfield, M. & Maddock, L. (ed.). *Light and Life in the Sea.*
© 1990. Cambridge University Press.

of marine or algal photosynthesis were written in the early to mid-1980s (Kirk, 1983; Larkum & Barrett, 1983; Prezelin & Boczar, 1986), much of the fine detail already requires revision. This account concentrates on the major developments of the current decade, attempting to bring these reviews up to date, without repeating the well established detail covered by them.

BIOCHEMISTRY OF LIGHT-HARVESTING

Pigments involved in light-harvesting

Not all of the pigments found in plants, or even in chloroplasts, are involved in light-harvesting for photosynthesis. Light-harvesting pigments are those which can contribute the energy that they absorb to the photosynthetic process. Chlorophyll a is the only pigment that can do this directly in oxygen-evolving photosynthetic organisms; all others - the 'accessory pigments' - transfer their absorbed energy to chlorophyll a, which in turn transfers it to the reaction centre of one of the two photosystems (PS I or PS II). In order to establish that a pigment is involved in photosynthesis, therefore, it is necessary to demonstrate either that light absorbed by the pigment results in photosynthesis (measured by oxygen-evolution or carbon-fixation) or, at least, that energy transfer occurs from the pigment to chlorophyll a. The latter process is best shown by measuring the intensity of fluorescence from chlorophyll a (at 682 nm) induced by wavelengths which are maximally absorbed by the pigment. Two types of action spectra may thus be presented as evidence that a pigment acts as an accessory in photosynthesis: an action spectrum for photosynthetic oxygen-evolution, or an action spectrum for excitation of chlorophyll a fluorescence. These two types of evidence will be distinguished in the following discussion of 'new' accessory pigments.

Chlorophyll

Chlorophylls a and b are well known as, respectively, the primary photosynthetic pigment in all photosynthetic plants, and the major accessory pigment of higher plants and green algae. Their chemical structures and absorption properties appear in most elementary text-books and need not be discussed here. Energy transfer from chlorophyll b to chlorophyll a is 100% efficient, and the effectiveness of chlorophyll b as an accessory pigment has been demonstrated in numerous action spectra for photosynthesis in green plants, including marine green algae (e.g. Ulva, Haxo & Blinks, 1950; Lüning & Dring, 1985).

One or more forms of chlorophyll c replace chlorophyll b in all of the various groups of 'chromophyte' algae. These pigments - three spectrally distinct forms have now been identified (Jeffrey & Wright, 1987) - differ from chlorophylls a and b in that the phytol 'tail' of the molecule is absent and the porphyrin ring has two less hydrogen atoms (Kirk, 1983). As a result, the Soret (blue) band in the absorption spectrum is more intense, and the alpha (red) band is less intense, so that chlorophyll c absorbs blue light far more strongly than red light. Chlorophylls c_1, c_2 and c_3 show only minor differences in absorption properties (Jeffrey & Wright, 1987) and there appear to be no physiological differences between plants with different forms of chlorophyll c, but the distribution of these forms does show some correlation with the taxonomy of the chromophyte algae (see below and Table 5.1). Action spectra for photosynthesis in the diatom *Phaeodactylum tricornutum* (Mann & Myers, 1968) and the dinoflagellate *Glenodinium* (Prezelin *et al.*, 1976) show clear peaks at around 630 nm, as do the fluorescence excitation spectra for light-harvesting pigment-protein complexes isolated from the full range of chromophyte algae (brown algae, Barrett & Anderson, 1980; diatoms, Owens & Wold, 1986b; dinoflagellates, Prezelin & Haxo, 1976; cryptophytes, Lichtlé *et al.*, 1987; see also Neori *et al.*, 1988). A light-harvesting role has, therefore, been demonstrated for chlorophyll c in the broad sense, although it is impossible to demonstrate that any individual pigment within this complex is a light-harvesting pigment.

Table 5.1. *Light-harvesting pigments in different groups of marine algae (based on Kirk, 1983; Prezelin & Boczar, 1986; Jeffrey & Wright, 1987; Stauber & Jeffrey, 1988; Hooks et al., 1988; Kohata & Watanabe, 1988).*

+   light-harvesting function demonstrated, occurs in majority of group
(+)   light-harvesting pigment occurs only in some members of group
-     occurs in majority of group, but light-harvesting function not demonstrated

	Chlorophytes				Chromophytes								Red	Prokar.	
	Vol	Prs	Ulv	Sph	Phe	Cry	Prm	Bac	Eus	Rap	Pyr	Crp	algae	Cyn	Pcl
Chlorophylls:															
a	+	+	+	+	+	+	+	+	+	+	+	+	+	+	+
b	+	+	+	+							(+)a				+
c_1		(+)b			+	(+)	(+)	(+)			(+)a				
c_2					+	+c	+	+		+	+	+			
c_3						(+)	(+)	(+)							
Mg 2,4-D		(+)													
Carotenoids:															
fucoxanthin					+	+	+	+		+	(+)a				
hexanoyloxyfx.							(+)	(+)							
butanoyloxyfx.						(+)	(+)	(+)							
siphonaxanthin	(+)	(+)	+												
peridinin												+			
violaxanthin	-	-	-	-	-	-				+	-				
prasinoxanthin		+d													
antheraxanthin						(+)									
Phycobilins:															
phycocyanobilin												(+)	+	+	
phycoerythrobilin												(+)	+	(+)	
phycourobilin													+	(+)	
phycobiliviolin												(+)			

Chlorophytes: *Vol*=Volvocales+other 'mainstream' green algae, *Prs*=prasinophytes, *Ulv*=Ulvales, *Sph*=Siphonales
Chromophytes: *Phe*=Phaeophyta, *Cry*=Chrysophyceae, *Prm*=Prymnesiophyceae, *Bac*=Bacillariophyceae, *Eus*=Eustigmatophyceae, *Rap*=Raphidophyceae, *Pyr*=Pyrrhophyta, *Crp*=Cryptophyta
Red algae: Rhodophyta
Prokaryotes: *Cyn*=Cyanophyta (Cyanobacteria), *Pcl*=Prochloron

[a] Thought to be due to endosymbiotic green algae (Watanabe *et al.*, 1987) or chrysophytes (Jeffrey & Wright, 1987)
[b] Wilhelm (1987, 1988)
[c] Chlorophyll c_2 may be absent from freshwater chrysophytes (Andersen & Mulkey, 1983)
[d] Wilhelm & Lenartz-Weiler (1987)

Another chlorophyll c-like pigment has been detected in some of the green flagellates which have been referred to the class Prasinophyceae (Ricketts, 1966). Until recently, this pigment was thought to be magnesium 2,4-divinylpheoporphyrin a_5 monomethyl ester (Mg 2,4-D), which is an intermediate in the biosynthesis of chlorophyll, but Wilhelm (1987, 1988) has shown that the third chlorophyll of *Mantoniella* is identical to chlorophyll c_1. This suggests that prasinophytes are the only plants to contain all three chlorophylls (a, b and c), although it is not yet clear that all prasinophytes contain chlorophyll c rather than Mg 2,4-D (see Table 5.1). The fluorescence excitation spectra for whole cells and for chloroplast particles of *Mantoniella* show peaks at 633 nm, as well as at 648 nm (chlorophyll b) and 670-678 nm (chlorophyll a), so that the extra chlorophyll appears to act as a second accessory pigment in these flagellates (Brown, 1985).

Carotenoids

All photosynthetic plants contain several carotenoid pigments with similar chemical structures (C40 isoprenoid compounds) and - at least when they have been extracted from the plants into an organic solvent - similar absorption properties confined to the 400-500 nm waveband. β-carotene is often regarded as the type specimen and, indeed, this pigment is almost as widely distributed as chlorophyll a, since it is absent only from cryptophytes. None of the carotenoids found in higher plants, however, has been shown to transfer significant amounts of energy to chlorophyll a, or to contribute directly to photosynthesis. The role of these pigments is thought to lie in the protection of the photosynthetic apparatus from excessive irradiance, rather than in light-harvesting.

Among marine algae, carotenoids are more important for photosynthesis and are frequently present in higher concentrations relative to chlorophyll (Kirk, 1983). The characteristic brown coloration of diatoms and chrysophytes, and of the brown algae themselves, is due to fucoxanthin, while most dinoflagellates are coloured by peridinin. Many macroscopic marine green algae, together with a few prasinophytes, contain siphonaxanthin. All three of these carotenoids have been shown to act as accessory pigments through action spectra for photosynthesis and for excitation of chlorophyll a fluorescence (see Prezelin & Boczar, 1986). These three pigments are also unusual among carotenoids in that their *in vivo* absorption extends well into the green region of the spectrum, although they appear as typical carotenoids in organic solvents (fucoxanthin, Barrett & Anderson, 1980; peridinin, Prezelin & Haxo, 1976; siphonaxanthin, Anderson, 1985). Another common feature which may relate to their light-harvesting capacity is their relatively high degree of oxygenation and their ability to form linear dimers which may extend their fluorescence lifetimes and thus facilitate energy transfer to chlorophyll a (Prezelin & Boczar, 1986).

In addition to these major accessory carotenoids, a number of related pigments have recently been shown to contribute to photosynthesis in specific algal groups. HPLC techniques have revealed two derivatives of fucoxanthin that either replace or occur alongside the commoner pigment. These are 19'-hexanoyloxyfucoxanthin and 19'-butanoyloxyfucoxanthin, which differ from fucoxanthin simply by the substitution of a hexanoyl or a butanoyl group for a hydrogen at the 19' C (Haxo, 1985, Wright & Jeffrey,

1987). This chemical substitution has little effect on absorption, and action spectra for oxygen-evolution and fluorescence excitation in the coccolithophorid *Emiliania*, showed that hexanoyloxyfucoxanthin apparently functioned in photosynthesis as efficiently as fucoxanthin (Haxo, 1985). As yet, there has not been a similar demonstration of the photosynthetic action of butanoyloxyfucoxanthin, but it seems probable that this pigment too will soon have to be added to our list of light-harvesting pigments.

Violaxanthin must be added to it already on the strength of work with one species in a rather unusual group of chromophytes - the Eustigmatophyceae. This is the only group of eukaryotic algae, apart from the red algae, which does not contain an accessory chlorophyll in addition to chlorophyll *a*. The major carotenoid in the marine species *Nannochloropsis* is violaxanthin, which is also found (but in lower concentrations) in most green algae and higher plants, brown algae and some other chromophyte groups (see Table 5.1). In *Nannochloropsis*, however, violaxanthin constitutes 60% of the total carotenoids and the excitation spectrum for chlorophyll *a* fluorescence in intact cells peaks at 490 nm, which corresponds to the peak absorption of violaxanthin *in vivo* (Owens *et al.*, 1987). It is surprising that violaxanthin should act as a light-harvesting pigment in this species but not in the many others in which it occurs and the time may be ripe for a reassessment of this pigment in other plants. Its possible contribution to photosynthesis in other plants may, however, be hard to demonstrate because of its low concentration and because it is masked by other carotenoids with similar absorption spectra.

Two other carotenoids of more limited distribution are also able to transfer absorbed energy to chlorophyll *a*. The major carotenoid of prasinophytes is prasinoxanthin, and energy transfer has been demonstrated in *Mantoniella* (Wilhelm & Lenartz-Weiler, 1987). Antheraxanthin occurs in addition to fucoxanthin in the freshwater chrysophyte *Chrysosphaera*, and also contributes to photosynthetic light-harvesting (Alberte & Andersen, 1986).

Phycobilins

The term phycobilin strictly refers to a set of four chromophores with similar open-chain tetrapyrrole structures. These chromophores are always covalently bound to proteins, and it is the resulting pigment-protein complexes - the phycobiliproteins - which are extracted when red algal and cyanobacterial cells are disrupted in water. There are three basic types of phycobiliprotein - phycoerythrin (red), phycocyanin (blue) and allophycocyanin (violet) - but, for a long time, these names have had to be qualified because the phycoerythrin and phycocyanin extracted from red algae (R-phycoerythrin and R-phycocyanin) had rather different absorption spectra from the superficially similar pigments which were extracted from cyanobacteria (C-phycoerythrin and C-phycocyanin). It is now clear that the various biliproteins of red algae and cyanobacteria (at least nine have been described) are all derived from just three chromophores (phycoerythrobilin, phycocyanobilin, phycourobilin), and that their different spectral properties are due to different combinations of the chromophores, and the different proteins to which the chromophores are bound (Glazer, 1981). For example, R-phycoerythrin contains phycoerythrobilin + phycourobilin, whereas C-phycoerythrin contains phycoerythrobilin alone, and R-phycocyanin contains phycocyanobilin + phyco-erythrobilin, whereas C-phycocyanin contains phycocyanobilin alone.

The fourth phycobilin chromophore, phycobiliviolin, is mainly found in the only other algal group which possesses biliproteins, the cryptophytes. These unicellular flagellates contain either phycoerythrin or phycocyanin but, as might be expected, they have their own distinctive types of phycoerythrin and phycocyanin. There are three cryptomonad phycoerythrins, with slightly different absorption maxima, all of which resemble C-phycoerythrin in that they contain phycoerythrobilin alone but presumably differ from the cyanobacterial pigment and from each other in the fine detail of their protein structures. Similarly, there are three cryptomonad phycocyanins, but these are more strongly differentiated from C-phycocyanin because they contain phycobiliviolin in addition to phycocyanobilin. The phycobiliproteins of cryptophytes also differ from those of red algae and cyanobacteria in that they are located in the intrathylakoidal space (Lichtlé *et al.*, 1987), rather than in spherical phycobilisomes attached to the surface of the thylakoids.

The classic photosynthetic action spectra of Haxo & Blinks (1950) demonstrated clearly that both phycoerythrin and phycocyanin contributed to photosynthetic oxygen evolution in a variety of marine red algae, and this conclusion has been confirmed by several subsequent investigations (*e.g.* Larkum & Weyrauch, 1977; Lüning & Dring, 1985). Similar results confirming the light-harvesting role of biliproteins in cyanobacteria and cryptomonads are reported by Prezelin & Boczar (1986).

Distribution of light-harvesting pigments among marine algae

Table 5.1 summarises the algal groups in which the various light-harvesting chlorophylls, carotenoids and phycobilins are found. Most of this information is familiar and has been presented in a similar way in previous reviews. However, the recent discovery of chlorophyll c_3 and of the derivatives of fucoxanthin (hexanoyloxyfucoxanthin and butanolyoxyfucoxanthin) has already resulted in a revision of what have been regarded as the diagnostic pigment complements for two chromophyte groups, and it is possible that further revisions will be needed as other groups are systematically examined for these 'new' pigments. A survey of 51 species of marine diatoms and a review of recent pigment data for 13 strains of coccolithophorids (Stauber & Jeffrey, 1988) has shown that chlorophylls c_1 and c_2 are not found throughout these groups, as previously thought (Jeffrey, 1976), but that three distinct patterns of pigments may be found in both diatoms and coccolithophorids:

(a) chlorophylls $c_1 + c_2$ and fucoxanthin;
(b) chlorophylls $c_2 + c_3$ and fucoxanthin;
(c) chlorophylls $c_2 + c_3$ and fucoxanthin + fucoxanthin-derivatives.

The first pattern was commonest among marine diatoms, occurring in 88% of those tested, whereas the third pattern was found in only one diatom species. The three patterns were more evenly represented among coccolithophorids, however, and the third pattern was found in such important and common phytoplankton species as *Emiliania huxleyi* (Wright & Jeffrey, 1987), *Phaeocystis pouchetii* (Jeffrey & Wright, 1987) and *Corymbellus aureus* (Gieskes & Kraay, 1986). The picoplanktonic coccoid chrysophyte *Pelagococcus subviridis* also contains chlorophylls c_2 and c_3, together with fucoxanthin and butanoyloxyfucoxanthin (Vesk & Jeffrey, 1987), so that pattern (c) could be even more widespread among chromophyte algae. All brown algae examined so far contain chlorophylls c_1 and c_2 whereas the peridinin-containing dinoflagellates and most crypto-monads contain chlorophyll c_2 only (Jeffrey & Wright, 1987), but none of these groups has yet been examined with the techniques that have revealed chlorophyll c_3 and the fucoxanthin derivatives. In freshwater, however, certain chrysophytes contain chlorophyll c_1 alone (Andersen & Mulkey, 1983).

The intense interest in picoplanktonic cyanobacteria during the past decade (see Joint, Chapter 6) has forced a further revision to the established picture of how photosynthetic pigments are distributed among the algae. The chromophore phycourobilin was thought to occur only in the phycoerythrins found in red algae (see above), but oceanic strains of *Synechococcus* have been found to contain phycoerythrins with phycourobilin (Alberte *et al.*, 1984). More recent work using flow cytometry (Olson *et al.*, 1988), immunofluores-cence and fluorescence excitation spectroscopy, which enables individual cells to be examined (Campbell & Itturiaga, 1988), has shown that much higher phycourobilin:phycoerythrobilin (PUB:PEB) ratios are found in natural populations of *Synechococcus* from oceanic sites than are found in populations from coastal waters or in any of the strains in laboratory culture.

Use of pigments as markers for specific groups in the phytoplankton

The increasing variety of pigments detected in the phytoplankton, coupled with knowledge about their distribution in the various groups of algae, has led several workers to suggest that the taxonomic composition of natural populations of phyto-plankton could be assessed on the basis of pigment analyses (Gieskes & Kraay, 1983, 1986; Guillard *et al.*, 1985; Hooks *et al.*, 1988; Klein & Sournia, 1987; Smith *et al.*, 1987; Stauber & Jeffrey, 1988; Wright & Jeffrey, 1987). This approach is particularly valuable for pico- and ultraplankton which are almost impossible to identify on the basis of morphological characters visible by light microscopy. Hooks *et al.* (1988) were able to separate 16 clones of coccoid eukaryotic ultraplankton into four groups using their chlorophyll *b*, prasinoxanthin and butanoyloxyfucoxanthin contents:

Type I : chlorophyll *b* - chlorophyte-like;

Type IIA: chlorophyll *b* + prasinoxanthin (px:chl *b*>0.23) - prasinophyte-like (*cf. Micromonas*)

Type IIB: chlorophyll *b* + prasinoxanthin (px:chl *b*<0.15) - prasinophyte-like

Type III: butanoyloxyfucoxanthin - chrysophyte-like (*cf. Pelagococcus*).

Analysis of natural water samples for these pigments and chlorophyll *a* enabled the contribution of green algae as a whole and of prasinophyte-like species to the total chlorophyll *a* biomass to be estimated. Hexanoyloxyfucoxanthin was the dominant carotenoid at the station sampled, which suggested that prymnesiophytes were a significant component of the phytoplankton, and the relatively low ratio of chlorophyll *b* to chlorophyll *a* also suggested that a large proportion of species containing chlorophylls *a* and *c* were present. As yet, the techniques for separating the three forms of chlorophyll *c* and Mg 2,4-D have insufficient resolution for use on natural water samples but, if improved separation can be developed, Mg 2,4-D may provide an additional marker for prasinophytes, while chlorophylls c_1 and c_3 will distinguish particular groups of chromophytes in the plankton (Table 5.1). Other less widely distributed carotenoids, including some which may not be involved in light-harvesting, can be used at present to identify particular groups, *e.g.* peridinin for dinoflagellates, zeaxanthin for cyanobacteria (Guillard *et al.*, 1985), alloxanthin for cryptomonads (Gieskes & Kraay, 1983).

Phycobilin pigments cannot be so readily extracted as carotenoids and chlorophylls, but they have the useful characteristic of fluorescing *in vivo*, and either the fluorescence spectrum or the excitation spectrum for this fluorescence can be used to estimate the abundance of cyanobacteria in a fresh sample of picoplankton. It is also possible to distinguish between strains of *Synechococcus* with different chromophore ratios. For example, phycoerythrin with a high proportion of PUB fluoresces at 560 nm when excited with 488 nm light, but phycoerythrin with more PEB fluoresces at 575 nm. The ratio of 560:575 nm fluorescence, therefore, indicates the proportion of high PUB strains (Olson *et al.*, 1988). Alternatively, the ratio of the fluorescence induced by 490 nm to that induced by 545 nm provides a similar assessment of the PUB:PEB ratio in a natural population of *Synechococcus* (Campbell & Itturiaga, 1988).

Organisation of chlorophylls and carotenoids in pigment-protein complexes

All of the chlorophyll and carotenoids associated with the photosynthetic apparatus of plants is complexed to protein in the cells, and gentle extraction of photosynthetic membranes with detergents leads to the isolation of intact pigment-protein complexes. A characteristic feature of these complexes is that the absorption properties of many of the component pigments are modified by the protein environment, and the absorption peaks of all chlorophylls and the major light-harvesting carotenoids (fucoxanthin, peridinin, siphonaxanthin) are shifted to longer wavelengths relative to their value in organic solvents. There are a number of different types of pigment-protein complex in most plants, and all algae appear to resemble higher plants in possessing at least two complexes which are associated with the reaction centres of PSI and PSII, and which contain chlorophyll *a* and β-carotene but none of the other accessory pigments (Kirk, 1983; Larkum & Barrett, 1983; Prezelin & Boczar, 1986).

The bulk of the chlorophyll *a* plus any accessory chlorophylls and the light-harvesting carotenoids are associated with one or more light-harvesting complexes (LHC). The absorption spectra of these complexes reflects the sum of the *in vivo* absorption spectra of the component pigments, but energy transfer occurs within the complex from the accessory pigments to chlorophyll *a*, so that the spectrum of the fluorescence emitted from the complex is that characteristic of chlorophyll *a*. The energy transferred in this way to chlorophyll *a* may be contributed to either of the photosystems. The pigment composition of the LHCs varies considerably from one algal group to another, because it is here that the different light-harvesting pigments are found. All LHCs contain chlorophyll *a* plus one or more accessory pigments, but the different accessory pigments in any one plant may be associated with different LHCs. For example, the major LHC of the diatom *Phaeodactylum* contains chlorophylls a, c_1 and c_2 and fucoxanthin, while another minor LHC is composed only of the three chlorophylls (Owens & Wold, 1986a). Table 5.2 lists the bulk of the marine algae whose pigment-protein complexes have been isolated and characterised.

Table 5.2. *Genera of marine algae from which photosynthetic pigment-protein complexes have been isolated and characterised*

Division Class	Order	Genera (reference no. - see key below)
CHLOROPHYTA		
Chlorophyceae		
	Ulvales	*Enteromorpha, Ulva* (23)
	Siphonales	*Acetabularia* (5, 18)
		Bryopsis (21, 23, 25)
		Codium (2-4, 13, 23)
Prasinophyceae		*Mantoniella, Micromonas* (29-31)
PROCHLOROPHYTA		*Prochloron* (19)
PHAEOPHYTA		*Acrocarpia* (7), *Chorda, Leathesia, Pylaellia* (1)
		Cystoseira, Dictyota (12), *Ecklonia* (6)
		Fucus (11-12), *Hormosira* (22)
		Laminaria (1, 12), *Macrocystis* (9)
CHRYSOPHYTA		
Prymnesiophyceae		*Pavlova* (15, 20)
Bacillariophyceae		*Chaetoceros, Nitzschia, Thalassiosira* (9)
		Phaeodactylum (9-10, 12, 14-17, 26-27)
		Skeletonema (1)
PYRRHOPHYTA		*Glenodinium, Gonyaulax* (8, 28)
CRYPTOPHYTA		*Cryptomonas* (24)

References: 1, Alberte *et al.*, 1981; 2-3, Anderson, 1983, 1985; 4, Anderson *et al.*, 1987; 5, Apel, 1977; 6-7, Barrett & Anderson, 1977, 1980; 8, Boczar & Prezelin, 1987; 9, Brown, 1988; 10, Caron & Brown, 1987; 11-12, Caron *et al.*, 1985, 1988; 13, Chu & Anderson, 1985; 14, Fawley & Grossman, 1986; 15, Fawley *et al.*, 1987; 16-17, Friedman & Alberte, 1986, 1987; 18, Green *et al.*, 1982; 19, Hiller & Larkum, 1984; 20, Hiller *et al.*, 1988; 21, Itagaki *et al.*, 1986; 22, Kirk, 1977; 23, Levavasseur, 1989; 24, Lichtlé *et al.*, 1987; 25, Nakayama *et al.*, 1986; 26-27, Owens & Wold, 1986a, b; 28, Prezelin & Haxo, 1976; 29, Wilhelm, 1988; 30, Wilhelm & Lenartz-Weiler, 1987; 31, Wilhelm *et al.*, 1986.

Recent investigations have started to explore the molecular biology of algal LHCs. The individual polypeptides have been separated and characterised, and the similarity between the polypeptides from different species has been tested by immunological techniques. The results of such studies indicate that the LHCs of green algae (including Siphonales; Anderson *et al.*, 1987) and land plants are highly conserved (Larkum & Barrett, 1983), but that the LHCs of the various groups of chromophyte algae may differ from each other as well as from those of green algae (Friedman & Alberte, 1987). Most of the diatoms investigated have similar complexes, suggesting that they belong to a monophyletic group, possibly of relatively recent origin, but there was evidence for a closer relationship between diatoms and brown algae than between diatoms and other chromophyte groups (Friedman & Alberte, 1987; Brown, 1988). Other results have indicated, however, that the fucoxanthin-chlorophyll a/c-protein complex from *Phaeodactylum* is almost identical to that from the prymnesiophyte *Pavlova* (Fawley *et al.*, 1987). It is clearly too early to expect definitive results from this new approach to the pigment-protein complexes of marine plants, but it promises to open up the evolutionary biology of light-harvesting as well as increasing our knowledge of the biochemistry.

PHYSICS OF LIGHT-HARVESTING (PACKAGE EFFECT)

The quality and quantity of pigments in an algal population are the primary chemical factors which influence the amount of light harvested for photosynthesis by the plants. The efficiency with which these pigments absorb the available radiation, however, as measured by the absorbance of a given amount of pigment, is affected by the way in which the pigments are distributed in the population. It can be shown from physical theory that, when measured under identical conditions, the absorbance of a suspension of particles containing a pigment will always be less than the absorbance of a solution containing the same amount of free pigment (Kirk, 1975, 1983). In effect, this means that concentrating the available pigment into discrete packages, such as individual cells or colonies in the population or individual chloroplasts in the cells, reduces the ability of the pigment to absorb light. This has been called the 'package effect'. The intensity of the package effect varies with the following factors:

Wavelength of incident light. The package effect is greater at wavelengths that are absorbed more strongly. Consequently, the absorbance spectrum for a cell suspension has less differentiation between the peaks and the troughs than the spectrum for an equivalent amount of pigment in solution.

Absorbance of individual particles. If the intracellular pigment concentration is increased, but the size and shape of the cells and the total pigment content of the population remains constant (*i.e.* a given amount of pigment is concentrated into a smaller number of cells), the package effect also intensifies. This intensification will be most pronounced at the most strongly absorbed wavelengths, and the end result is a flattening of the absorbance spectrum as the individual cells become optically black. This is analogous to changes in absorbance in a macroalgal thallus as thickness increases (Lüning & Dring, 1985). Conversely, a larger number of less strongly pigmented cells will show a reduced package effect, and will thus absorb more light per unit of pigment concentration. An effect of this type has recently been demonstrated in the green flagellate *Dunaliella* (Berner *et al.*, 1989). Cells grown at 70 μE m^{-2}s^{-1} had more chlorophyll *a* per cell than cells grown at 700 μE m^{-2}s^{-1}, but absorbed red light less strongly per unit of chlorophyll *a*. Part of this effect was shown to be due to changes in the ratio of chlorophyll *a* to the accessory pigments, but over half was attributed to the package effect. There was no significant change in cell size, but intracellular pigment concentration clearly increased on transfer to low light. This was reflected in an increase in thylakoid stacking, together with a decrease in the lipid:protein ratio of the thylakoids (an index of membrane transparency). These two factors appeared to contribute more or less equally to the overall increase in the package effect.

Size of particles. An increase in the size of individual algal cells (or colonies) without any change in shape or in the total pigment content of the population, also results in a concentration of the pigment into a smaller number of particles. There is, therefore, a greater chance that photons will be transmitted through the suspension without hitting an absorbing particle, and the package effect is again increased. The relationship between the specific absorption coefficient of chlorophyll *a* at the red absorption maximum and the diameter of cells or colonies has been calculated by Kirk (1976, 1983). Cells greater than about 50 μm in diameter suffer a reduction of over 70% in specific absorption, relative to a solution of chlorophyll, but the reduction for nanoplankton (20 μm diameter) with the same amount of chlorophyll per unit cell volume is about 25%, while that for picoplankton (2 μm diameter) is as little as 5%.

Shape of particles. A spherical shape appears to be the least efficient for light absorption. Elongated cells exhibit a smaller package effect than spherical cells of the same volume and pigment content, and long thin cylinders are almost as efficient in harvesting light as are a much greater number of smaller spherical particles of the same diameter (*e.g.* a single cylinder 3537 x 6 μm would have a similar absorbance to 884 spherical particles with a diameter of 6 μm; Kirk, 1976).

The implications of the package effect for the ecology of phytoplankton will be considered after discussing other aspects of the environmental biology of light-harvesting.

ENVIRONMENTAL BIOLOGY OF LIGHT-HARVESTING

The currently fashionable subject of environmental biology includes studies of the effects of the environment on organisms, and of the distribution of organisms in relation to the variation of different factors in the environment. In the present context, these two approaches translate into two questions: how do environmental factors affect the light-harvesting apparatus of marine plants? and how do variations in the light-harvesting apparatus affect the ability of plants to live in different marine environments?

Environmental effects on light-harvesting apparatus

Two aspects of the light climate will be the only factors considered here, but nutrient supply clearly interacts with pigment composition, particularly through the availability of nitrogen for the synthesis of phycobiliproteins (see Kana & Glibert, 1987).

Growth irradiance

The ways in which the photosynthetic apparatus of plants changes in response to variations in the irradiance in which plants are grown have attracted a lot of attention in recent years, and aquatic plants have provided particularly attractive subjects for such studies because of the wide range of irradiances which they may experience in natural conditions. Richardson *et al.* (1983) have analysed the various 'strategies' that unicellular algae show in their adaptation to irradiance, and Kirk (1983) includes the more disparate studies of macroalgae in a discussion of 'ecological strategies' for photosynthesis in the aquatic environment. The basic picture has not changed radically since these reviews were written, although a few more examples can be added to reinforce the conclusions which were drawn. Most plants respond to decreases in irradiance which last for at least several hours by an increase in the concentration of photosynthetic pigments. The different photosynthetic pigments in any one plant rarely increase at the same rate and, in general, accessory pigments are found to increase more rapidly than chlorophyll *a* (see table 3.4 in Dring, 1982). The commonest exception to this generalisation is fucoxanthin, which often appears to increase less rapidly than chlorophyll *a* as either irradiance decreases or depth increases (*e.g. Phaeodactylum, Ascophyllum, Dictyota, Fucus, Laminaria, Sphacelaria* - see references in Kirk, 1983 and Dring, 1986; but *Macrocystis* shows the opposite trend - Wheeler, 1980; Smith & Melis, 1987).

   This general increase in the ratio of the accessory pigments to chlorophyll *a* in response to a decrease in growth irradiance is often interpreted as an increase in the concentration of LHCs, which contain the bulk of the cells' accessory pigments, while the reaction centre pigment-protein complexes, consisting largely of chlorophyll *a* alone, remain relatively constant. By comparing graphs of photosynthesis per cell against irradiance with those of photosynthesis per unit chlorophyll against irradiance for plants grown in different irradiances, it is possible to distinguish between two possible ways in which the pigment content of a plant could be increased in response to declining irradiance: increasing the number of photosynthetic units (PSU) per cell, or increasing the size of existing units. Most results for marine plants suggest that PSU size increases (*e.g. Glenodinium*, Prezelin, 1976; *Gracilaria*, Beer & Levy, 1983), and this supports the earlier conclusion that the LHCs increase, since this would result in a larger antenna being associated with each reaction centre.

   The relatively recent discovery of the importance of picoplanktonic cyanobacteria in oceanic waters has meant that studies with such organisms were initiated only after this consensus was reached. One investigation of the response of *Synechococcus* to changes in growth irradiance suggested that marine cyanobacteria reacted differently from most other marine algae (and freshwater cyanobacteria), since PSU size appeared to decrease as irradiance decreased (Barlow & Alberte, 1985). A more recent study of the same strain, however, showed clearly that PSU size increased (by the addition of phycoerythrin to the phycobilisome) with declining irradiance when growth was close to light-saturation, but that phycobilisome numbers increased when irradiance was reduced further and growth became light-limited (Kana & Glibert, 1987). These results are more compatible with ecological observations, and the disagreement with the earlier results may mean that the photosynthetic apparatus of cyanobacteria changes rather slowly in response to variations in irradiance (see Joint, Chapter 6).

Light quality

Plants may respond to changes in the spectral composition of the light field either by altering the overall pigment content of the cells, or by altering the balance between different photosynthetic pigments. The first type of response is analogous to intensity adaptation, and may indicate which wavelengths are most effective in inducing the plant's response to changes in irradiance. The second type of response is generally described as 'chromatic adaptation', even though the effect of the changes may not be adaptive (*i.e.* they may not increase the total absorption of the plant in the light field).

Several species of marine diatoms, one dinoflagellate and a cryptomonad formed more chlorophyll in weak blue-green light than in the same irradiance of white light (Vesk & Jeffrey, 1977), which suggests that, as in more laboratory-based green algae (*Chlorella, Scenedesmus*; Senger, 1987b), blue light stimulates chlorophyll formation. However, other species in the same survey showed a weaker, or no, response to blue-green light (Vesk & Jeffrey, 1977), and the dinoflagellate *Prorocentrum* and the red alga *Gracilaria* had reduced chlorophyll *a* concentrations after growth in blue light (Faust *et al.*, 1982; Beer & Levy, 1983). Sporophytes of *Laminaria* showed an increase in chlorophyll content as irradiance decreased in blue and red light, but little response to changing irradiance in green light (Dring, 1986). There is, therefore, insufficient evidence to conclude that blue light is always seen as dim light by marine plants, but so-called 'intensity adaptations' are certainly not independent of light quality.

Chromatic adaptation which increases the ability of a plant to absorb radiation in a specific light field was described as "complementary chromatic adaptation" by Engelmann (1883) because the resulting colour of the plants (*i.e.* those wavelengths not absorbed) is complementary to that of the light field. This term has persisted in the literature, even though any other response of pigment composition to a change in light quality (sometimes described as 'inverse chromatic adaptation') cannot really be regarded as adaptive. Some cyanobacteria from freshwater habitats clearly exhibit complementary chromatic adaptation, since the phycoerythrin content is higher in green light than in red light, and phycocyanin content may also vary in a complementary manner (de Marsac, 1977), but there is as yet no clear evidence that any marine species of cyanobacterium shows a similar response. Two strains of *Synechococcus* grown in green light had a higher photosynthetic efficiency in green light than cells grown in blue-green light (Glover *et al.*, 1987), but no pigment analyses are available to demonstrate the mechanism of this response. Growth in green light has, however, been shown to increase the content of phycoerythrin in *Cryptomonas* (Kamiya & Miyachi, 1984) and of peridinin in *Prorocentrum* (Faust *et al.*, 1982), but other investigations of a wide range of marine algae have shown that light quality has no significant effect on pigment composition. These investigations have included surveys of pigment ratios in 18 unicellular algae and seven seaweed species after growth in blue-green or green light compared with white light (Vesk & Jeffrey, 1977; Ramus, 1983), and more detailed examinations of *Gracilaria* (Beer & Levy, 1983) and *Laminaria* (Dring, 1986) after growth in red, green and blue light. Adaptation to a particular light quality also failed to enhance the efficiency with which five ultraplanktonic eukaryotes (a prasinophyte, three prymnesiophytes and a diatom) were able to photosynthesise in the same light quality (Glover *et al.*, 1987). Few marine plants, therefore, appear to be able to adjust their pigment composition in response to changes in light quality.

Light-harvesting and the depth distribution of marine algae

Failure to demonstrate ontogenetic (or phenotypic) chromatic adaptation does not invalidate the hypothesis that the distribution of marine plants is controlled by the match between light-harvesting ability and the spectral quality of the light in different sites (*i.e.* that plants are phylogenetically - or genotypically - adapted to particular light qualities). It was this hypothesis that was originally proposed by Engelmann (1883) with reference to benthic marine algae. In its popular form, the hypothesis of 'complementary chromatic adaptation' is used to explain why blue-green and green algae occur mainly in the upper intertidal, brown algae dominate the mid- to lower intertidal and the upper subtidal zones, and red algae become increasingly dominant in deeper water. As underwater light becomes greener with depth, it is argued, so the possession of, first, fucoxanthin and then phycoerythrin becomes more advantageous.

Recently, this hypothesis has been criticised for a number of reasons. Light quality can influence the competitive ability of plants only if their growth is light-limited, so that chromatic adaptation cannot explain the distribution of plants in the intertidal or upper subtidal zones because most marine species would be light-saturated for at least some of the time in these sites. The thick thalli of many of the more familiar brown algae (*e.g.* kelps, fucoids) and several siphonaceous green algae (*e.g. Codium, Caulerpa*) are optically black and the photosynthesis of such species is almost independent of wavelength (Lüning & Dring, 1985). The ecological success of plants of this type also, therefore, cannot be attributed to chromatic adaptation. Logically, the only plants whose distribution could be influenced by chromatic adaptation are species with thin thalli growing near the bottom of the photic zone, where light is the main growth-limiting factor.

Examination of the theoretical (Dring, 1981) or actual (Dring & Lüning, 1985) photosynthetic efficiency of different species in light qualities which prevail at the limits of plant growth in the full range of oceanic and coastal water types has shown that, as predicted by Engelmann (1883), red algae are best adapted for photosynthesis in the green light of most deep coastal waters, but that brown and green algae are considerably more efficient in the blue light of clearer oceanic waters. If chromatic adaptation was the primary factor controlling distribution, therefore, brown and green algae should dominate the flora at depth in oceanic waters. However, red algae penetrate to greater depths than brown and green algae in most oceanic sites, and the deepest record for a benthic alga (268 m) is held by a red algal crust (Littler *et al.*, 1985). At the other end of the water type range, in the most turbid coastal waters, the light at the bottom of the photic zone is orange in colour (peak at 575 nm). Chromatic adaptation would predict that phycocyanin-containing red algae or cyanobacteria should dominate here, but few benthic plants with a high phycocyanin content are found in any subtidal sites. Red algae with a high phycoerythrin content occupy all of the benthic marine habitats with lowest irradiance, regardless of the colour of the light, and most recent investigators have concluded that the success of red algae in these sites depends on their tolerance of low irradiance conditions rather than on the precise wavelengths which are absorbed (Dring, 1981; Ramus, 1983; Larkum & Barrett, 1983).

There has been some resistance among marine ecologists to the abandonment of such a long-standing and attractive hypothesis, but most of the recent arguments in favour of chromatic adaptation have been based on the distribution of phytoplankton, rather than of benthic algae. The hypothesis had not previously been applied to planktonic algae because, since they do not grow at fixed depths, they were thought to be under little selection pressure to adapt to light quality. The photosynthesis and growth of a range of ultra- and pico- phytoplankton has recently been measured at different depths in oceanic waters (Wood, 1985) and in various colours which simulate those at different ocean depths (Glover *et al.*, 1986, 1987). As Engelmann might have predicted, eukaryotic algae (mainly prasinophytes and prymnesiophytes) utilised blue-violet and blue light more efficiently than *Synechococcus* strains, and strains of *Synechococcus* with a high PUB content were more efficient than those with little or no PUB. These results suggest that eukaryotic ultraplankton should out-compete cyanobacteria at the bottom of the photic zone in oceanic waters, and that strains of *Synechococcus* with high PUB:PEB ratios should penetrate such waters more deeply than those with lower ratios (see Joint, Chapter 6).

The ecological evidence available so far does not provide convincing support for either of these predictions. Although the densest populations of small eukaryotic algae are often said to occur at greater depths than those of picoplanktonic cyanobacteria, the most comprehensive survey of the distribution of these two groups (Murphy & Haugen, 1985) showed that the biomass of eukaryotes rarely exceeded that of the cyanobacteria at any depth. In addition, the abundance of both groups at most stations was fairly uniform from the surface down to the thermocline, below which the populations of both declined, and it was difficult to identify a depth at which a marked population maximum occurred. Kana & Glibert (1987) emphasise also that there is little ecological evidence for the common assertion that *Synechococcus* shows its maximal growth at low irradiances. As for the pigment composition of *Synechococcus*, high PUB:PEB ratios were found in most oceanic populations and there was no detectable increase in this ratio for populations from deeper waters (Campbell & Itturiaga, 1988).

What does seem clear, however, is that the smaller photosynthetic plankton penetrate more deeply in oceanic waters than do larger forms (*e.g.* diatoms, dinoflagellates), and this may be due to the package effect. Small cells are more efficient light-harvesters than larger cells (see above), so that they will be better able to utilise low irradiances. Thus, the conclusion that irradiance is more important than light quality in controlling the depth distribution of marine plants seems to apply to both phytoplankton and benthic algae, although the physiological mechanism by which this control is effected is quite different for the two groups.

Chapter 6

THE RESPONSE OF PICOPHYTOPLANKTON TO LIGHT

IAN JOINT

Plymouth Marine Laboratory, Prospect Place, West Hoe, Plymouth, PL1 3DH

Oceanic picoplankton are often considered to be shade organisms, with a preference, if not an absolute requirement to grow at low irradiance. This conclusion is based on two types of observation. Firstly, cultures of marine *Synechococcus* grow well at low irradiance; growth saturates at irradiance of less than 50 $\mu E\,m^{-2}\,s^{-1}$ and photoinhibition occurs at high light levels. Secondly, many vertical distribution studies of natural assemblages suggest that *Synechococcus* is abundant towards the base of the euphotic zone; eukaryotic picoplankton appear to be more dark adapted, since they show a biomass peak below the cyanobacterial maximum. However, other data exist which suggest that picoplankton does not have a preference for low-light conditions. Studies from a variety of marine regions demonstrate that cyanobacteria can be more abundant and more productive in the near surface waters than at the base of the euphotic zone. Although cyanobacteria possess phycoerythrin which absorbs the blue/green wavelength light found at depth in the euphotic zone, there is no evidence that this pigment confers any significant advantage to the cyanobacteria. Evidence from fine structure studies shows increases in thylakoid membranes in response to low-light conditions. However, physiological measurements and determinations of photosynthetic parameters show that picoplankton taken a few metres below the sea surface, where irradiance is high, are not shade organisms.

INTRODUCTION

Oceanic picoplankton were first described a decade ago when epifluorescence microscopy demonstrated the presence of small unicellular cyanobacteria in oceanic coastal and estuarine waters (Waterbury *et al.*, 1979; Johnson & Sieburth, 1979). In addition to cyanobacteria, eukaryotic picophytoplankton cells, with diameters of about 1 μm, were also present in natural assemblages (Johnson & Sieburth, 1982; Joint & Pipe, 1984). It is now clear that picoplankton make a significant contribution to primary production in the world's oceans (see reviews by Fogg, 1986; Joint, 1986) and in some marine provinces picoplankton may account for more than 50% of the annual primary production.

Picoplankton are commonly considered to be 'shade' organisms, in the sense described by Sournia (1982); that is, "some species are preferentially, if not exclusively, found in the deeper levels of the euphotic layer or even in the oligophotic layer of the sea". The evidence for picophytoplankton as shade organisms comes from observations on both natural assemblages and cultures of picoplankton.

Herring, P.J., Campbell, A.K., Whitfield, M. & Maddock, L. (ed.). *Light and Life in the Sea.*
© 1990. Cambridge University Press.

In warm oligotrophic regions of the oceans a distinct chlorophyll maximum is usually associated with the pycnocline and chlorophyll concentrations are often an order of magnitude greater in this layer than in the surface mixed layer. Bienfang & Szyper (1981) found that picoplankton accounted for more than 80% of the biomass in the chlorophyll maximum in water off Hawaii. Takahashi & Bienfang (1983) also found about 80% of the sub-surface chlorophyll peak was associated with picoplankton. Therefore, in tropical and sub-tropical waters, it appears that the subsurface chlorophyll maximum is often dominated by picoplankton. In a study of North Atlantic waters, Glover et al. (1985) reported that the greatest abundance of cyanobacteria was towards the base of the euphotic zone but that there was an additional peak of eukaryotic algae below the cyanobacterial peak. Since picoplankton cells are less than 1 μm in diameter, sinking rates are negligible (Takahashi & Bienfang, 1983) and any accumulation of biomass at a particular depth must be the result of in situ growth rather than accumulation as the result of sinking. These observations suggest that picoplankton grow best under conditions of greatly reduced irradiance.

Other evidence from cultures of cyanobacteria support the suggestion that these are shade organisms. All cultures of *Synechococcus* grow at low irradiances and enrichment cultures to isolate cyanobacteria require growth at 10-20 $\mu E\ m^{-2}\ s^{-1}$ (Waterbury et al., 1986). Optimum growth rate of *Synechococcus* cultures occurs at low irradiance; Morris & Glover (1981) working with a laboratory culture of *Synechococcus*, strain DC-2 (renamed WH7803), found optimal growth rate at 45 $\mu E\ m^{-2}\ s^{-1}$ and Richardson et al. (1983) found that maximum growth rate of cyanobacteria occurred at an average irradiance of 39 $\mu E\ m^{-2}\ s^{-1}$. These authors also determined that a minimum irradiance of 5 $\mu E\ m^{-2}\ s^{-1}$ was required to support autotrophic growth of cyanobacteria.

These observations suggest that the optimum depth of the water column for the growth of cyanobacteria must be where irradiance is low. However, in order for picoplankton to be considered as shade organisms, it must be demonstrated that these species are preferentially, if not exclusively, found in the deeper levels of the euphotic layer. An alternative hypothesis is that the picoplankton are not shade organisms but that they are adaptable and can exist in a variety of light conditions. The purpose of this paper is to examine the evidence for obligate growth at low irradiances or, alternatively, for adaptability, to consider the factors which influence photosynthesis at depths in the water column and to examine those features of picoplankton photosynthetic apparatus which might explain the observed growth and distribution in the water column at low irradiance values.

Light quality at the base of the euphotic zone

The attenuation of light in the sea is the result of absorption by four components of the aquatic system; the water itself, dissolved yellow pigments, phytoplankton, and inorganic particulate matter. Pure water absorbs only very weakly in the blue and green regions of the spectrum but red light is significantly attenuated. At 680 nm, the absorption maximum of chlorophyll, a one metre thick later of pure water would absorb about 35% of the incident light (Kirk, 1983). In coastal waters, the presence of dissolved organic matter, frequently referred to as 'Gelbstoff', results in the rapid attenuation of blue light. Particulate matter does not absorb light strongly but it scatters quite intensely; the direct result of this scattering is that absorption is low or absent in the red end of the spectrum but is quite significant at blue and ultra-violet wavelengths. In coastal waters, the result of absorption by all these processes is that blue and red light are significantly attenuated with depth but green light shows the greatest transmission. In oceanic waters, there may still be significant transmission of short wavelength blue light, which could be absorbed by the Soret band of chlorophyll. But, at any significant depth in the water column, the direct absorption of light by chlorophyll *a* may be greatly reduced and accessory pigments become more important in harvesting light.

Morphological response

If marine cyanobacteria are obligate shade organisms, then they should show a constant cellular morphology, whatever the growth conditions. Figure 6.1 shows that this is not the case. The photosynthetic membranes (thylakoids) within the cyanobacteria are clearly visible in these transmission electron micrographs. Figure 6.1A shows a cyanobacterium taken from the surface 10 m in the Celtic Sea and Figure 6.1B shows *Synechococcus*, strain WH7803, grown in the laboratory at an irradiance of *ca* 50 µE m^{-2} s^{-1}. It is clear that the thylakoid membrane is much larger in the cultured cyanobacterium than in the cell taken from the surface 10 m of the sea. This strongly suggests that *Synechococcus* can adapt to different growth irradiances by increasing the total photosynthetic membrane per cell. Similar changes in thylakoid content with growth irradiance were shown by Kana & Glibert (1987), who found increased stacking of the thylakoids when *Synechococcus*, strain WH7803, was grown at decreasing irradiances. This evidence for adaptation to light is not a feature to be expected from an obligate shade organism.

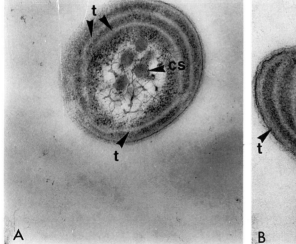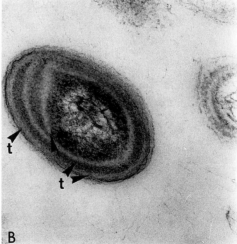

Figure 6.1. Transmission electron micrographs of *Synechococcus*. (A) From the surface mixed layer of the Celtic Sea and (B) laboratory culture of *Synechococcus*, strain WH 7803, grown at an irradiance of *ca* 50 µE m^{-2} s^{-1}. The thylakoid membranes (t) and carboxysome (cs) are indicated.

Photosynthetic pigments of cyanobacterial picoplankton

If picoplankton assemblages are successful at the base of the euphotic zone this must be a consequence of the pigment content of the cells. Light energy is collected by three basic types of photosynthetic pigments; chlorophylls, carotenoids and biliproteins. All photosynthetic organisms contain chlorophyll but cyanobacteria also contain phycobiliproteins, which are either red (phycoerythrins) or blue (phycocyanins) in colour. Oceanic strains of cyanobacteria are characterised by high concentrations of phycoerythrin, which has an absorption maximum in the green region of the spectrum at about 546 nm (Waterbury *et al.*, 1986); this part of the spectrum has high light transmission in sea water.

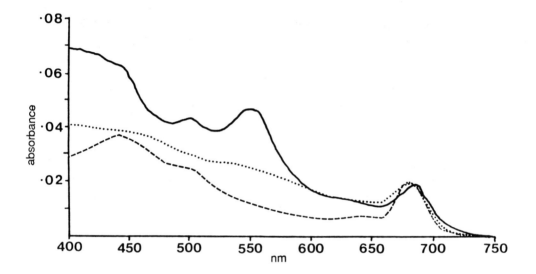

Figure 6.2. Spectrum of light absorbance by dilute cultures of *Synechococcus* strain WH 7803 (——), *Thalassiosira weissflogii* (- - -) and *Gonyaulax tamarensis* (····).

It might be expected that algal cells possessing accessory pigments which absorb light at green wavelengths would grow preferentially at depth in the sea; the presence of phycoerythrin might explain why cyanobacteria are most abundant towards the base of the euphotic zone in oceanic waters. However, this would only be a conclusive argument if no other phytoplankton species was capable of utilizing green light. This is clearly not the case; Yentsch & Yentsch (1984) show that, whilst not overlapping exactly, the absorption spectra of fucoxanthin (a characteristic accessory pigment of diatoms) and peridinin (characteristic of dinoflagellates) are close to that of phycoerythrin. The absorption spectra of three phytoplankton cultures, *Synechococcus*, strain WH7803, *Thalassiosira weissflogii* and *Gonyaulax tamarensis* are shown in Figure 6.2. As a means of comparing the light absorption of these different species, cell concentrations were adjusted to have the same absorption at 680 nm. The light absorption at 550 nm by phycoerythrin in *Synechococcus* is clearly significantly more than that by *T. weissflogii* but not very much greater than that by *G. tamarensis*. But all three organisms absorb light at the wavelengths of maximum transmission in oceanic waters. These results must obviously be interpreted with caution since this analysis has taken no account of the effect of cell size, scattering and culture conditions; Figure 6.2 serves only as an illustration that cyanobacteria are not the only organisms capable of absorbing light at the absorption maximum of phycoerythrin. Diatoms, dinoflagellates and indeed many other phytoplankton groups possess pigments which will harvest light of wavelengths found at the base of the euphotic zone. The pigment complement of *Synechococcus* is, therefore, not strong evidence for the hypothesis that picoplankton are obligate shade organisms.

However, evidence for adaptation comes from studies of the cellular concentration of photosynthetic pigments in natural assemblages. Glover *et al.* (1988) showed that there was a maximum in phycoerythrin concentration at a depth corresponding to 1-5% of surface irradiance. This peak was not the result of increased numbers of *Synechococcus* at this depth, but rather of an increase in phycoerythrin per cell; *Synechococcus* from the surface contained *ca* 20 fg phycoerythrin per cell but, at the 1% light level, the intracellular concentration had increased to *ca* 100 fg phycoerythrin per cell. This increase could have been a response to low irradiance but it might equally have been a consequence of higher nutrient concentrations in the deeper water. Wyman *et al.* (1985) suggested that *Synechococcus* might use phycoerythrin as a nitrogen reserve, as well as a photosynthetic pigment, when growing in the presence of increased nitrogen supply. Nutrients are continually advected across the pycnocline from deeper water and the response observed by Glover *et al.* (1988) might have been a consequence of the increased availability of nutrients towards the base of the euphotic zone. Whatever the reason for the intracellular increase in phycoerythrin, these data do show adaptation by picoplankton in response to their growth conditions.

There also appears to be little evidence from the optical properties of oceanic waters to support the hypothesis that the phycobiliproteins of cyanobacteria confer any advantage over eukaryotic picoplankton. Lewis *et al.* (1986) state "Of the thousands of measurements that have been made of the spectral characteristics of submarine light, none that we have seen to date show any evidence of the presence of phycobilipigments". These authors conclude that phycobiliproteins are not important light-harvesting pigments in the ocean, relative to chlorophyll and carotenoids, and they suggest that more attention should focus on the eukaryotic picoplankton. It is now known that eukaryotic picoplankton, which do not contain phycobiliproteins, can become the dominant pico-plankton biomass at the base of the euphotic zone (Glover *et al.*, 1988; Wood, 1985) and the presence of phycobiliproteins does not therefore appear to confer a significant advantage on the cyanobacteria at these depths. The recent discovery of *Prochloron*-like picoplankton in the deep euphotic zone (Chisholm *et al.*, 1988) again demonstrates the importance of chlorophyll and carotenoids in adapting phytoplankton for efficient light harvesting at very low irradiances.

Photosynthetic reaction centres

One other indicator of the ability of a phytoplankton cell to adapt to different irradiances involves changes in the ratio of the two reaction centres of photosynthesis; photosystem II is associated with the production of oxygen from water and photosystem I is responsible for the reduction of NADP. Each photosystem consists of chlorophyll and accessory pigments bound in a protein complex. The light harvesting complexes for photosystem II in cyanobacteria are large protein complexes known as phycobilisomes, which deliver >90% of the light absorbed to the reaction centres of photosystem II (Manodori & Melis, 1984). Phycoerythrin associated with photosystem II is also present in phycobilisomes. In most cyanobacteria, phycobilisomes can be observed by transmission electron microscopy as electron-opaque granules, evenly spaced on the thylakoid membrane; Cohen-Bazire & Bryant (1982) show a good example from *Pseudanabaena*. In freshwater *Synechocystis*, the phycobilisomes show chromatic adaptation and are significantly smaller when grown in red, rather than green light (Cohen-Bazire & Bryant (1983). In contrast, there is no evidence for similar changes in phycobilisome structure in marine *Synechococcus*. Figure 6.3 shows transmission electron micrographs of *Synechococcus*, strain WH7803, grown in the laboratory under low irradiance. No identifiable phycobilisomes were found within these cells and it was, therefore, not possible to identify any effect of changes of irradiance on the membrane structure of this marine cyanobacterium.

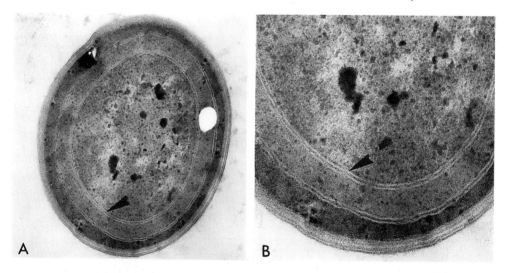

Figure 6.3. Transmission electron micrographs of the thylakoids of *Synechococcus*, strain WH7803 grown at *ca* 50 μE m^{-2} s^{-1}.

Photosynthesis/irradiance curves of marine cyanobacteria

One of the earliest measurements of photosynthetic characteristics of marine picoplankton was by Platt *et al.* (1983), working in the sub-tropical north Atlantic. The photosynthetic parameters of picoplankton, taken from the chlorophyll maximum at 65-89 m, were very different from those of the >1 μm size fractions; these authors concluded that the photosynthetic picoplankton were adapted to photosynthesis at lower irradiance than larger phytoplankton. Similar data were obtained in the tropical Pacific by Li *et al.* (1983) and by Glover *et al.* (1985) in the Gulf of Maine.

These observations confirm that, like other phytoplankton, picoplankton can adapt the photosynthetic apparatus to utilise the low irradiance and wavelengths of light found towards the base of the euphotic zone. However, there are suggestions that *Synechococcus* may show different physiological responses to other phytoplankton. Barlow & Alberte (1985) found that the response of *Synechococcus* to increasing light was an increase in photosynthetic unit size; these observations are exactly opposite to the response of larger phytoplankton, such as diatoms and dinoflagellates, which have been shown to increase photosynthetic unit size in response to *decreasing* irradiance. Barlow & Alberte suggest that *Synechococcus* responds to low irradiance by increasing the numbers of photosynthetic units per cell; coupled with this increase in number of photosynthetic unit was an increase in the absorption of light by phycoerythrin. These observations confirm a high degree of photoadaptation in marine *Synechococcus*.

Surface distributions

The evidence discussed above supports the hypothesis that *Synechococcus* shows adaptation to different light fields but does not support the hypothesis that *Synechococcus* is an obligate shade organism. If natural assemblages of *Synechococcus* show the same level of adaptation to different irradiance that are observed in culture, then there should be examples of cyanobacteria growing at high irradiances near to the surface of the euphotic zone.

In the initial descriptions of marine *Synechococcus*, both Waterbury *et al.* (1979) and Johnson & Sieburth (1979) found highest numbers of cyanobacteria in the surface waters. Even in those cases where picoplankton are considered to account for most of the biomass in the sub-surface chlorophyll maximum (Takahashi & Bienfang, 1983), picoplankton are also found in the surface water. In European shelf seas, we have found that the biomass of the picoplankton is highest in the surface 15 m (Joint & Pomroy, 1986). In a recent study of the North Sea, Howard & Joint (1989) found as many as 2×10^8 cells l⁻¹ in the surface waters (Figure 6.4). Therefore, cyanobacteria are not found exclusively at the base of the euphotic zone; indeed, it is common to find the highest biomass close to the surface (Howard & Joint, 1989).

In addition, we have found that the rate of fixation of ¹⁴C by picoplankton, determined by *in situ* incubations, is always highest at about 10 m, in the near surface waters (Joint & Pomroy, 1983, 1986; Joint *et al.*, 1986). Populations of picoplankton are not only abundant under conditions of high irradiance, but are also capable of active photosynthesis. On occasion, photoinhibition was present in the surface 5 m, but the irradiance at these depths was considerably higher than reported by Barlow & Alberte (1985) for laboratory cultures.

Measurements of photosynthetic characteristics of natural assemblages from the surface mixed layer that we have made are very different from the data obtained by Platt *et al.* (1983) for the subsurface chlorophyll maximum in the sub-tropical Atlantic. Joint &

Pomroy (1986) showed that, although values of assimilation number (P_m^B) were generally higher for the picoplankton fraction, there was no evidence of lower values of the adaptation parameter (I_k). The concept of the irradiance at which photosynthesis saturates (I_k), was introduced by Talling (1957a) and the adaptation parameter (equivalent to P_m^B / α^B) is a measure of the degree to which phytoplankton is adapted to low irradiance.

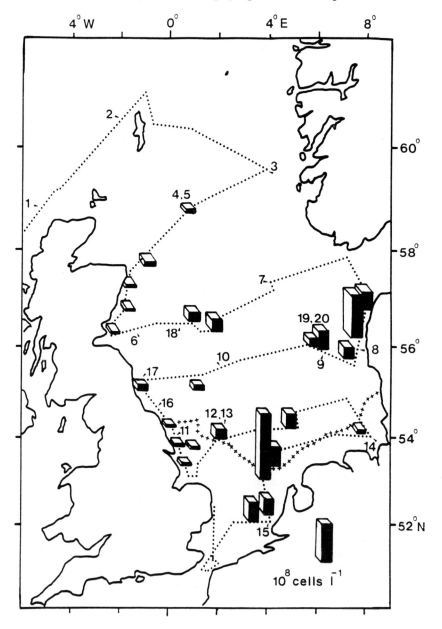

Figure 6.4. Distribution and abundance of cyanobacteria in the water taken from 2 m depth in the North Sea during July 1987; the cruise track (····) and position of the frontal boundary (+++) are indicated. The data are plotted on a linear scale.

Joint & Pomroy (1986) also found no evidence of photoinhibition of either picoplankton or larger phytoplankton in these experiments. However, the maximum irradiance used was only 500 μE m^{-2} s^{-1}; subsequent experiments have used higher irradiance of up to 1100 μE m^{-2} s^{-1}; Figure 6.5 again shows no evidence of photoinhibition in the picoplankton fractions of these natural assemblages taken from near surface of the North Sea (Howard & Joint, 1989).

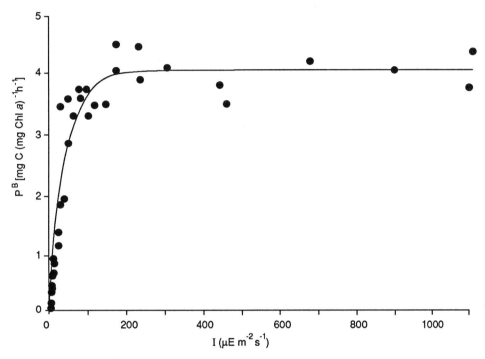

Figure 6.5. Typical photosynthesis/irradiance curves for the picoplankton fraction taken from the surface 5 m in eastern North Sea (station 20, from Figure 6.4) on 30 July 1987.

Cultures grown at high irradiance

The evidence presented so far supports the hypothesis that picoplankton photosynthesis can adapt to changes in irradiance, rather than the alternative that picoplankton are obligate shade organisms. The only remaining evidence in support of the shade hypothesis is the observation that *Synechococcus* cultures do not appear to tolerate high irradiance conditions (Waterbury *et al.*, 1986).

Kana & Glibert (1987) have recently demonstrated that it is possible to culture *Synechococcus* in the laboratory at high irradiance. *Synechococcus* strain WH7803 was adapted to high irradiance levels by allowing a 3-4 day period for each two-fold increase in irradiance. This procedure enabled the cultures to grow successfully, with a doubling time of *ca* 10 h, at irradiances equivalent to full sunlight (2000 μE m^{-2} s^{-1}). Growth rate of *Synechococcus* in this study saturated at 200 μE m^{-2} s^{-1}; this contrasts with the earlier experiments of Morris & Glover (1981), who measured maximum growth of *Synechococcus* strain WH7803 at between 45 and 70 µE m^{-2} s^{-1}, and of Barlow & Alberte (1985), who found light saturation of growth of strain WH8018 at 25-50 µE m^{-2} s^{-1}.

Kana & Glibert (1987) found a change in colour of the cultures in response to increasing irradiance; the cells were red at low irradiances but became pale yellow at high irradiance. These pigment changes have been investigated by Kana *et al.* (1988) who have now demonstrated that *Synechococcus* strain WH7803 responds to increasing irradiance by regulating the cellular content of phycoerythrin, chlorophyll and zeaxanthin. Light absorption by low-light cultures was dominated by phycoerythrin; at high growth irradiances there was a significant decrease in phycoerythrin and chlorophyll content and zeaxanthin became the most abundant pigment in the cell. However, carotenoids appear to play only a minor role in light absorption in *Synechococcus* and Kana *et al.* (1988) suggest that variations in phycoerythrin alone could explain the changes in photosynthetic characteristics that occur with altered growth irradiance. These authors also show that there were a number of similarities between cultures of *Synechococcus* grown at high light and natural assemblages of cyanobacteria taken from the surface of the ocean; this is further evidence to suggest that photoadaptation does occur in marine *Synechococcus*.

CONCLUSION

The evidence summarised in the paper shows that many of the assumptions made about an obligate requirement for low light do not stand up to examination. It is true that picoplankton are frequently found at the base of the euphotic zone and may be the major biomass in the sub-surface chlorophyll maximum; but there are also many examples where picoplankton are more abundant in near surface waters than at depth. Evidence from natural assemblages shows that photosynthetic rates of picoplankton are high in near surface water; photoinhibition does occur in the surface few metres, but larger phytoplankton, as well as picoplankton, show photoinhibition under these circumstances. In short-term incubations to determine photosynthesis/irradiance parameters, picoplankton do not appear to be more adapted to low light than other phytoplankton. Finally, the demonstration that *Synechococcus* cultures are indeed capable of growth at 2000 µE m^{-2} s^{-1}, and are not restricted to irradiances of <50 µE m^{-2} s^{-1} is conclusive proof that *Synechococcus* is not a 'shade' organism.

Picoplankton do not have an obligate requirement for low light conditions and we must look for a reason, other than light, to explain the observations of high picoplankton abundance at the base of the euphotic zone.

This work forms part of the research of the Plymouth Marine Laboratory, a component body of the Natural Environment Research Council.

Chapter 7

LIGHT AND DEVELOPMENT: CELLULAR AND MOLECULAR ASPECTS OF PHOTOMORPHOGENESIS IN BROWN ALGAE

C. BROWNLEE

Marine Biological Association, The Laboratory, Citadel Hill, Plymouth, PL1 2PB

Responses to light play an essential role in the development of all photosynthetic multicellular plants. Plants can respond to light quality, irradiance, duration and direction. Two broad classes of response have been identified. These are phototropism in which the direction of cell growth or elongation is influenced by the light direction, and photomorphogenesis in which the patterns of cell differentiation and division respond to the incident light. The major pigments involved in these processes will be summarized. The special role of cell growth and polarity in plant morphogenesis is discussed. In zygotes of the Fucales, directional light can be considered to have both phototropic and photomorphogenic effects, *i.e.* the light direction not only dictates the direction of growth of the emerging rhizoid, but influences the differentiation of the single cell zygote into distinct rhizoid and thallus cells. Evidence for the types of pigments involved in this response and their location plays a key role in translating a directional light signal to spatial information in the cytoplasm, which is ultimately transmitted to the genome. Progress has now been made in identifying the second messenger systems involved. Increase in our knowledge of how polarity is established and maintained at a molecular level in plant cells is leading to increased understanding of how a directional light signal is perceived and translated. A model is presented to describe the likely molecular mechanisms involved.

INTRODUCTION

Light is the single most important environmental parameter affecting the growth and development of all plants. The quality and direction of incident light are key factors influencing the polarity and differentiation of many plant cell types. Other factors include gravity, external chemical and ionic gradients and growth regulators. The acquisition of polarity at the cell level is a fundamental process in the growth and development of plants and animals. Since cell migration cannot play a role in the development of multicellular plants, the directions of growth and planes of division of individual cells dictate the pattern of morphogenesis of the whole plant. In fertilized zygotes, the initial polarization of a single cell sets the plane of division and subsequent polarity.

Much of what we know about how polarity is established and maintained in a plant cell comes from work with zygotes of brown algae (Phaeophyceae), particularly the

Herring, P.J., Campbell, A.K., Whitfield, M. & Maddock, L. (ed.). *Light and Life in the Sea.*
© 1990. Cambridge University Press.

Fucales and species of *Fucus* and *Pelvetia*. There are several reasons for this. Large numbers of eggs and sperm can easily be obtained. Soon after fertilization, the zygote begins to secrete a cell wall and can be dramatically polarized within a few hours by a variety of directional external signals including light. This polarity is manifest by the appearance of a rhizoid under natural conditions, which begins to grow from the side of the zygote away from the direction of incident light (Figure 7.1). The general events associated with polarization in the Fucales and other plant systems have been reviewed elsewhere (Evans *et al.*, 1982; Schnepf, 1986). To summarise, sperm entry causes the egg to become activated so that further sperm entry is prevented and a well defined sequence of developmental events is set in motion. A few minutes after fertilization, the egg begins to secrete a cell wall and dramatic changes in the ionic conductivity, ion fluxes and membrane potential begin. These are associated with the development of internal turgor (Allen *et al.*, 1972). In unilateral light, a positive current can be detected entering the shaded hemisphere. This precedes any detectable structural changes. The polar axis becomes fixed around 12 hours after fertilization and marked polarization in the distribution of cytoplasmic inclusions can be detected around this time. This is followed by visible germination of the zygote.

The purpose of this chapter is to bring together current knowledge on the mechanism of action of the photoreceptor(s) involved in sensing the direction of incident light and the cellular mechanisms thought to be involved in translating a unidirectional light signal into positional information leading to the development of a polarized plant.

PERCEPTION OF THE LIGHT STIMULUS

Photomorphogenic sensory systems in plants

Plants have evolved the capacity to sense and respond to the light direction, quality, irradiance, duration and photoperiod, without the development of specialised organs and often using apparently the same photoreceptor pigment(s) (for general reviews see Smith, 1981; Senger, 1987*a*). Photomorphogenic and phototropic pigments have been classified into two major types, phytochrome and cryptochrome (excluding the fungal blue-light receptor, mycochrome, which is now thought to be a complex of more than one pigment type (Corrachano *et al.*, 1988)). The phycobiliprotein, phytochrome, is the ubiquitous red/far-red photoreversible green plant pigment and the one for which there is most information concerning the types of responses which it controls and its mode of action (Quail, 1983; Colbert, 1988). Phytochrome has been implicated in responses as diverse as greening, flowering and phototropism. It is involved in sensing the quality, irradiance and photoperiod. In addition to the red/far-red sensitivity, it is probably involved in the blue-light responses of several tissues (Mohr, 1987). Much evidence now exists for a primary effect of phytochrome at a membrane location, involving second messengers such as calcium and protein phosphorylation and a role in regulating gene expression (Pooviah & Reedy, 1987; Roux *et al.*, 1987). Though there are reports of the presence of phytochrome in red algae (Dring, 1967), and red/far-red-reversible responses have been described for other non-green algae, including brown algae (Dring,

1988; Duncan & Foreman, 1980), there is doubt whether true phytochrome responses exist in non-green algae (Dring, 1988). Phytochrome has not been extracted from non-green algae.

Another major pigment (or class of pigments) involved in photomorphogenesis and phototropism is the blue-light receptor, cryptochrome (Thomas, 1981; Senger, 1987a). This is found in a wide range of species from green plants to brown algae. Much less is known about the mode of action of cryptochrome. Its distribution, likely identity and possible mode of action have been extensively discussed (Briggs & Iino, 1983; Senger, 1987a). Though the identity of the blue-light receptor is still not firmly established there are two main candidates, based on action and absorption spectra (Briggs & Iino, 1983; Galland, 1987; Schmidt, 1987; Song, 1987). The first is a flavoprotein, as described for the first positive phototropic curvature of *Avena* coleoptiles (Thimann & Curry, 1960), with action spectrum peaks between 450 and 460 nm. Evidence for the involvement of a flavoprotein comes from light-induced absorbance changes as well as action spectra (Briggs & Iino, 1983; Song, 1987; Widell, 1987). Light-induced absorbance changes have been shown to reflect the reduction of a *b*-type cytochrome (Muñoz *et al.*, 1974; Muñoz & Butler, 1975). The second major candidate for the blue-light photoreceptor is a protein-bound carotenoid (Briggs & Iino, 1983) However, the function of carotenoids other than in screening (Vierstra & Poff, 1981) and as energy transfer pigments (Song, 1980) has been seriously questioned (Song, 1987).

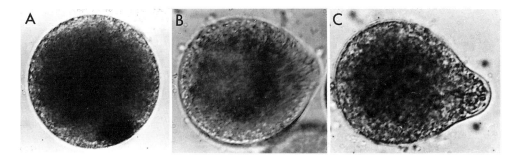

Figure 7.1. Sequence of events in the polarization of a *Fucus* zygote by a unidirectional light stimulus. Arrow indicates the direction of incident light. (A) Unpolarized zygote up to 12 h after the onset of polarizing stimulus. (B) Germination from the shaded side of a zygote, occurring after 12 h. (C) Cell division with the plane of division at 90° to the polar axis. first cell division occurs around 30 h after fertilization.

The nature of the photoreceptor in the polarization of Fucus

The majority of light responses in brown algae are blue-light-sensitive and crypto-chrome appears to be the photoreceptor in most cases (Dring, 1987, 1988), though action spectra similar to the fungal blue-light response have been found in photoperiodic responses in brown algae (Dring, 1987). These and other algal light responses have recently been discussed in considerable detail (Dring, 1987, 1988). Examples of the blue-light response in brown algae include the formation of hairs and the onset of two-dimensional growth in *Scitosyphon* (Dring & Lünning, 1975). This is a true photomorpho-genic response and not simply a phototropic response since it involves changes in the patterns of cell division and differentiation. The action spectrum shows a close similarity to those of other typical cryptochrome responses, particularly those involving a flavo-protein. Rudimentary action spectra and fluence response curves have been obtained for the polarization response in *Fucus* (Jaffe, 1958; Bentrup, 1963). Blue light is most effective, though Bentrup (1963) demonstrated additional high activity in the UV region, around 250 nm. The data of Bentrup (1963) showed a fluence-response curve with two peaks. This, together with action spectra at different irradiances led to the suggestion that two pigments are involved in the response, probably corresponding to a flavoprotein and a carotenoid. Desensitisation of the response was observed as irradiance increased. More detailed action spectra and fluence response curves are required to confirm these observations and the conclusions drawn from them. Schmidt (1983) has argued strongly against the the unequivocal interpretation of action spectra in the identification of photoreceptor pigments, and the involvement of carotenoids as primary photoreceptors in the blue-light response is still questioned (see above).

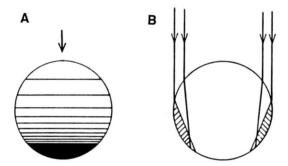

Figure 7.2. Effects of screening and refraction (lens effect) on the light distribution in a spherical cell. (A) System dominated by the presence of screening pigment. In this case the surface facing away from the incident light (distal surface) receives least light. (B) System in which screening pigments are absent and with a refractive index significantly higher than the surrounding medium. A sub-equatorial band exists which receives very low light levels. The distal region receives highest light due to the focussing effect. Arrows indicate the direction of incident light.

Location of the photoreceptor and mechanism of perception of unidirectional light

The mode of action of a photoreceptor will be highly dependent on its location. Clearly, a close relationship between the photoreceptor and a fixed cellular structure will facilitate the translation of a light gradient into spatial information inside the cell. The obvious candidate for the location of a photoreceptor which relays information from the outside to the inside of a cell is the plasma membrane. This will enable interaction with other molecules and complexes in the plasma membrane, such as those now known to be involved in the transduction of a wide variety of signals. These include the phospho-inositol pathway (Berridge & Irvine, 1984; Berridge, 1987), G protein complexes (Gilman, 1987) and ionic channels, especially calcium channels (Reuter, 1986). Raven (1981, 1983) has critically discussed evidence for and against the action of photoreceptors at a membrane location.

There are two general mechanisms by which the direction of incident light can be sensed by a spherical cell. These are shading by the pigments contained in the cell and focusing by the lens effect. Figure 7.2 illustrates the relative importance of shading and the lens effect caused by differences in the refractive index between the cell and the medium. In a transparent cell, the lens effect predominates and the sub-equatorial region (region b) receives least light. In a cell containing screening pigments, region b or the side of the cell facing away from the direction of incident light (region c) may receive least light, depending on the degree of shading and refractive index. The relative importance of screening and refraction in selected systems has been discussed by Poff (1983).

In considering the perception of unidirectional light by *Fucus* zygotes, the ability to sense the light direction over a very wide range of irradiance values must be taken into consideration. Unlike specialised organs such as animal visual organs, the receptor pigments are almost certainly uniformly distributed. In a classic study, Jaffe (1958) examined the polarization response of *Fucus* zygotes to plane polarized and unpolarized unilateral light. Both the response to unpolarized and polarized light showed the involvement of the blue photoreceptor. Three response types were identified (Figure 7.3). In plane polarised light, zygotes germinated in the plane of polarisation and 'subequatorially', *i.e.* around 90° to the the direction of the incident light (Figure 7.3A).

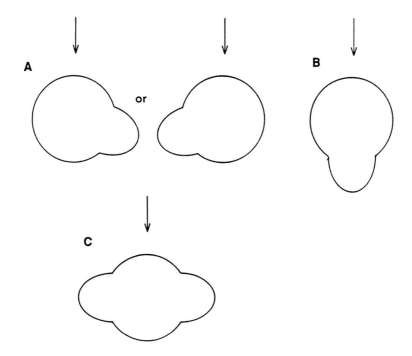

Figure 7.3. Germination responses to polarized and unpolarized light. (A) Sub-equatorial response to polarized light or low irradiance unpolarized light. (B) 'Rear pole' response to high irradiance unpolarized light. (C) Double rhizoid germination at 180° occasionally observed in polarized light or low irradiance unpolarized light.

This response was also observed in a significant number of zygotes germinating at relatively low irradiances of unpolarized light (10^{15}-10^{16} quanta s^{-1}). At higher irradiances (10^{17}-10^{19} quanta s^{-1}), the rhizoid germinated from the rear pole (*i.e.* the pole furthest from the incident light) (Figure 7.3B). A third type of response in which zygotes produced two rhizoids at 180° to each other was found in a significant number (up to 50%) of zygotes germinating in polarized light (Figure 7.3C). The subequatorial response in plane polarized light has been convincingly explained in terms of the location of the photoreceptor in the plasma membrane with the photoreceptor molecules oriented parallel to the membrane. Germination occurs where the photoreceptor molecules are excited least, *i.e.* in the plane of vibration and subequatorially (Jaffe, 1958). Subequatorial germination in low level unpolarized light was explained by the lens effect with subequatorial regions receiving least light. Different optical properties of the egg had to be invoked to explain rear pole germination in high irradiance. These might include altered scattering and absorption but so far there is little experimental evidence to support this view. It is also possible that different pigment systems are involved in the polarization response, one acting at low irradiance and the other at high irradiance. Recent work has revealed the presence of blue-light desensitisation of the blue-light response in other systems. For example, the blue-light-mediated production of hair whorls in *Acetabularia* can be desensitised by a prior pulse of blue light (Schmidt *et al.*, 1988). The light requirement of the desensitisation response was around 50 times less than for the photomorphogenic response. The desensitisation was concluded to be a result of inactivation of the photoreceptor or some early intermediate in the transduction chain, rather than a decreased ability of the cells to respond to an active photoproduct. In other systems, such as phototropism in maize coleoptiles, inactivation of an active photoproduct is likely to be involved in the desensitization process (Iino, 1987).

It may be possible to explain the observations of Jaffe (1958) by a mechanism (Figure 7.4) involving one photoreceptor which requires a critical low irradiance for the production of an active photoproduct. The lens effect and shading combine to affect the optical properties of the egg. At high irradiance, desensitisation occurs. The *Fucus* egg is heavily pigmented with chlorophyll, carotenes and xanthophylls, all of which will strongly absorb blue light. Consequently, the side of the egg facing away from the incident light will receive least light, since light reaching that region of the cell will have the longest path length. The subequatorial membrane region will receive significantly less light than the region in the lighted hemisphere due to the lens effect, assuming that the egg refractive index is significantly higher than that of sea water (Jaffe, 1958 calculated that a refractive index of 1·4 would produce a significant lens effect in *Fucus*). However, due to scattering and a shorter path length, the subequatorial region will receive more light than the distal region. In addition, light will be focussed in a region a few microns inside the egg in the subequatorial region (Jaffe, 1958). Internal scattering from the subequatorial region may also increase the light level at the adjacent membrane. Photoreceptor pigment will be inactivated at all times in the lighted hemisphere. At low irradiance levels (see Figure 7.4A), the shaded hemisphere will receive suboptimal light for production of active photoproduct. Active photoproduct concentration will be highest

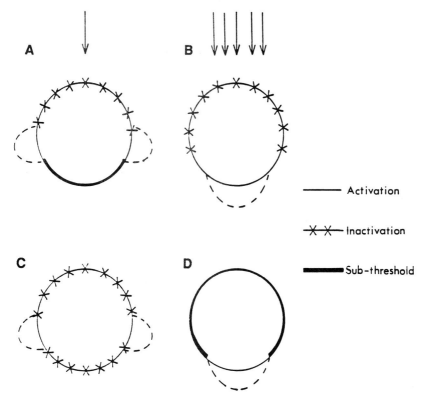

Figure 7.4. Possible mechanisms to explain the sub-equatorial and rear pole germination responses. Broken lines indicate the site of rhizoid germination. (A) & (B) Model based on a single photoreceptor and a combination of shading and the lens effect (see text for details). In low light (A), inactivation occurs in the proximal region and shading results in no activation in the distal region. The subequatorial region receives significantly less light than the proximal region but more light than the shaded distal region. Photoreceptor is activated only in the sub-equatorial region. In high light (B), all regions are photoinactivated except the shaded distal region. (C) & (D) Model based on two photoreceptor pigments. In low light (C), a high irradiance photoreceptor receives below threshold irradiance. The low irradiance photoreceptor is light-inactivated except in the sub-equatorial regions (the lens effect dominates in this model). At high irradiance (D) the low light photoreceptor is inactivated and the high irradiance photoreceptor is activated in the distal region which receives most light.

in the subequatorial regions. At high irradiance, the photoreceptor will be inactivated in all but the most densely shaded region (Figure 7.4B). Two sub-equatorial rhizoids might be expected since two regions in the horizontal plane will receive identical light. It is difficult to explain why only one rhizoid is often produced even in low light. Slight asymmetries in the unpolarized egg may result in one side becoming polarized marginally earlier than the other. Quickly developing positive feedback mechanisms involving asymmetric current fluxes would then suppress growth of the opposite pole (see below). Germination vertically up or down may be suppressed by the presence of the substrate (Jaffe, 1958) and the possible effect of gravity. To test these models, measurements of the light gradient around the surface of eggs in different light levels are required. The technology is available for such measurements to be made, using optical fibres and photon counting or video image analysis of unilaterally illuminated eggs. If the lens effect is shown to dominate the optical properties of the egg (*i.e.* if the distal region receives more light at all irradiances) then it may be necessary to invoke the involvement of two photoreceptor systems, one which is inactivated even at low irradiance, giving rise to the subequatorial response (see Figure 7.4C) and the other which is activated at high irradiance, giving the rear pole response (Figure 7.4D).

TRANSLATION OF THE LIGHT STIMULUS

Mechanism of action of the photoreceptor

Blue light has been shown to cause absorbance changes in several systems (Poff & Butler, 1974; Muñoz & Butler, 1975). These have been found to be related to reduction of a *b*-type cytochrome in plasma-membrane-enriched fractions. It is likely that a flavin-cytochrome complex is directly associated with a membrane protein (Leong *et al.*, 1981; Leong & Briggs, 1981). Thus, the primary action of the blue-light receptor involves a light-activated redox reaction. The next step in the signal transduction chain is less clear and is discussed in detail by Schmidt (1987) and Horwitz & Gressel (1987). Excitation and relaxation of the flavin may cause conformational changes in the protein component of the complex. There may be several or many different types of flavoproteins, each controlling a different signal transduction chain. These might include direct modulation of specific enzymes such as the blue-light-activated cAMP phosphodiesterase in *Phycomyces* (Cohen & Atkinson, 1978).

Other responses exist where the modulation of transport processes is the first measurable step in the transduction chain. These include increased calcium influx and elevated intraflagellar calcium in the blue-light reversal of flagellar beat in *Haematococcus* (Litvin *et al.*, 1978) and *Chlamydomonas* (Bessen *et al.*, 1980). These changes have not yet been correlated with any light-induced absorbance changes which may occur. Other systems, however, do not appear to involve changes in calcium fluxes or cytoplasmic concentrations as the initial step. These include modulation of flagellar beat in *Euglena,* where changes in intraflagellar free calcium are preceded by changes in potassium and sodium fluxes (Doughty *et al.*, 1980). Apical growth in *Vaucheria terrestris* is promoted by blue light. This is preceded by increased entry of positive current at the apex (Kataoka & Weisenseel, 1988). This response does not require calcium and is actually increased by the calcium channel-blocker, verapamil. Recently, Morse *et al.* (1987, 1989) showed that white light stimulates inositol phospholipid turnover in *Samanea saman* Pulvini, leading to increased levels of diacylglycerol and inositol polyphosphates, which act as important second messengers (Berridge, 1987; Berridge & Irvine 1984). It is likely that this is a blue-light response (Satter *et al.*, 1981).

Formation of the polar axis

Assuming that a unidirectional light stimulus results in spatial differences in the excitation of the photoreceptor and that this results in different rates of production of a photoproduct (see above), it is possible to speculate on the molecular mechanisms involved in translation of this asymmetry to the cytoplasmic machinery involved in segregation of cytoplasmic components and growth. To stimulate further experimentation, Figure 7.5 describes the possible sequence of events in the polarization of a *Fucus* zygote from the perception of the light stimulus to the fixation of the polar axis and rhizoid growth. Where experimental data for *Fucus* are absent, information from other systems has been applied, where appropriate. Only the 'normal' rear-pole response is considered, with the rhizoid growing directly away from the light source. Excitation of the receptor causes the stimulation of phosphoinositide breakdown via interaction with G-protein (Cockcroft & Gomperts, 1985). The breakdown products, inositol phosphates and diacylglycerol cause the phosphorylation of proteins via protein kinase C (Berridge, 1987) and the release of calcium from intracellular stores in the endoplasmic reticulum (Figure 7.5A). Calcium release may in turn stimulate reversible protein phosphorylation (Kikkawa *et al.*, 1986; Blowers & Trewavas, 1988; Ranjeva & Boudet, 1987). To test this hypothesis it will be necessary to characterize the G-proteins in the *Fucus* plasma membrane and to monitor changes in phosphoinositide turnover, particularly increased levels of inositol polyphosphates, during light treatment. Similarly, localised changes in the levels of free cytoplasmic calcium following blue-light absorption need to be demonstrated.

Polarization in *Fucus* and *Pelvetia* involves the formation of an initially labile axis, *i.e.* one which can be altered in unilateral light by rotating the zygote by 90°. Approximately 10 h following the onset of polarizing light in *Fucus* (6-8 h in the case of *Pelvetia*), the axis becomes fixed. Formation of the labile axis must involve the establishment of a reversible gradient. The first recorded asymmetric event associated with polarization is the entry of positive current at the site of future rhizoid outgrowth (Robinson & Jaffe, 1975). This current is well established within one hour after the onset of a polarizing stimulus and several hours before axis fixation. During axis formation and fixation, calcium ions constitute a major component of the current (Robinson & Jaffe, 1975). This has led to the hypothesis that a gradient of cytoplasmic calcium is established in the polarizing zygote with higher concentrations at the growing apex and that this gradient contains the information required by the cytoplasmic machinery to enable mechanisms associated with fixation, rhizoid growth and cell division to proceed. Direct evidence for the existence of such a gradient is scarce. Work so far in this laboratory with the fluorescent calcium indicator, fura-2, has provided preliminary evidence that elevated calcium levels exist around the future rhizoid growth zone just before and during axis formation (Brownlee, 1989), though this observation needs to be confirmed and further characterized. Physiological evidence for the involvement of a calcium gradient in axis formation comes from the work of Speksnijder *et al.* (1989) who showed that injection of polarizing *Fucus* zygotes with analogues of the calcium buffer BAPTA with different binding constants and at varying concentrations could arrest polarization and growth. The most effective concentrations of various analogues indicated that BAPTA was acting as a mobile calcium chelator and could be buffering a gradient with maximal concentration up to 10 µM. Evidence against the involvement of calcium in polarization has, however been provided by Kropf & Quatrano (1987). Here it was shown that zygotes could fix a

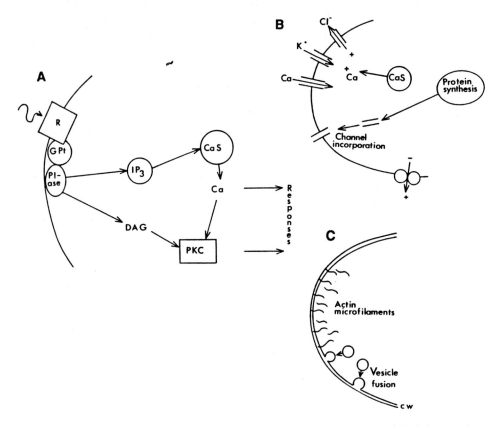

Figure 7.5. Summary of putative molecular events involved in the perception of the light stimulus through to the fixation of a polar axis. (A) Activated photoreceptor (R) interacts with G protein (GPt) on the inner face of the plasma membrane. This stimulates phosphoinositidase (PIase) activity and the hydrolysis of phospholipids to form inositol trisphosphate (IP$_3$) and diacylglycerol (DAG). Localized IP$_3$ formation causes localised calcium release from internal stores (CaS). DAG and calcium stimulate protein kinase C (PKC) activity. PKC and calcium activate a cascade of responses. (B) These responses include stimulation of channel activity and localized increased positive current entry. This, together with increased calcium entry, serves to amplify the the signal. (C) Axis fixation involves calcium-activated microfilament localization and vesicle fusion.

polar axis in the absence of external calcium and in the presence of verapamil which may block calcium entry. These apparently conflicting pieces of evidence could be reconciled if calcium influx serves to amplify an already existing calcium gradient established by localised intracellular calcium release (Figure 7.5A & B). The assumption here is that in the absence of calcium influx, the (IP$_3$-mediated?) intracellular release can be maintained without localised depletion of the intracellular stores, and that sustained production of IP$_3$ is possible. There is as yet little or no evidence to support this assumption. Responses involving IP$_3$ almost exclusively involve only transient changes in the levels of IP$_3$ (Berridge, 1987). However, there is evidence that diacyl glycerol levels can remain elevated during a protracted stimulation in some animal systems (Griendling *et al.*, 1986).

Qualitative changes in protein synthesis during axis formation have recently been described (Kropf *et al.*, 1989*b*). The first 12 hours following fertilization are characterised by qualitative differences in the synthesis of four proteins. It has been suggested that these proteins play an important role in the formation of the polar axis. Their precise identity and function remain to be elucidated.

Fixation of the polar axis

Fixation of the polar axis probably involves irreversible structural changes and has recently been shown to involve the localisation of actin microfilaments and the formation of transmembrane bridges between the cell wall and the microfilament skeleton (Kropf *et al.*, 1988, 1989*a*) (Figure 7.5C). This can be regarded as an irreversible structural change. Axis fixation is followed by germination of the zygote and rhizoid growth. Rhizoid growth involves the influx of calcium ions at the growing tip and the generation of a gradient of free cytoplasmic calcium with elevated levels in the tip region (Brownlee & Wood, 1986; Brownlee & Pulsford, 1988). The elevated calcium levels in the tip region may play a role in stimulating the transport and fusion of vesicles with the plasma membrane to supply new cell wall material. Calcium may also be involved in the regulation of transport processes via channel activity and ATPase activity, so providing the reinforcing mechanisms and positive feedback for sustained growth. Rhizoid elongation is strongly dependent on the influx of calcium ions. Removing calcium or blocking calcium entry with verapamil inhibits elongation. Germinating rhizoids show a typically negative phototropic response. The mechanism of this phototropic response and the relations between this photoreceptor system and that involved in the acquisition of polarity are not clear. Calcium influx may be involved since treatment of rhizoids with verapamil or removal of external calcium blocks the ability to perceive the light stimulus (Brownlee, unpublished).

Following the fixation of the polar axis and plane of first cell division, further divisions result in the multicellular plant. Major unanswered questions at the whole plant level include the identity of the extracellular morphogenic gradients, their establishment and maintenance, and the cellular mechanisms by which they are perceived. At the molecular level, differentiation ultimately relies on differential gene expression in response to cytoplasmic signalling. Though light regulation of gene expression has been demonstrated in many plant systems (Tobin & Silverthorne, 1985) there is little concrete evidence to suggest the mechanisms involved in the transfer of information from the cytoplasm to the genome. Some indication that changes in the levels of cytoplasmic calcium can control chromatin condensation and breakdown of the nuclear envelope has recently been obtained with sea urchin embryos (Twigg *et al.*, 1988). It is only when we know more about the molecular mechanisms involved at the membrane and in the cytoplasm which are controlling polarity and cell growth, and their interaction with the genome that we will be able to understand how a membrane-associated light receptor can function to shape the growth of a whole plant.

SUMMARY

The acquisition of polarity by zygotes of the Fucales is a blue-light response sharing several features with the blue-light cryptochrome response which is widespread throughout the plant kingdom. A membrane-bound flavinoid is likely to be the photoreceptor. Much knowledge has accumulated on the cell and molecular mechanisms by which plant cells acquire and fix a polar axis from work with the Fucales. Mechanisms by which a zygote can detect and respond to a unidirectional light gradient are discussed. Information is accumulating on the cell-molecular mechanisms of polarization. Major unanswered problems include the interaction of the photoreceptor pigment with the cell's second messenger systems, and the nature of the labile gradient established during polarization. In addition, the feedback mechanisms involved in amplification of the initial gradient need to be elucidated. The identity and characterisation of specific membrane channels which serve to regulate asymmetric ionic fluxes will greatly increase our understanding of the molecular mechanisms of polarization. Much information is also required on the mechanism of transfer of information to the genome. Work with zygotes of the Fucales is likely to continue to provide answers to the above problems.

Chapter 8

LIGHT AND VISION AT DEPTHS GREATER THAN 200 METRES

E.J. DENTON

Marine Biological Association, The Laboratory, Citadel Hill, Plymouth, PL1 2PB

In oceanic water below about 200 m depth, daylight is blue and almost constant in spectral composition and in angular distribution. However, the intensity of daylight falls tenfold for approximately every 75 m increase in depth. Bioluminescent lights are mostly blue or blue-green. They generally have wider spectral bands than the penetrating daylight. They differ greatly from each other in intensity, duration, frequency of flashing and angular distribution. The absolute threshold of vision for daylight for a man looking upwards will be reached at a depth of about 850 m. It is likely that some deep-sea animals can detect daylight down to about 1000 m. Above the depth of the absolute threshold visual acuity will increase rapidly with decreasing depth becoming less and less limited by the number of light quanta available. We should expect the possible acuity to improve more than 100-fold on going upwards by about 300 m. The rate of bleaching of visual pigments will vary enormously with depth. The time for half bleaching by daylight being measured in thousands of years at 800 m and in minutes at 200 m. As background daylight increases with decreasing depth the visibility of bioluminescent lights falls - by at least 1000 times on going between 100 and 600 m above the depth where the absolute threshold of vision is reached. Because sea water scatters and absorbs light a disproportionate extra intensity of bioluminescent light would be needed to extend its visibility much above 100 m even in the dark. Some deep-sea fish have two visual pigments and probably have colour vision. The possession of multibank retinae makes colour vision possible with only one visual pigment. For most animals the most useful kind of colour vision is one that gives good hue discrimination in the blue-green part of the spectrum. Some highly specialised fish eyes are described and the possible functions of the specialisations are discussed.

INTRODUCTION

Murray & Hjort (1912) in their famous book 'The Depths of the Ocean' write: "Nothing has appeared more hopeless in biological oceanography than the attempt to explain the connection between the development of the eyes and the intensity of light at different depths in the ocean. In a trawling from abyssal depths in the ocean we may find fishes with large eyes along with others with very small eyes or totally blind. Nowhere would a more perfect uniformity be expected than in the dark and quiet depths of the ocean. Brauer, who has given a valuable contribution to our knowledge of the eyes of deep-sea fishes, remarks in his treatise on the fish collections of the 'Valdivia' Expedition: 'If the surroundings really acted directly on the organisms, and were the only agents which

Herring, P.J., Campbell, A.K., Whitfield, M. & Maddock, L. (ed.). *Light and Life in the Sea.*
© 1990. Cambridge University Press.

could produce alterations, their influence would be much more uniform and general. Instead of this we find the greatest variation. Thus we find the eyes now altered or permutated, now highly differentiated even in closely related forms.'"

Since the publication of 'The Depths of the Ocean', which does in fact give good reasons for some of the variations found, a good deal has been discovered about the properties of light in the deep sea and about the eyes and luminescent organs of the animals that live there. However, very little indeed is known about the behaviour of deep-sea creatures and we are far from a full understanding of the functions even of the most specialized structures found in the eyes of deep-sea animals. Here, since the properties of light in the upper layers of the sea are well described by Sathyendranath & Platt (Chapter 1), I give only a brief account of the properties of light below 200 m depth. I then discuss, with reference to the human eye, some visual functions in deep-sea animals that depend on the limitations set to vision by the quantal nature of light and by signal-to-noise considerations. Since the optics of the eyes of deep-sea animals are described by Land (Chapter 9), I devote this chapter mainly to other topics. Finally I describe some features of a few very specialized eyes and speculate on their functions.

DAYLIGHT AND BIOLUMINESCENCE

As daylight penetrates into the sea it is absorbed and scattered and its intensity diminishes rapidly with depth. Even in clear oceanic water the intensity, as measured by an upward facing irradiance meter, falls to one tenth for about every 75 m increase in depth even for those wavelengths that penetrate best. When the sun is high in the sky the

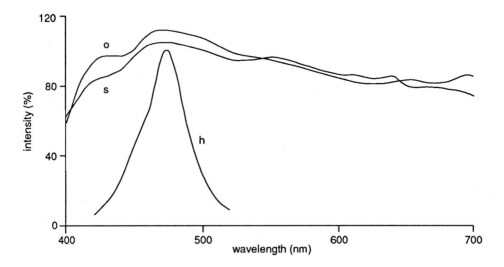

Figure 8.1. A comparison of spectral distributions of light intensities above the sea (o and s) and at depths of around 500 m in clear waters (h). o, overcast sky and s, sun and sky; both after Taylor & Kerr (1941) cited by Le Grand (1952). After Kampa (1970). Curves o, s, h have been made equal to 100% at the wavelengths 520, 505 and 475 nm respectively.

intensity of light on the surface of the sea is more than 1 kW m^{-2} whilst at about 800 m depth the intensity of light has fallen to about 10^{-9} W m^{-2}. The intensity of daylight above the sea varies little with wavelength over our visual spectrum. But, because the sea transmits some wavelengths very much better than others, the penetrating light, even at moderate depths, becomes confined to a relatively narrow waveband of wavelengths between about 430 nm and 530 nm centred in the blue at a wavelength of about 475 nm. However the deeper layers of the sea have different optical properties from those near the surface because phytoplankton is largely absent from the deeper layers and, for these deeper layers, extinction coefficients vary little for wavelengths between 430 and 530 nm. The consequence is that at depths below about 200 m the spectral distribution of light changes very slowly with increasing depth. On Figure 8.1 I show (after Kampa, 1970) this spectral distribution of intensities and compare it with curves for daylight above the surface of the sea.

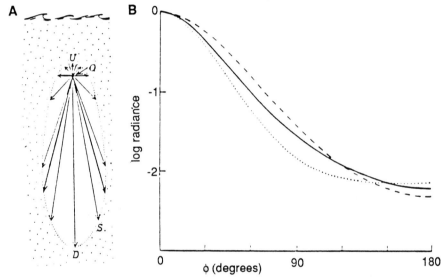

Figure 8.2. (A) Diagram of the kind of distribution of radiance found in the sea. For a given point *e.g.* O, the distribution in three dimensions is given by the surface formed by revolving the dotted area S around the axis UD. The relative radiance in a given direction is the length of line in that direction joining the point O to a point on the surface S. (B) The log radiance of light is plotted against θ, the angle between a given direction and the downwards vertical axis. The log radiance for θ=0 has been made equal to zero. ——Lake Pend Oreille (after Tyler 1960); ---- from an equation given by Tyler (see Denton *et al.*, 1985); .... for oceanic Mediterranean waters (after Lundgren, 1976, cited by Jerlov, 1976).

In discussions on the visibility of objects in the deep sea it is evidently more useful at each depth to have radiance values for all possible directions than merely to have downward irradiances. At the depths that we are considering the angular distribution of daylight is symmetrical about the vertical axis and changes little with depth (Figure 8.2). (Discussions on this phenomenon and references to original work on this topic are given in the general accounts of marine optics by Jerlov, 1976 and Kirk, 1983.) In the deep sea scattering is relatively low and mostly forward and there are very great changes of radiance with direction, thus to give examples the upward radiance is only about 0·5% of the vertically downwards radiance and the horizontal radiance is about 3% of the downward radiance. A useful fact is that at these depths the vertically downward radiance in W sr^{-1} m^{-2} is about equal numerically to the value of the downward irradiance measured in W m^{-2}.

Bioluminescent light is found at all depths in the deep sea. The luminescent organs found amongst deep-sea animals are of a variety quite unequalled in nature (see *e.g.* Herring, Chapter 16; Nicol, 1960; Marshall, 1954) and they are, in the sophistication of their structures and chemistry, of quite exceptional interest. Some animals produce orange and red lights but the great majority emit lights that are mainly confined to the blue, green and greenish-yellow part of the spectrum. Herring (1983) gives an account of his own extensive measurements and those of other authors. He concludes that the wavelengths of maximum emission for most specimens fall within the range 450-490 m, and that species found in the pelagic environment are mostly blue-emitting. In general, apart from the light emitted by some ventral photophores, the emission curves for bioluminescent pelagic animals are broader than those for submarine daylight. Many animals produce bioluminescent light in flashes. These flashes vary greatly in intensity and duration (Nicol, 1958*a*; Herring, 1983; Widder *et al.*, 1983). (For an account of the kind of flashes of bioluminescence measured in the sea with bathyphotometers see Clarke & Hubbard, 1959; Clarke & Denton, 1962.)

At dusk and dawn daylight changes greatly - at the equator by a factor of about 10^8 and by as much as 6×10^6 in one hour. At higher latitudes the changes are less great and less quick. In relation to photosynthesis moonlight is of no importance. It can, however, play a significant role in vision because the eyes of animals are so extremely sensitive. The level of bright moonlight is about 3×10^{-6} of that of sunlight. This means that on bright moonlit nights its contribution to submarine 'daylight' at about 400 m is close to that found in full sunlight at about 800 m. The intensity of daylight varies with the sky conditions. When the sun is obstructed by thin clouds its value falls by a factor of about two; for average cloud conditions obstructing the sun's rays the value falls about threefold. Occasionally for dark clouds preceding a heavy thunderstorm the value for a clear day is reduced about tenfold (Brown, 1952).

Apart from colour, penetrating daylight and bioluminescent lights are very different in their properties and the adaptations of animals to them are correspondingly very different. Near the surface of the sea in the day-time, daylight is dominant. Below about 900 m daylight is of trivial importance. Between about 200 m and 900 m depth we have to take account of both.

REFLECTED LIGHTS

The reflective properties of animals are evidently of the highest importance in determining their visibility (Partridge, Chapter 10) and the ideal properties for camouflage are different for daylight and bioluminescent lights. Thus the very highly reflective surfaces found in the vertical sides of some fish that give effective camouflage in penetrating daylight are less good against predators looking for their prey with their own 'headlights' or reflected bioluminescence from other animals. Curiously, however, the more perfect the mirror the safer it could be in these circumstances. This is because a perfect mirror would reflect a given light to such a predator over only a very restricted range of angles. Although the silvery sides of fish are not completely specular they often scatter light over only a small range of angles (see Denton & Nicol, 1965). In clear oceanic waters below the depths at which daylight is significant, or on dark nights everywhere, good camouflage against being seen by bioluminescence is given by having body surfaces that have a very low reflectivity for blue light and many oceanic animals living deeper in the sea are either black or dark red. In general pigments that absorb and so do not reflect blue light will appear dark no matter what other colours they reflect (Newton, 1730). To give perfect camouflage the reflectivity would have to be very low because clear oceanic waters scatter so little light back towards a source of light that any reflected light will be highly visible.

VISUAL PROBLEMS OF DEEP-SEA ANIMALS

The animals living in the light environment described above have to find their food, avoid predators and find their mates and reproduce. Amongst the visual problems they have to solve are those of:

(1) holding an appropriate depth station in the day-time. The vertical stratification of many deep-sea animals in the daylight is a most remarkable feature in their lives. It demands that these animals can measure something approximating to absolute light levels.

(2) camouflaging themselves in the day-time. Here, we may consider the problem of camouflaging a fish of square cross-section. The back of the fish will be invisible to an observer above the fish if the reflected light equals the light passing upwards on either side of the fish. In the ocean this demands that the back be very dark. The vertical sides of the fish will be invisible if they are perfect mirrors for they would then always reflect to any observer an intensity of light which he would see if the 'fish' were not there at all. The ventral surface will be invisible only if it emits light of an intensity, angular distribution and spectral distribution which will match that of the sea around the fish. At depths greater than about 250 m the angular and spectral distributions of light are constant but to match the intensity of the penetrating daylight demands that the animal 'measure' the intensity of light striking the dorsal surface and match it by the light emitted from its ventral surfaces (Figure 8.3). Nearer the surface, as Young (1977) and Young & Roper (1977) have shown, cephalopods can change both intensity and spectral composition of the light they emit so as to match the ambient light.

In squid this process is mediated by special extra-ocular receptors (Young, 1977; Young & Arnold, 1982) but in some fish the eyes seem to be involved. It appears that light from special photophores that shine into the eyes is coupled, in the intensity of light that they produce, to the ventral photophores. This allows the animal to match the output of its ventral photophores, which the fish cannot see, to the daylight from above which it can see.

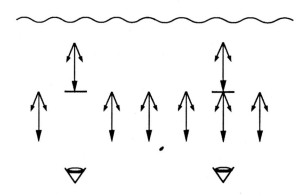

Figure 8.3. Diagrams showing on the left an observer in the deep sea viewing an object against a background of submarine daylight; on the right the object is camouflaged because it emits light from its lower surface to replace that which its upper surface absorbs or reflects.

(3) detecting and identifying other animals either by the daylight and bioluminescent lights that they reflect in particular directions or, in the day-time, by the 'shadows' that they cast below themselves. Since upward light is so much less intense than downward light the lower surfaces of animals cannot reflect light of sufficient intensity to match the downward light and these surfaces will appear dark to observers below them.
(4) seeing bioluminescent light:-
    (a) emitted by other organisms including the lights used to distinguish members of an animal's own species - sometimes the patterns of bioluminescence are different between the sexes of a given species. Beebe (cited by Marshall, 1954) writes that he could tell at a glance what and how many of each species of lantern fishes were represented in a catch, from their "luminous hieroglyphics".
    (b) reflected by other animals. Examples of fish that hunt and communicate with their lights are found in some of the astronesthids, melanostomiatid and malacosteid fishes. One such animal was seen by Dr E.R. Gunther emitting a beam of blue light directly forwards and using this light to catch its prey (Marshall, 1954).
(5) as a special case of (3) and (4) locating the exact position of prey in front of their jaws.

INTENSITY LEVELS AND VISUAL ACUITY

A convenient starting place in discussions on intensity levels and visual acuity is the depth at which daylight could just be detected by a dark-adapted human looking upwards in clear oceanic waters. For such a broad field of light this will be when the flux of quanta absorbed by the retina is about one quantum per second per 5000 retinal rods, where a human rod has a diameter of about 2·6 μm (Denton & Pirenne, 1954. They used a field subtending about 40° at the eye and 5 second flashes.). This flux will be found at about 850 m. At the absolute threshold of his vision an observer can detect only that light is present and cannot recognize shapes. However, at intensity levels only a few times above the absolute threshold significant form vision is possible and if an observer extends his hand towards the light he can easily distinguish his fingers. In the sea such an increase in intensity would be given by moving upwards by about 30-40 m say from 850 m to 810 m. Visual acuity rises very quickly at first as intensity increases above the absolute threshold (Figure 8.4). Thus at intensity levels of 3, 10 and 100 times the absolute threshold the minimum angles that a black disc would have to subtend at the eye to be detected are about 2°, 1° and 0·25° respectively. For such an observer looking straight upwards a luminance level of 100 times the absolute threshold would be found at depths of around 700 m and here the acuity would be such that a black object about the size of a housefly could be detected at a range of about 1 m. By a similar kind of argument we should, neglecting scattering, expect an animal with an eye like that of the human, but with a spectral sensitivity more appropriate to the light available, to be able to detect objects subtending about 10' at 600 m depth and about 1' at 400 m depth.

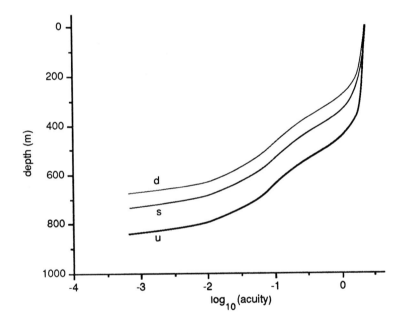

Figure 8.4. Log acuity for a human observer plotted against depth in the sea (daytime), u, upward viewing, s, sideways viewing and d, downwards viewing. The curves are calculated from data which can be found in Pirenne and Denton (1952) and Denton et al. (1985). Visual acuity is defined as the reciprocal of the angle, in minutes of arc, subtended at the eye by a test object when it just ceases to be detectable. If acuity is limited by the numbers of quanta absorbed by the photosensitive retinal pigments then the corresponding curves for the eye of a deep-sea fish (with the same size of pupil) would be displaced downwards by about 120 m.

With respect to directions other than looking almost directly upwards we may note that if a given downward radiance is to be found at a depth of h metres the same radiance (and acuity) can be found for directly downwards viewing and for sideways viewing at depths that are about $(h - 170)$ m and $(h - 110)$ m respectively. As Figure 8.4 shows, at some depths acuity can be very different in different directions.

A fish (or squid) eye with the same area of pupil as the dark adapted human eye (*i.e.* about 0·5 cm^2) has a focal length (*f*) of about 1·0 cm (the human eye's object focal length is about 1·7 cm). We find then that *i*, the focal length divided by linear aperture, is smaller, the ocular media are more transparent, and the absorption of light by the retinal visual pigments much higher for the deep-sea fish than for the human. This means that when facing a given extended source of light, such as that given by daylight in the sea, the number of quanta absorbed by the visual pigments per unit area of retina per second will be about 50 times greater. Now Pirenne & Denton (1952) show that, for the human eye, over the large range of acuities corresponding to 100-1 in minutes of arc subtended by the gap which can just be discriminated in a Landolt C test object, the number of quanta absorbed by the retinal receptors which receive light from the gap will be between about 7 and 45 s^{-1}. These numbers are not very different from the value of 5-6 quanta absorbed for a just-detectable short flash on an otherwise completely dark field. Similar and probably even lower numbers for given acuity will almost certainly apply to a fish provided that it has the appropriate nervous organization.

With a given external field of light the levels of retinal illumination (number of quanta per unit area of retina per second) for the fish eye will be independent of the size of the eye and the linear sizes of the images will be proportional to the size of the eye. If acuity is limited by the levels of retinal illumination, then it follows that acuity will also be proportional to size. We may note, however, that if predator and prey are of similar form (especially size of eye to overall size of animal) they will see each other at exactly the same distance apart. We also conclude that for the fish eye the range of depth over which acuity changes appreciably is likely to be roughly the same as for the human but that the absolute depths will be different. Thus (for given acuities) a fish eye for which $f = 1·0$ cm and which has a high absorption of light by the retinal visual pigments, the corresponding depths will be about 125 m deeper, whilst for a fish eye for which $f = 0·1$ cm the corresponding depths will all be about 20 m less deep than those for man. We may note that when acuity is changing quickly with depth, the acuity for sideways or downwards viewing will be much less good than for upward viewing. Here we have not taken account of the fact that the common visual pigments of deep-sea fish are almost certainly more stable than rhodopsin in man and hence the dark noise (without external lights) will be less. This will have a significant effect in lowering the absolute threshold of vision.

For the human eye visual acuity falls by a factor of only about two as the intensity of light drops from about 10^{10} to about 10^6 times the absolute threshold of vision. The former value corresponds to the luminance levels found for a white object under direct sunlight, the latter to the values of luminance found at about 300 m depth in the ocean by an observer with an eye similar to the human eye, but with a spectral sensitivity curve with maximal sensitivity at a wavelength of about 475 nm and looking upwards. The conclusion we reach, therefore, is that even in the mesopelagic zone the intensities of daylight available for visual discriminations are not as limiting as might be supposed. In particular we may note that, over much of this zone, animals should be able to detect animals above themselves relatively easily, unless these animals are camouflaged by light emitted from their ventral surfaces.

However, the rapid changes in acuity with depth at lower levels of intensity indicate that below say 400 m great advantages in acuity could be given here by having eyes that are even a few times more sensitive than those of competitors. Just as the vision of cats and owls is so much more effective than our own when the light is dim - even though their sensitivity to light is only a few times (3-10) greater than ours - a small factor in the enormous range of intensities over which vision can be useful (Hecht & Pirenne, 1940; Martin, 1982).

Of course animals which are searching for prey by the 'shadows' that these prey cast beneath themselves or by the light they reflect will not always have the necessary retinal and nervous organizations needed to have high acuity of vision even when this is not limited by the number of quanta available. In humans in bright daylight very high acuity is given only for images formed on the foveal region of the retina.

THE USEFUL WORKING RANGES OF VISION

The ranges over which bioluminescence can be useful for vision are greatly affected by the transparency of the sea and the background light against which it is viewed. We shall give two examples :

(1) Let L be the intensity of the bioluminescent source needed to produce a given illumination, A, which can just be seen by an observer at a distance of r metres then $L = Ar^2 e^{cr}$ where c is the total attenuation coefficient of light for the sea-water. On Figure 8.5 we show as a function of r the extra intensity of the source needed to produce unit

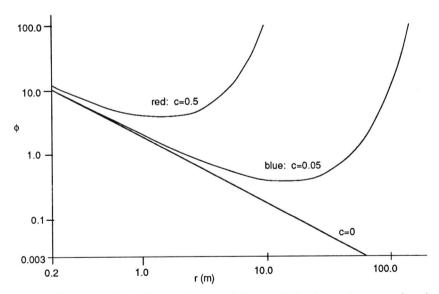

Figure 8.5. The abscissa is the distance r in metres between a bioluminescent source and an observer looking towards this source. The ordinate (ϕ) is the extra intensity, on an arbitrary scale, needed to produce unit increase in the volume within which the source can be seen. The three curves are for c = 0, 0.05 and 0.5 (where c is the total attenuation coefficient) for a transparent medium, for blue, and for red light in oceanic sea water. Both r and ϕ are plotted on logarithmic scales.

increase in the volume within which the source can be seen. Curves are given for $c = 0$ to which very clear air will approximate and for which A falls as the inverse square of r; for $c = 0.05$ which is for blue light ($\lambda = 475$ nm); and for $c = 0.5$ for red light ($\lambda = 700$ nm). If light were only absorbed and not scattered the values for the blue and red lights would be about 0.025 and 0.5. For small values of r the rate of decline of A with distance is dominated by inverse square law attenuation. However, it may be seen that, even for blue light, the price to be paid, in increases in value of L, the light emitted, for gains in the volume within which the source can be seen, rises very quickly as r rises above 100 m. This does, in effect, limit the useful range of bioluminescence for communication in the oceans. To give a numerical example, let us consider a very bright flash of blue bioluminescence measured by Nicol (1958a). This gave 2×10^{-5} µJ of blue light cm^{-2} at a distance of 1 m. This would just be seen by a human eye sensitive to 100 quanta entering the pupil at a distance of around 70 m or at about 90 m for an eye with the same size of pupil but sensitive to five quanta entering the eye. Five quanta of blue light is a more likely figure for the eye of a deep-sea fish or squid. (For blue light 1 Joule is equivalent to about 2.4×10^{18} quanta.)

(2) Let us consider a source giving a flash of light of short duration and subtending a small angle at the eye of an observer. The number of quanta ΔN entering the eye for the source to be seen will rise with the intensity of the steady background daylight I (of blue light) against which it is viewed. The way in which ΔN varies with I (measured in quanta per square degree per second entering the eye) was explained by Barlow (1957a) in an analysis based on signal to noise arguments where the noise is given by the inevitable fluctuation in the number of quanta delivered to the eye by flashes and background fields that are nominally of constant values. Since similar considerations will apply to all eyes we would expect, in a general way, other eyes to behave like the human eye. On Figure 8.6 we show the values of ΔN needed for various depths for the human eye.

It may be seen that the value of ΔN has to be increased by a thousandfold as the observer moves upwards from a depth of about 800 m to depths of 200-300 m.

To give a numerical example of what this means we may note that a source that will just be visible at a range of 5 m at a depth of 800 m will not be detected at distances greater than about 17 cm at either 300 or 200 m (depending on whether it is viewed looking upwards or sideways). The value of I found at the depths where ΔN equals 10^5 quanta are around 10^9 quanta per square degree per second. These are levels at which human scotopic vision (i.e. vision based on rods) begins to break down (saturate) and ΔN has to rise very quickly with I for the source to be seen at all (Aguilar & Stiles, 1954). These are also the levels at which the photosensitive visual pigment of the rods begins to bleach quickly. We estimate that the time needed for daylight to bleach about half of the retinal pigment in an eye looking upwards would be greater than 5000 years at 850 m and only a few minutes at 200 m.

So far we have mainly discussed the ways in which vision of man would be limited at various depths by the special conditions found in the sea. However, the cases chosen for discussion are ones in which the performance of the eye is limited either by the number of quanta available or by signal-to-noise considerations.

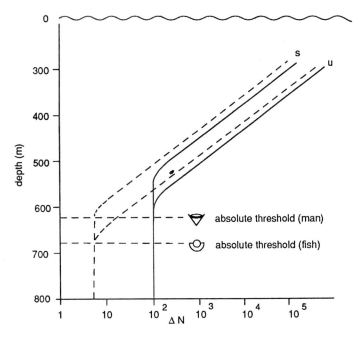

Figure 8.6. For clear oceanic waters in the daytime. The abscissa ΔN is the threshold number of quanta entering the eye of an observer from a short-lasting bioluminescent source that subtends a small angle at the eye of the observer. The ordinate is depth. —— a human observer looking straight upwards, u, and sideways, s, and - - - for a deep-sea fish looking upwards and sideways.

For a fish with the same linear aperture as a human eye we should expect ΔN without background illumination to be about five instead of about 100 quanta yet I for a given depth will be the same for the fish and the man. However, for the background field of daylight, the number of quanta absorbed per unit area of retina will be about 50 times higher (see above) for fish. This will mean that, at higher levels of I, the noise (statistical variations in the number of quanta for a given value of I) will be higher and the number of quanta absorbed from the flash will have to be increased by a factor of about $\sqrt{50}$ i.e. about seven times for the flash to be seen. A line giving ΔN against depth for a fish eye of this size looking upwards is given in Figure 8.6. The same line will in fact apply approximately to fish eyes which are larger or smaller. However, this coincidence hides the fact that to attain a given ΔN the intensity of the source of the flash will have to be decreased or increased by the square of the ratio of the linear apertures. This can be a large effect reflecting the fact that to detect a small flashing light a large eye, which can collect light over a large area of pupil and concentrate it onto a small area of retina (within which the effects of the stimulation of the rods are summated), is clearly better than a small eye, just as a large telescope is better than a small one at detecting distant stars. In the sea we do find animals, particularly squid, with very large eyes indeed. The eye of one giant squid measured 37 cm in diameter (Cooke cited by Beer, 1897) and the eyes of deep-sea animals, particularly those in the 'twilight zone' are generally very large with respect to the size of the animal.

For the human eye Barlow (1958) has shown that with larger test spots (corresponding to bioluminescent animals whose lights subtend appreciable angles at the eye of an observer) the increment thresholds (ΔN) lie above the limits set by quantal fluctuations of the background light (corresponding here to penetrating daylight). If the eyes of a deep-sea animal resembled those of man in this respect then, above depths of around 700 m, the effect would be to increase the relative importance of daylight over luminescent light.

It now seems certain that animals emit light from ventral photophores to obscure the shadows they would otherwise cast below themselves. Since animals like *Argyropelecus olfersii* and *A. hemigymnus* in which the ventral photophores are extremely large are only rarely found at depths less than about 200 m in the day-time; this must be the depth above which animals can no longer emit continuous light of the intensity needed to match the ambient daylight (Dr J. Swinney, personal communication; see also Young *et al.*, 1980).

We shall now briefly recall the main conclusions given above for clear oceanic waters. Below 900-1000 m daylight has no significance for vision. The effective light is bio-luminescent and the properties of sea-water limit its effective visual range to about 100-150 m.

The absolute threshold at which a human observer could just detect daylight will be about 850 m when looking upwards and about 740 m looking sideways. We should expect that deep-sea animals with large sensitive eyes will detect daylight at depths greater by 100-200 m than man; animals with small eyes at lesser depths than man. In so far as acuity is limited by the number of quanta available we should expect this to increase at first rapidly and then steadily with decreasing depth over a range of about 450 m and then to become almost independent of depth. In this range the advantage of having a larger and more sensitive eye could be particularly great.

In about the top 200 m the acuity possible will vary little with the direction of viewing whilst at greater depths acuity will vary greatly with direction, being best for upward viewing at all depths. It is impossible by reflecting light to avoid casting a shadow downwards. However at depths greater than about 200 m animals can emit light sufficiently intense to obscure the shadows that would otherwise make them very visible.

For say 100-200 m above the absolute threshold the background daylight will not be sufficiently intense to affect the visibility of bioluminescent lights. As we go further towards the surface, daylight becomes progressively more dominant but its effect will vary greatly with the direction of viewing. In this range of depths a small flashing light will have to be at least 16 times more intense to be seen at a given distance when viewed from below than when viewed from above. The corresponding figure for sideways viewing is about x6. The acuity possible rises very quickly with decrease in depth. At a depth of around 200-300 m daylight is sufficiently strong to bleach the visual photosensitive pigment quickly and, in fish, rod vision will probably saturate and become less useful for vision.

LENSES AND TAPETA

The lenses found in the eyes of deep-sea fish and squid are mostly spherical in shape. They are, like the lenses of animals living in shallow water, of very good quality optically (Land, Chapter 9). Apart from the lens from the larger eye of an oceanic squid of the family Histioteuthidae the few lenses examined by the author had the expected ratio of about 2·5 for the focal length divided by the radius of the lens (Matthiessen's ratio) and the optical properties of these lenses were not dependent on the direction of the light. Some of the lenses of deep-sea fish and squid are markedly yellow.

Some deep-sea fish have highly reflective tapeta based on guanine crystals. Many of these tapeta are silvery (particularly those of elasmobranchs), but others are markedly blue. In accord with the red light that it produces from its sub-orbital photophore the eye of *Malacosteus* has a spectacularly red tapetum. Its structure is not crystalline, the reflecting material being small globules containing a red pigment.

VISUAL PIGMENTS

It is generally found that in the eyes of vertebrates the retinal rod receptors are concerned with vision in dim lights and the cone receptors with bright lights. In man colour vision is entirely mediated by the cones, a fact that has sometimes led people to suppose, wrongly, that cones are essential for colour vision in all vertebrates. Here we are dealing almost entirely with rod receptors (Brauer, 1908; Locket, 1977).

Until the late 1950s it was widely thought that only two photosensitive pigments were to be found in the retinal rods of vertebrates. One, rhodopsin, had an absorption maximum at a wavelength of about 500 nm, the other, porphyropsin, an absorption maximum at about 520 nm: the first being based on vitamin A1 and found in terrestrial animals including man, and in marine fish; the second, based on vitamin A2, and found in fresh water fish. We now know that there are a large number of rod pigments with a wide range of absorption maxima (Partridge, Chapter 10). In the late 1950s the visual pigments of the rods of many deep-sea fish were shown to absorb maximally at wavelengths around 480 nm *i.e.* close to the wavelength of light that penetrates best into the sea (Denton & Warren, 1957; Munz, 1958). These pigments were named chrysopsins because they were golden in colour. Here in the deep sea there was a close correspondence between the absorption of light and the spectral quality of the light available. Such straightforward agreement is not found in the eyes of most animals, certainly not in night vision in man (Partridge, Chapter 10). In deep-sea fish the retinal rods are often over 100 μm long as against 26 μm for the human rod and they contain proportionately more photosensitive pigment. Because they have so much pigment the spectral absorption curves for intact retina of deep-sea fish are usually very flat. In Figure 8.7 we give a retinal spectral absorption curve for a deep-sea fish, *Xenodermichthys*, and compare it with a similar curve for man. For animals with a great deal of pigment the exact position of the absorption maximum does not, at first sight, seem to be very important. Nevertheless, in the upper layers of the sea there does seem to be a significant correlation between the depth at which fish live and the wavelengths of the maxima of the absorption spectra of their visual pigments (Wald *et al., 1957*). Again in the fish *Diretmus* the photosensitive pigments in the parts of the retina that receive light from above the fish differ from those found in the parts that 'look' sideways. Some deep-sea fish have highly reflective tapeta behind the layer of rods that contain the visual pigments. This allows a high absorption of light with shorter rods and less visual pigment (Denton & Shaw, 1963).

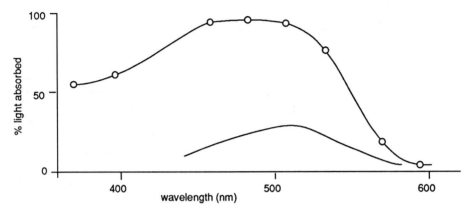

Figure 8.7. *Xenodermichthys copei* (a fish) - the upper curve shows the percentage of incident light absorbed in one passage through the retina plotted against wavelength. The lower curve is calculated for the human rods from the maximum density of 0·15 given by Rushton (1956). After Denton (1959).

COLOUR VISION

Since the lights available to deep-sea fish were found to be so overwhelmingly in the blue-green part of the spectrum and, since they mostly had only rod receptors, there seemed, at first, little reason to believe that they would have colour vision. The conclusion that should have been drawn was that they might possibly have colour vision and, if so, this would usually be specialized to give good hue discrimination in the blue-green part of the spectrum. In this connection we may consider the function of the yellow lenses found in the eyes of some deep-sea fish and squid. As soon as they were discovered it was clear that the functions ascribed to such lenses in terrestrial animals were not appropriate for deep-sea animals but no likely hypothesis presented itself. A very interesting suggestion was then made by Muntz (1976), and Somiya (1976), who independently advanced the plausible hypothesis that, by reducing the relative intensity of blue daylight reaching the retina, the yellow lenses of deep-sea fish made the detection of luminescent animals whose lights were, for example, greener, much easier. Muntz argued that a function of the yellow lens might often be that of 'breaking' the camouflaging effect of lights emitted by the ventral photophores of other animals.

When an animal has two or more visual pigments, there must be a possibility that it has colour vision and some deep-sea animals are known to have more than one retinal photosensitive pigment. This was first shown by Munz in 1958 and a general account of later work including their own is given by Partridge *et al.* (1988). Some fishes have retinae in which there are several banks of rods (Vilter, 1953; Munk, 1966b; Locket, 1977) and Partridge *et al.* (1988) have shown that one such fish *Bathylagus bericoides* has two photosensitive pigments and that these pigments are in different rods. Bowmaker *et al.* (1988) have found that three species of deep-sea fish *Aristostomias grimaldi*, *Malacosteus niger* and *Pachystomias microdon* have rhodopsin-porphyropsin systems of paired pigments and that each pigment is restricted to a single class of rods. With such a retina it would, for colour vision, be advantageous to have these pigments in different banks of rods.

It is evident that here discrimination in a given part of the spectrum will be easier if the ratios of stimulation of the different sets of rods changes quickly with wavelength. In

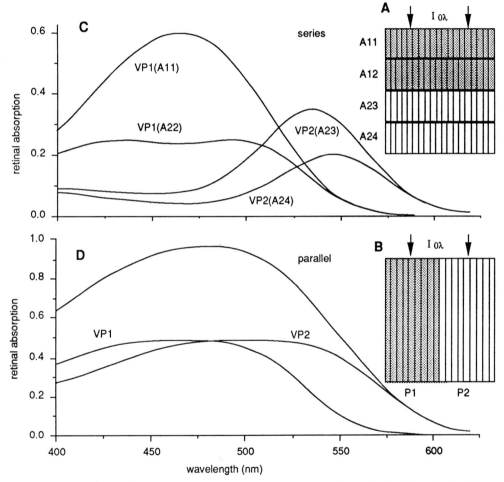

Figure 8.8. (A) & (B) Diagrams of series and parallel arrangements of retinal rods. (A) the banks A11 and A12 contain the visual pigment VP1 (λ_{max} = 465 nm), the banks A23 and A24 contain the visual pigment VP2 (λ_{max} =500 nm). (B) the rods in B1 contain VP1, the rods in B2 contain VP2. The areas of B1 and B2 are equal. (C) & (D) The ordinates are the absorptions of light in one passage through the retina. The abscissae are wavelengths. (C) For a retina with four banks. (D) The parallel case. The total quantity of each visual pigment per unit area of retina is the same for both series and parallel cases. For the series case the optical density of each layer for the wavelengths of maximum absorption is taken as 0·4. For the parallel case the optical density for each set of rods is taken as 1·6.

Figure 8.8A & B we show diagrammatically what may be described as a 'series' arrangement of rods in multibank retinae and also the 'parallel' arrangement of rods for a more 'conventional' retina in which the receptors (here rods) containing different pigments lie side by side. In Figure 8.8C & D we compare spectral absorption curves for different sets of rods given by these series and parallel arrangements. Clearly the series one is that which will give the better hue discrimination in the blue-green part of the spectrum (Denton & Locket, 1989).

When a retina does have several banks of rods it could have colour vision even with only one photosensitive pigment. This is because the first bank of rods that the light reaches on entering the eye changes the spectral composition of light reaching the second bank and this effectively gives the first and second banks different spectral sensitivities. The spectral composition of light reaching a third bank will, in turn, be affected by the first and second banks and so on for later banks. A multibank retina with only one pigment absorbing maximally at about $\lambda = 480$ nm could give very good hue discrimination in the blue-yellow part of the spectrum.

Since some deep-sea animals emit orange and red lights it was not surprising that such animals have retinal photosensitive pigments sensitive to such lights (Denton *et al.*, 1970; O'Day & Fernandez, 1974; Bowmaker *et al.*, 1988). An account of the properties of red emitting photophores and of the retinae of such animals is given later.

SOME SPECIAL CASES

An account of the optical properties of the eyes of deep-sea animals is given by Land (Chapter 9). General accounts of the eyes of deep-sea fishes with many references to earlier work are given by Marshall (1954), Munk (1966*a*) and Locket (1977). Here we shall only give, as examples, an account of the eyes of some of the deep-sea fish and squid chosen to illustrate particular specializations to the light conditions (daylight and bioluminescence) described above.

Diretmus argenteus. This fish lives in the day-time at depths between 100-600 m and so can be found at depths where daylight is much more important for vision than bioluminescence. Diagrams of the eye of *Diretmus* are given in Figure 8.9. Its retina is multibank and divided into distinct regions. This seems appropriate in view of the great changes of intensity found on moving from above to below the fish (Figure 8.2). In the lower region (L) of the eye there is a patch (Lc) containing cones. This might be expected in a fish found sometimes at depths less than 200 m. The whole of part L of the retina has a bank of exceptionally long (up to 600 µm long) rods. This bank is the most vitread one so that the light must pass through this layer before reaching the more sclerad banks. It seems certain that these long rods contain a stable yellow pigment. This part of the retina receives light from above the fish and it contains a different photosensitive pigment from the part of the retina (S) that receives light from below the fish. The lower part contains a photosensitive pigment (or pigments) absorbing maximally at $\lambda = 500$ nm. This, together with the stable yellow pigment contained in the long rods will give a system that will enhance the contrast between the blue daylight and bioluminescent lights that are

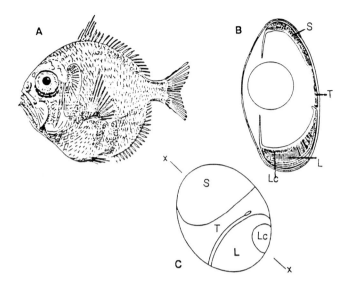

Figure 8.9. *Diretmus argenteus*. (A) Intact animal. (B) Section through the eye in a plane perpendicular to its front surface along the line X--X of C. (C) Retina viewed from its vitread side. L is a region that contains very long rods, up to 600 μm in length. This region subserves vision upwards and forwards. In part of this region there is an area Lc containing cones. Sclerad to the long rods there are several banks of short rods. S is the rostrodorsal part of the retina. This has one bank of moderately long rods and several banks of short rods sclerad to to these rods. T is the central part of the retina. This contains only short rods. Diagrams B and C, partly after Munk (1966).

'greener' than daylight. The upper part's photosensitive pigment absorbs maximally at λ = 485 nm (Denton & Locket, 1989). It is particularly interesting that similar but much more striking results have been obtained on a deep-sea cephalopod, *Watasenia scintillans* by Matsui *et al.* (1988a). They showed that this squid's retina has three visual pigments. The photoreceptor cells are exceptionally long in the small area of the ventral retina receiving downwelling daylight whereas they are short in the other regions of the retina. The short rods contain a visual pigment with an absorption maximum at wavelength 484 nm. The outer segment of the long photoreceptor cells have two strata: a yellow stratum containing a pigment absorbing maximally at a wavelength of about 471 nm through which light must pass before reaching a pinkish stratum containing a pigment absorbing maximally at a wavelength close to 500 nm. Matsui *et al.* (1988b) have made the significant further discovery that the three visual pigments found in *Watasenia* have different chromophore groups.

BAJACALIFORNIA AND OTHER FISH WITH FOVEAS

The eyes of *Bajacalifornia drakei* have been studied by Locket (1985). Although this fish lives at depths between 700 and 1600 m it has highly specialized foveas. The axes of vision to the foveas of the two eyes converge in front of the jaws, presumably to give binocular vision, yet these foveas must be concerned entirely with bioluminescent lights even in the day-time. The fovea of the eye of *Bajacalifornia* is a most remarkable structure having up to 28 superposed banks of rods (Figure 8.10). It is easy to show that, provided the various banks of rods are functional, this arrangement could give this animal good colour vision in the blue to yellow part of the spectrum (Denton & Locket, 1989). We may note that a fovea of this kind would probably not be useful in an animal living near the surface of the sea in day-time because of the delays involved in regenerating so much visual pigment once it has been bleached. At the depths where *Bajacalifornia* lives, the time daylight would take to bleach the pigment would be measured in thousands of years. The fovea of this animal is also convexiclivate. Various optical functions have been ascribed to such foveas. These are discussed by Steenstrup & Munk (1980) in their study of the foveas of notosudid deep-sea teleosts. These animals possess pure cone foveas. Steenstrup & Munk suggest that the main function of the notosudid fovea is to break the camouflage of mesopelagic animals that use small photophores for counter-illumination

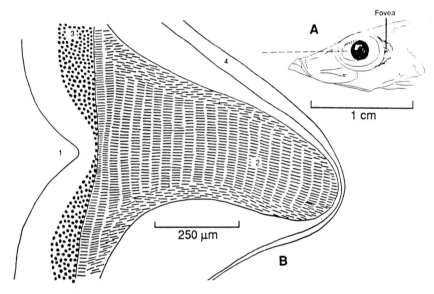

Figure 8.10. *Bajacalifornia drakei*. (A) Head with the tissues behind the eye cut away to show the location of the fovea. The axes through the centre of the lens and the foveas cross in front of the fish. (B) Fovea diagram. There is a steep-sided pit, 1, in the retina, lined by radial fibre. The fovea externa, 2, contains up to 28 banks of ellipsoid-outer-segment complexes, compared with one or two poorly defined banks in peripheral retina. Rod nuclei, 3 are piled up in the foveal shoulders. The sclera, 4, is displaced outwards by the fovea externa (from Locket, 1985).

and that they make use of the very different angular distributions of photophore lights and penetrating daylight. On their theory a two-spot image of a single small photophore is produced by the optical properties of the convexiclivate fovea. For *Bajacalifornia* perhaps the theories of Pumphrey (1948) or Harkness & Bennet-Clark (1978) and Locket himself may be more likely to be true. Pumphrey suggests that this type of fovea could help with the accurate focussing of objects and a better appreciation of their angular movements, whilst Harkness & Bennet-Clark argue that the convexiclivate fovea functions as a focus indicator and Locket that the two foveas allow the fish to find the directions and distances of prey. It would certainly be interesting with these various theories in mind to analyse the optical properties of the fovea of *Bajacalifornia* bearing in mind the ways in which light can be channelled along rods. Here, we merely note that, although the average illumination of these retinae from bioluminescent lights will be very low, the sources of light will generally subtend small angles at the eyes of the predator and the number of quanta of light absorbed in small areas of retina will be quite sufficient to allow an accurate determination of the positions and hues of the sources. Since bioluminescent light does appear in flashes there must be a premium on systems which, like those suggested by Harkness & Bennet-Clark and Locket, allow distance to be found without depending upon focussing an image by moving the lens. The eye of *Bajacalifornia* is markedly aphakic (a property discussed in a later section) with what Locket describes as sighting grooves to allow the foveas to receive light from just in front of the jaws.

FISH THAT EMIT AND ARE SENSITIVE TO RED LIGHT

At least three genera of fish have photophores beneath their eyes that emit orange or red lights and, as we have already noted, some such animals possess red-sensitive retinal pigments.

The light emitted by suborbital photophores of *Malacosteus* has been studied in some detail. The light generated inside this photophore contains a wide band of wavelength but such light has to pass through a colour filter before being emitted. This filter absorbs a large fraction of the generated light so that the emitted light contains only light of a narrow waveband centred at about 700 nm (Denton *et al.*, 1985). At first sight this is surprising because red light of this waveband is absorbed heavily by sea-water (see Figure 8.5). However, it is easy to show that for such a waveband retinae containing the red-sensitive retinal pigments found in the eyes of similar fish, *i.e.* in *Aristostomias* (O'Day & Fernandez, 1974) and in *Pachystomias* (Denton *et al.*, 1970) would be 10^5 and 10^6 times more sensitive than would a retina containing a typical chrysopsin (λ_{max} at about 475 nm). This will give such red emitting and red-sensitive fish the possibility of signalling to each other and seeking their prey with little risk of their lights being seen by other animals. The range of signalling with this waveband of light will be only a few metres. However, these animals also possess photophores that emit blue-green light and retinal pigments sensitive to blue-green light. Recently Bowmaker *et al.* (1988) have studied the retina of *Pachystomias*. They found two photosensitive pigments absorbing, respectively, maximally at $\lambda_{max}=544$ and 513 nm. They also gave good evidence that some rods from the eyes of *Malacosteus* contain chorin-like pigment which acts as a photosensitizer increasing long-wave sensitivity.

The retina of *Malacosteus*, which has been studied by Locket (1977) and Somiya (1982), has been shown to be multibank. The more vitread rods differ in staining properties from those that lie closer to the sclera. There certainly would be an advantage to having red-sensitive pigment in the more sclerad rods and blue-sensitive pigment in the more vitread rods. The filtering action of the blue-sensitive pigment on light reaching the red-sensitive pigment would be high in the blue and trivial in the red. This would make colour discrimination easier. *Malacosteus* also has a bright red tapetum into which the most sclerad rods are embedded. This would certainly confine the sensitivity of these rods to the red.

TUBULAR AND APHAKIC EYES

Although the eyes of many deep-sea fishes have a form found in many common shallow water fishes (*i.e.* an approximately hemispherical shape with a spherical lens set with its centre approximately at the centre of the hemisphere) others have tubular eyes. Drawings of such eyes are given *e.g.* by Brauer (1908), Munk (1966a) and Marshall (1954). These eyes either look upwards or in front of the jaws. The main retina often covers only the base of the tube and sometimes there are no retinal receptors on the sides of the tubes. The lenses of at least some of these fish have the usual properties of fish lenses so that the only region of the retina on which a reasonable image can be formed corresponds to a small range of angles of viewing of about $\pm 30°$ around the axis of the tube. Possessing eyes that 'look' directly upwards and which cover only a few per cent of the space around the fish must indicate a concentration of visual effort on searching for animals above themselves by the shadows these animals cast in the penetrating daylight. The eyes that 'look' in front of the jaws must be mainly useful for catching prey when this is very close.

Many deep-sea fish have aphakic eyes in which the pupil is very much larger than the lens so that light can pass around the lens and strike the retina. This light will give a background illumination on part of the retina against which images formed by the lens have to be detected. These aphakic eyes are particularly common amongst the deeper-living animals which deal almost entirely with bioluminescent lights. Such lights will, unlike daylight, usually be small sources. Whilst such a source will usually be significant for vision only when it is focussed onto a small area of retina, the background retinal illumination it produces by passing around the lens will be far from this area and will be weak. This background can, moreover, only rarely be significant either in timing or intensity, in diminishing the visibility of other flashing sources of luminescence coming from other directions. The merit of the aphakic eye is that it will increase the capture of light for the retina from obliquely placed sources.

ON THE VARIETY OF EYES

This chapter opened with a quotation from Murray & Hjort (1912) expressing astonishment at the very great variety of eyes found among deep-sea animals. Let us consider this variety in the light of modern knowledge.

Some of the limits to visual discriminations at different depths set by physical factors - the intensities, durations, and the spectral and angular distributions of daylight and bioluminescent lights - have been discussed above using arguments based on the numbers of light quanta available for vision and the inevitable quantum 'noise' of flashes and backgrounds of light. Land (Chapter 9) has described limits set to visual discriminations by optical factors. These accounts detail what *could* be achieved visually in particular situations with respect, for example, to thresholds and to acuities in different directions. However these *potentialities* for visual discriminations will usually far exceed what any given animal can achieve. This is, of course, because there are always 'physiological limitations' such as the ability of the nervous system to transmit, analyse and use the information the eyes produce. A simple example may be given relevant to acuity. To give an acuity of $1/60$ ($1°$) over the whole possible (spherical) field of vision, would, assuming the simplest possible retinal mosaic consisting of groups of receptors side-by-side each with its own nerve fibre, demand about 5×10^4 nerve fibres which would be a large number for a small animal, but only a small fraction of the number possessed by man. For an acuity everywhere of $1/10$ ($10'$) about 2×10^6 nerve fibres would be needed. This is the number possessed by man for all purposes. For an acuity of 1 ($1'$) everywhere about 2×10^8 nerve fibres would be needed together with an appropriate nervous system to use the information provided. In practice, although animals do approach physical limits a particular animal can do so only in a few respects or for a small part of the visual field. Even in a relatively uncomplicated light environment different animals will be specialised to approach different limits. Thus we find, for example, at the same depths, or more exactly at the same light levels, (i) animals with tubular eyes specialised for the detection of animals above themselves but effectively blind to over 90% of the possible visual field, and (ii) animals with aphakic eyes specialised to detect bioluminescent lights coming from all directions.

A particular type of animal with given 'physiological' resources can be more perfectly adapted to its environment if this environment is less variable. Now many deep-sea animals, by swimming upwards and downwards keep themselves in the daytime under almost constant conditions with respect to intensity as well as to spectral and angular distributions of light. We have seen that the visual possibilities with respect, for example, to acuity and ease of detecting bioluminescent animals, will vary very appreciably with 'depth'. Thus, as Figure 8.4 shows, at a given time a difference in depth of as little as 70 m could give a tenfold change in the possible acuities. For animals that keep themselves under constant illuminations we should, therefore, expect the optimal visual adaptations, *e.g.* in retinal organisation or in the resources devoted to regenerating visual pigments, to be appreciably different for depths that differ relatively little. The advantages of such visual adaptations (which would not be seen in gross anatomical studies) probably explains why closely related species of animals are found in distinct layers always with the same relationship in depth (*i.e.* light) to each other.

Above all it must be remembered that the bioluminescence found in the deep sea has been shown to be very varied in its properties (Herring, Chapter 16). This variety leads to specialisations in the eyes of animals, such as those that confer the ability to detect bioluminescence in the far red, and to match bioluminescent lights shone forwards from the head of a fish possessing tubular eyes 'designed' to give good visual acuity just in front of the jaws.

In summary, we see that increased knowledge has made the variety of eyes found in deep-sea animals that astonished Murray & Hjort (1912) readily understandable. However extraordinary structures in the eyes of deep-sea animals have been described by anatomists to which no function can be ascribed and, as Herring (Chapter 16) has emphasised, we know very little about the behaviour of deep-sea creatures.

In this chapter we have mostly discussed light and vision in the range of depths between 200 and 1000 m. A good account of the kinds of eyes found at greater depths, including those of benthic animals, is given by Marshall (1954). We have not discussed the ways in which animals might make use of polarised light. Here we refer the reader to the very lucid accounts given by Waterman (*e.g.* 1984, 1988), the acknowledged world authority on this subject.

The author has been generously supported by the Leverhulme Trust as one of the Trust's Emeritus Fellows.

Chapter 9

OPTICS OF THE EYES OF MARINE ANIMALS

MICHAEL F. LAND

School of Biological Sciences, University of Sussex, Brighton, BN1 9QG

The one type of eye that does not work under water is our own, where most refraction occurs at the curved air-cornea interface. Most marine animals with simple eyes make use of lenses instead, and the commonest design is the spherical 'Matthiessen lens' in which the refractive index decreases from centre to periphery. This has evolved at least 6 times. Alternatives are lenses with many components (some copepods), concave mirrors (some bivalves and ostracods), and in the unique and embarrassing case of *Nautilus*, a pinhole.

Several types of compound eye are found in marine animals, and the only one absent is the simple apposition type found in terrestrial insects, where the air-cornea interface of each facet produces the image. However, other apposition types that use inhomogeneous optics are common (*Limulus*, amphipods), and so are superposition eyes. In these eyes many facets contribute to the image at any one point, and they are of three kinds: those that use inhomogeneous (lens-cylinder) optics, those that use mirrors, and a recently discovered intermediate in crabs that uses a combination of a lens and a parabolic mirror.

INTRODUCTION

The main optical difference between the terrestrial and aquatic habitats is the refractive index of the surrounding medium: 1 in air, and between 1·333 and 1·339 in water, depending on the salinity. The consequence of this difference for eye design is that the commonest method of forming an image in terrestrial eyes - the use of a curved cornea separating air from fluid - will not work in water. This is because the curved interface has now almost the same refractive index on both faces, and hence causes little or no refraction. Whereas in humans two-thirds of the power of the optical system resides in the cornea and only one-third in the lens, in a fish all the power has to be in the lens. Thus, aquatic lenses tend to be spherical, and also to have a gradient of refractive index within them that allows ray-bending to occur through the structure and not just at its faces. Such lenses have evolved in several different phyla, and they have remarkable properties: f-numbers of about 1·25, and almost no spherical aberration. Much the same considerations apply to compound eyes. In the eye of a bee or a fly the image in each ommatidium is formed by refraction at the surface of each corneal facet, and again this interface is not available to aquatic arthropods. They also must obtain the necessary focussing power using either internal interfaces or inhomogeneous optics. In the first part of this chapter these, and some of the other alternatives to corneal optics are explored. The second part

Herring, P.J., Campbell, A.K., Whitfield, M. & Maddock, L. (ed.). *Light and Life in the Sea.*
© 1990. Cambridge University Press.

deals briefly with some of the special problems encountered by amphibious vertebrates - flying fish, diving birds and seals - which have to vary their optics to suit the medium they are in. More detailed treatments of some of these themes are given in Land (1981*a*) and Land (1987).

There are other more subtle optical differences between aquatic and terrestrial worlds that are also reflected in eye structure. Descent into the sea involves a progression of changes in the light environment. The light becomes dimmer, and restricted to a narrow waveband, which is in the blue in oceanic waters and green or even yellow in inshore and lake water. The distribution of light becomes symmetrical about the vertical, with the downwelling light being as much as 200 times brighter than the upwelling. For a deep-living animal a patch a few tens of degrees wide across the zenith is the only region where there is enough light for useful vision. The eyes of deep-living animals often point upwards rather than sideways or forwards, and elaborate camouflage strategies are often also employed to minimize an animal's detectability under these special conditions (see Lythgoe, 1979). The third part of the chapter explores some of these environmental adaptations.

OPTICAL MECHANISMS

1. Simple Eyes

a) Pits and pinholes

Amongst the less active invertebrates the commonest type of eye is a simple pit of pigment cells with a small number of receptors (1-100) within it. The resolution that such eyes can provide is inevitably poor, being limited to the angle that a single receptor subtends at the mouth of the pit, typically between 10 and 90°. Eyes like this are adequate for navigation towards or away from the general direction of a light source (tropotaxis) but not much more than this.

Such eyes were undoubtedly ancestral to all other types of simple eye, and in the course of evolution their performance has been improved along three different routes (Figure 9.1). First, enlargement of the eye and restriction of the aperture turns the structure into a pinhole camera, capable of providing a well-resolved, if dim, image. Second, the addition of a refractile lens of some kind in the mouth of the pit permits an image to be formed on the receptor layer behind. And thirdly, lining the back of the pit with a mirror will also produce an image, not at the back this time but towards the middle of the eye. First, we will consider the pinhole eye.

The eye of the ancient cephalopod mollusc *Nautilus* is the only example of a pinhole eye comparable in size and complexity with the lens eyes of other cephalopods and fish. There are a few other molluscs with more or less closed 'pit' eyes, the abalone *Haliotis* being one example. However, that eye is less than 1 mm deep whereas a *Nautilus* eye is close to 1 cm, and comparable with an *Octopus* eye. It has many of the attributes of optically more sophisticated eyes. For example, the pupil changes with light intensity from 0·4 to 2·8 mm diameter. The eye is equipped with a set of muscles, and these are used in a reflex mediated by the statocyst that keeps the eye-axis vertical as the animal rocks through the water (Hartline & Lange, 1984). There is also an optomotor response to motion of the environment, presumably concerned with preventing rotational slip of the image (Muntz & Raj, 1984). These are all features of advanced visual systems. The problem with the *Nautilus* eye is that to achieve a well-resolved image the pupil must be small, in which case the image brightness will be low (with a 0.4 mm pupil the f-number is about 25 compared with 1·25 for a fish eye, a difference in image brightness of 400 times). Opening the pupil, however, ruins the image. This invidious trade-off is obviated by a lens, which permits good resolution with high image brightness. It seems that from an optical point of view almost any lens would improve the situation in *Nautilus* and it is a genuine mystery as to how this unique, but unsatisfactory design has been maintained for 400 million years.

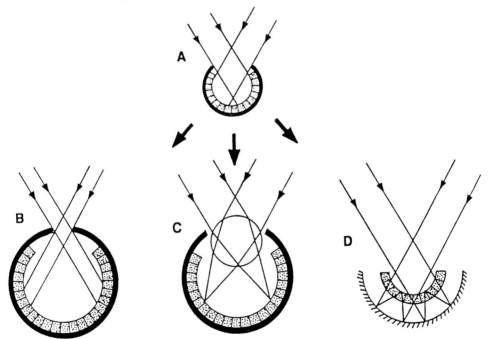

Figure 9.1. Three types of eye derived from the ancestral pigment-pit eye of the lower phyla (A). (B) pinhole eye (*Nautilus*); (C) lens eye (fish, cephalopods and others); (D) reflecting eye (*Pecten*).

b) Matthiessen lenses

As aquatic animals do not have an optically usable cornea the whole of the focussing power of the eye must be in the lens, and this tends to be spherical to keep the focal length short. This raises problems, however, because a spherical lens with the same focal length as that of a fish (about 2·5 radii) would have spherical aberration so bad that it would be virtually useless (Figure 9.2A). Furthermore, the refractive index would need to be about 1·66, and this is not attainable using normal biological materials like proteins. The upper limit to plausible refractive indices is probably 1·56, which would give the lens a focal length of about 4 radii, much longer than it actually is in either fish or squid. The solution to this conundrum was worked out by Matthiessen in the 1880s, although Maxwell had already reached similar conclusions in 1854 "while contemplating his breakfast herring" (Pumphrey, 1961). These lenses are not optically homogeneous, but contain a gradient of refractive index, highest in the centre and falling at the periphery to a value not much exceeding that of the surrounding water. The beauty of this arrangement is that it kills two birds with one stone. The presence of the gradient means that rays of light are bent continuously within the lens rather than just at the front and rear surfaces, and this results in a shorter focal length than could be provided by a homogeneous lens with the same central refractive index. Furthermore, provided the gradient has the correct form, the lens will be free from spherical aberration, since the rays furthest from the axis will be bent less relative to the central ones than would be the case in a homogeneous lens (Figure 9.2B). The exact form of the gradient was not known to Matthiessen - although he made an enlightened guess - and was not in fact worked out until many decades later (Fletcher *et al.*, 1954). A simple examination of a fish lens (cooked or raw) shows that there is a gradient: the outside is soft and easily squashed, but the centre is hard and crystalline. The texture reflects both the protein concentration and the refractive index.

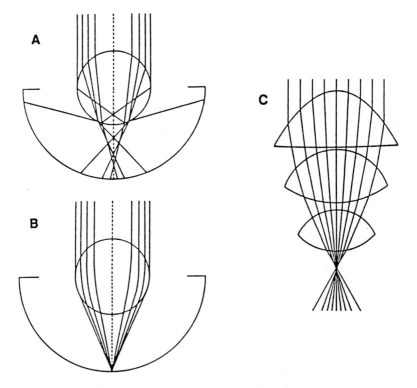

Figure 9.2. (A) Spherical aberration in a homogeneous lens with a focal length of 2·5 radii. (B) Continuous ray bending and perfect focus produced by a lens with an appropriate refractive index gradient. (C) Lenses of the eye of a male *Pontella*, seen from below. The parabolic front surface corrects the aberration of the others. The combination has an axial length of 0·35 mm. (A) & (B) after Pumphrey (1961), (C) from Land (1984).

This solution to the problem of spherical aberration has evolved independently at least six times. Besides fish, eyes with essentially the same geometry are found in the cephalopod molluscs, in the gastropod molluscs (conches and heteropods), once in the annelid worms (the Alciopidae) and once in the crustaceans (the copepod *Labidocera*). It is a simple matter to determine whether or not the lens is of this design, because the focal length/radius ratio is diagnostic. If it is around 2·5 ('Matthiessen's ratio') then the lens must be inhomogeneous. It seems that this is both the best and the most popular way of producing a well-corrected lens. It has the additional virtue that the spherical symmetry of the lens and eye ensures that the image is well-resolved over a very wide field; there is no single optical axis as there is in the human eye.

The eyes of the copepod crustacean *Labidocera* and the heteropod snail *Oxygyrus* deserve special mention not because of their optics but because they have linear scanning retinae (Land, 1982, 1988). In *Oxygyrus* the retina is only three receptors wide but it is 410 receptors long, giving a field of view that is a long strip. The eye tilts through 90° every second or so, along an arc at right angles to the retinal strip, thus scanning the surrounding water for food particles. *Labidocera* is similar, but on a much smaller scale. Its retina contains only 10 elongated receptors resembling slab waveguides, and this receptor line scans through an arc of about 35°. This remarkable behaviour was first described by G.H. Parker in 1891 (see Land, 1988).

c) Multi-element lenses

In spite of the excellence and ubiquity of spherical 'Matthiessen' lenses, there are lenses that are constructed along totally different lines. The optical system of the single ventral eye of the copepod crustacean *Pontella* consists of a number of components, not unlike a camera lens, or more precisely a microscope objective (Figure 9.2C). In the males the eyes have three elements, two outside the eye-cup in the rostrum, and one in the front of the eye-cup itself (Land, 1984). Curiously, the females have a doublet not a triplet, with only one component in the rostrum. The first lens of the triplet in the male has an interesting shape, its front surface being approximately parabolic. The reason for this became clear when rays were traced through it. It turns out that this one parabolic surface corrects the spherical aberration of all the other surfaces in the system, and if it is itself replaced by a spherical surface the image becomes messy.

It is not clear what this eye does for the animal. It contains only six receptors in a curious concentric array, and the field of view of the eye is quite small. Nevertheless, the sexual dimorphism of the eyes and the colourful appearance of the animals themselves strongly suggests a role in mate finding. So far, however, the only known behaviour of these small planktonic animals is that they jump out of dishes! It is worth noting that *Labidocera*, mentioned above as having spherical inhomogeneous lenses, is in the same copepod family as *Pontella*. Thus in closely related genera we find the two alternative solutions to the problem of spherical aberration: a gradient of refractive index in one, and a non-spherical surface in the other.

d) Concave mirror optics

Bivalve molluscs are perhaps not the kind of animal one would look to for optical surprises, nor even much in the way of eyesight. However, in one case at least this would be wrong. Scallops provide an almost unique example of an eye which forms an image using a concave mirror, rather than a lens. Scallops (*Pecten* and related genera) have 50 to 100 eyes around the edge of the mantle of the two shells. The eyes are quite large - 1 mm diameter - and really quite 'eye-like', but they usually pass unnoticed because they are on the inedible part of the animal. A section through a scallop's eye shows an overall layout not unlike that of a fish eye (Figure 9.3). It has a single chamber so is camera-like rather than compound, it has a lens of sorts, and behind this a retina filling the space between the lens and the back of the eye, which is lined with a reflective tapetum. The assumption until comparatively recently had been that this was a conventional lens eye. The first indication that this was not correct came from an observation I made in 1965. When I looked into the eye I saw an inverted image of the room, including a distorted picture of myself looking through the microscope. After some thought it became clear that this image could not have been formed by the lens (if that were the case its apparent position would not be in the eye but near infinity), which left only one candidate, the spherical silvery mirror at the back of the eye. It turned out that the lens was soft, with hardly any power, and that the mirror was the optical system. Thus the scallop eye is the optical analogue of a Newtonian telescope (Land, 1965). The reflected image lies on one of the two layers that make up the retina, and as early as 1938 the pioneer electrophysiologist H.K. Hartline (1938) had found that this layer gave responses to the offset rather than the onset of light. This 'off-responding' layer lies immediately behind the lens,

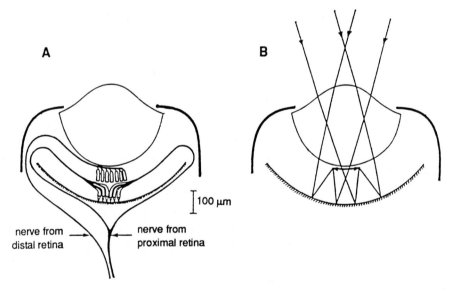

Figure 9.3. Optics of the scallop eye. (A) the locations of the two retinae (only the central regions are shown). (B) the image is formed mainly by the reflecting tapetum, with a very small contribution from the lens. It falls on ciliary receptor structures in the distal part of the distal retina. Based on Land(1965).

whereas the other 'on-responding' layer is in contact with the mirror, and receives no image. The 'off' layer is undoubtedly the one that enables the scallop to see trouble coming, and to shut its shells, but the function of the 'on' layer still remains a mystery.

Mirror eyes of this kind are rare, and the eyes of scallops are certainly the best of them from an optical point of view. A few other molluscs like the cockle *Cardium* have similar but much smaller eyes, and there are also examples in the Crustacea, in the nauplius eyes of some copepods and ostracods. There are even examples in the flat-worms and the rotifers (reviews in Ali, 1984). These, however, are all very small, and their image quality is unknown and probably poor. The only other large mirror eyes are in the big (10 mm) deep-sea ostracod *Gigantocypris*. These were described by Sir Alister Hardy (1956) as follows:

> "The paired eyes have huge metallic-looking reflectors behind them, making them appear like the headlamps of a large car; they look out through glass-like windows in the otherwise orange carapace and no doubt these concave mirrors behind serve instead of a lens in front."

Hardy was undoubtedly right, but these eyes have a very odd structure indicating enormous light-gathering power, but poor resolution (Land, 1978). This is consistent with life in the nearly lightless zone below 600 m. I suspect that the reason for the general unpopularity of mirror eyes is that the image contrast is necessarily poor, because focussed light reaching the retina has already passed through it once unfocussed (see Figure 9.3). The question of how biological mirrors are made is an intriguing one, but beyond the scope of this chapter, and the reader is referred to Denton (1970) or Land (1972).

2. *Compound eyes*

Compound eyes are usually associated with the arthropods, and in particular the insects and crustaceans. There are, however, odd examples from other phyla; some sabellid tube-worms (Annelida) have small compound eyes on modified tentacles (Kerneis, 1975), and so too do bivalve molluscs of the genera *Arca*, *Barbatia*, and *Pectunculus* (see Land, 1981*a*). These 'one-offs' evolved in isolation as predator detectors, but they do show that compound eyes are not found only on animals with exoskeletons.

In the last two decades knowledge of the optics of compound eyes has increased substantially. At the turn of the century, following Sigmund Exner's great study of 1891, compound eyes were divided into apposition and superposition types. In the former each element (ommatidium) had its own optics, and viewed a small angular field of space, adjacent to that of its neighbour; in the latter many facet lenses contributed to the image at each point in the deeper-lying receptor layer, their contributing ray-bundles superimposing on each other - hence the name superposition (Figure 9.4). For reasons documented elsewhere (Land, 1981*a*) the superposition principle suffered an eclipse in the 1960s, but re-emerged in the 1970s in two different forms: one based on the original refracting mechanism of Exner (Figure 9.5), and another based on mirrors (Vogt, 1975, 1980). More recently, Nilsson (1988) has found an intermediate between these two, known as 'parabolic' superposition, involving a combination of lenses and mirrors (see Figure 9.6). There are thus about three-and-a-half distinct types of compound eye.

a) Apposition optics

In apposition eyes each lens forms a separate image at the distal tip of a rod of photosensitive pigment called the rhabdom. This rod, which is made up of contributions from about eight receptor cells, behaves as a light-guide, so that light entering its tip is effectively scrambled; if there were any detail in the pattern of the entering light, this is lost. This solves a problem that goes back to Leeuwenhoek, namely how it is that each lens forms a small inverted image, when the geometry of the image as a whole is erect. In apposition eyes this is simply not a problem because the rhabdom does not 'know' that the scrambled image it receives is inverted (dipteran flies are exceptions here because the individual components of the rhabdom remain separate). For most apposition eyes the overall image is made up of the contiguous fields of view of the rhabdom tips in the image plane of each lens, and these fields are typically about one degree across (this compares pretty unfavourably with less than one minute of arc for the cones in our own eyes). Amongst marine arthropods, apposition eyes are found in the Xiphosura (*Limulus*), most of the lower crustacean sub-classes, and amongst the Malacostraca in the stomatopods, isopods, amphipods and the crabs. A characteristic feature of most apposition eyes with reasonably light pigmentation is the 'pseudopupil', a dark spot which appears to move around the eye as the observer changes position, and which indicates the ommatidia that are accepting light from the viewer's direction. It is particularly obvious in living hyperiid amphipods, and in many crabs.

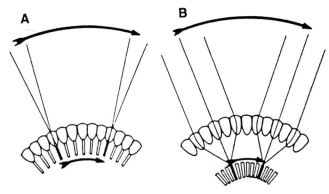

Figure 9.4. The two basic types of compound eye. (A) Apposition eye. Each facet forms a small inverted image at the rhabdom tip, but the overall image is erect. (B) Superposition eye. Many facets contribute to the erect image in the deep-lying retina.

Because the cornea itself has little power in water, the ommatidial image is generally produced by ray-bending within the lens. This was proposed for *Limulus* by Exner (1891), who put forward the idea of a 'lens-cylinder' whose refractive index decreased parabolically from the axis to the outside. It was shown much later that the required refractive index gradient was indeed present (Land, 1979). The hyperiid amphipods also have apposition eyes with lens-cylinders, but their construction is rather more complicated with two graded-index regions, one at each end of rather elongated trumpet-shaped lenses (Nilsson, 1982). They function, however, as ordinary apposition eyes.

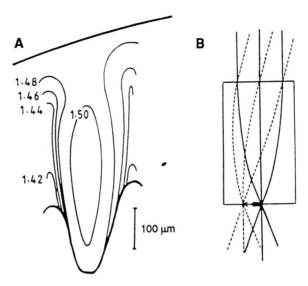

Figure 9.5. Lens-cylinder optics. (A) the refractive index distribution in a facet of a *Limulus* eye, which forms an image at its proximal tip; from data in Land (1979). (B) diagram from Exner (1891) showing the paths of image-forming rays in a lens-cylinder with a gradient of refractive index decreasing parabolically from the axis outwards.

b) Superposition optics: refracting, reflecting and parabolic (Figure 9.6)

Exner found that in the cleaned eye of the male glow-worm it was not possible to see multiple inverted images, as was the case with apposition eyes, but only a single, erect image. This image lay relatively deep in the eye, not immediately behind the optical systems as in apposition eyes. This obviously raised difficulties, because simple optical systems - single lenses and mirrors - produce inverted images, and an array of single lenses cannot be persuaded to produce a single erect image either. If one draws in the paths that rays of light would have to take in order to contribute to such an image, it becomes clear that what each optical element must do is to redirect rays so that they emerge from the back of the element at the same angle that the ray entered it, but reflected across the element's axis. Exner concluded that the simplest device that would do this was a two-lens combination: a simple inverting telescope with a magnification of -1. He therefore proposed that the optical elements in the glow-worm eye, the so-called crystalline cones, actually consisted of two lenses somehow embedded in the same physical structure. They would work in the same way as lens-cylinders in apposition eyes, like that of *Limulus* mentioned earlier, except that they have twice the effective length, so that they behave as a pair of lenses with the focus between them, rather than as a single lens with an external focus. Amongst marine invertebrates this type of superposition eye is typical of mysids and euphausiids (Land *et al.*, 1979).

The 'two-lens telescope' mechanism is not the only basis for a superposition optical system. In the eyes of crayfish and shrimp the optical elements are flat-faced, four-sided truncated pyramids of relatively low refractive index jelly. In 1975 Klaus Vogt found that the faces of these pyramids acted as plane mirrors, and indeed had a coating on them similar to the reflecting layer in the scallop eye (see Vogt, 1980). It is easy to show that a series of radially arranged plane mirrors produces almost the same ray-paths as the radial inverting telescopes in the lens-based superposition eyes proposed by Exner, light being reflected across the axis of each optical element in both cases. Clearly, superposition images can be produced by either kind of optical array. An intriguing feature of the decapod eyes with this kind of reflecting image formation is that they all have square facets, rather than the hexagonal ones seen in almost all other compound eyes. This turns out to be a necessary consequence of the reflecting optical system, since most rays bounce off two faces of each mirror box rather than one, and for proper image formation these must be reflected through 180° in the plane of their direction of travel. Only a 90° corner mirror can do this (Vogt, 1980; Land, 1981a). The presence of square facets provides a simple, though not entirely infallible, way of distinguishing *reflecting* superposition from *refracting* superposition eyes.

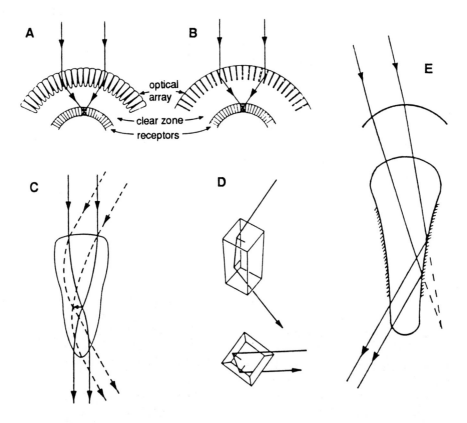

Figure 9.6. Superposition optics. (A) Refracting superposition eye. Rays are redirected to the focus by an 'afocal' double-length lens cylinder, as shown in (C). (B) Reflecting superposition eye, with mirrors replacing the lens-cylinders. Rays that do not fall exactly along a line of mirrors are reflected from two sides of each mirror 'box', as shown in (D). Notice that, viewed end on, the rays are reflected through 180°, as with a single mirror. (E) Parabolic superposition. Light is partly focussed by a lens then unfocussed by a parabolic mirror to give a parallel final beam, as in (C) and (D). Based on Nilsson (1988).

In the recent past there has been a degree of healthy controversy over the relations between the different kinds of superposition eye, and this has centred around the question of whether functioning intermediates between the mirror and telescope versions could exist. Could there be a hybrid system that used both lenses and mirrors? Last year Nilsson (1988) described a new type of superposition eye in a crab, which does indeed have features in common with the other two, as well as some of its own. He calls it 'parabolic' superposition, because each optical system is a combination of a short-focus lens and a convex parabolic mirror which intercepts the partially focussed beam and re-collimates it. A feature of this arrangement is that axial rays are focussed at the rear of the optical element, so that when the reflected rays are excluded - by screening pigment when the eye is light adapted - the eye effectively becomes an apposition eye. The significance of this optical system is that it forms a possible evolutionary link between apposition eyes and both the other types of superposition eye.

The principal merit of superposition eyes of whatever kind is their sensitivity, with receptors typically receiving from a hundred to a thousand times more light than in an apposition eye (see Land, 1981a for a discussion). For this reason, superposition eyes are common in both nocturnal insects (fireflies and moths) as well as the deep-water crustaceans already mentioned.

AMPHIBIOUS VISION

To be able to function in both air and water, an eye must be capable of dealing with large changes in external refractive index. There are basically two solutions to the problem. The first is to have an optically ineffective, flat, cornea that has no power in either medium; and the second is to retain a usable cornea in air, but to compensate for its absence in water by accommodation. Such a mechanism has to be powerful; we all know from trying to see underwater without a mask that our own accommodation mechanism is nothing like adequate.

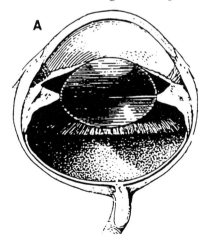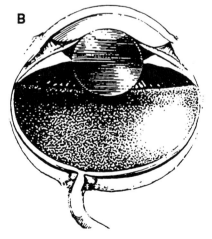

Figure 9.7. Hemi-sections of the eyes of a terrestrial and an aquatic carnivore: lynx (A) and seal (B). Notice the spherical lens and flattened cornea in the seal. From D.W. Soemmerring, (1818). *De oculorum hominis animalque.*

Seals and penguins both make partial use of the first method. The seal cornea is much flatter than in most terrestrial mammals, and the lens has reverted to the spherical fish design, presumably because it has taken over the function of providing nearly all the optical power (Figure 9.7). In fact the harp seal cornea is rather more strongly curved along the horizontal meridian than the vertical (Piggins, 1970), possibly as a result of the need for streamlining, and this may result in a somewhat astigmatic image in air. In general, however, the flattened cornea removes the necessity for a powerful accommo- dation mechanism to make up for the power loss in water. Sivak & Millodot (1977) found that penguin eyes had almost flat corneas, and that they were in focus (emmetropic) in air. They would thus be slightly hyperopic in water, but by an amount that should be well within the power of the accommodatory mechanism to deal with. The flat cornea solution to the problem of amphibious vision does have some difficulties. In air the peripheral parts of the field are compressed and distorted, and in water the field of view is reduced because the lens cannot protrude into the corneal bulge, as it does in most fish. Some amphibious fish have produced interesting solutions to these problems. The flying fish (*Cypselurus*) has a tent-shaped cornea with three almost flat triangular facets, and a peak containing the lens in the centre. Similarly, the clinid fish *Mnierpes*, which lives in rock pools and looks into air, has a pair of flat facets (Figure 9.8).These adaptations will improve the size and quality of the field of view, but at the expense of a disjointed image in air (Graham & Rosenblatt, 1970).

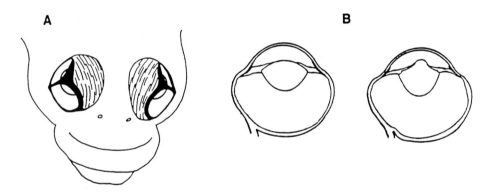

Figure 9.8. Alternative designs for amphibious eyes. (A) Flat corneal 'goggles' in the rock-pool fish *Mnierpes macrocephalus*. Drawn from Graham & Rosenblatt (1970). (B) Extreme accommodation achieved in the merganser by squeezing the lens through the iris. Based on Sivak (1978).

In compound eyes there is the same problem of the change in power of the corneal surface between media. Where it occurs, as for example in shore crabs, *Limulus*, and some insects of the water surface, the answer is always to make the cornea flat, and to make up the required power either using internal curved surfaces between different media, or by lens-cylinder optics (see Figure 9.5, and Land, 1987).

The only animals that have a strongly curved cornea, and try to make up the power deficit in water by accommodating, are certain diving birds. Levi & Sivak (1980) found that during accommodation the front surface of the lens of a merganser (*Mergus cucculatus*), was very strongly curved. The deformation was produced by an unusual method: the lens is squeezed into and partially through the rigid iris by the powerful ciliary muscle, creating a local 'blip' with a very high curvature (Figure 9.8). Sivak *et al.* (1985) studied a variety of ducks, and found that whereas in the majority accommodative changes were only of a few dioptres (similar to man), in the mergansers and goldeneyes, both divers, the changes were 80 and 67 dioptres respectively, and these are commensurate with the power lost from the cornea. A similar mechanism to this has been proposed for some other birds, reptiles and even otters, but with evidence of varying quality (Sivak, 1978).

A very few animals are able to see in air and water simultaneously, and of these the 'four-eyed fish' *Anableps* is perhaps the most famous. Here the eye is bisected by the meniscus, and has two pupils, one looking into air and the other into water. The lens is ovoid with the long axis pointing in the direction of the water. This means that light from the water below, which does not pass through an optically effective cornea, encounters the strongest curvatures in the lens, whereas light from above, already partially focussed by the aerial cornea, meets the weaker curvatures of the lens's short axis (Walls, 1942). This wonderful *ad hoc* arrangement does indeed work; Sivak (1976) measured the refractive state of the eye and found that in-focus images were produced in both air and water.

OPTICS AND THE UNDERWATER LIGHT ENVIRONMENT

The final section of this chapter deals with the ways that eyes are modified to deal with the rather special light conditions in the sea. The first part concerns structural changes that are related to the low light levels of the deep ocean; the second part with changes resulting from the narrow angular distribution of light in the sea; and the third with adaptations to conditions in shallow water and near the surface. These modifications take a variety of forms depending in part on the nature of the 'basic' eye design of the particular phylogenetic group, but they can ultimately all be understood as evolutionary adjustments tending to 'fine tune' eye structure to animal's environment.

1. Adaptations to dim light

In the clearest oceanic water, blue light attenuates by a factor of 10 for every 70 m (Jerlov, 1968). Thus at 700 m depth the intensity is a ten-billionth that at the surface, which is at or just below the absolute threshold of human vision when the surface is sunlit. Some animals with specialized eyes may be 10 or perhaps even 100 times more sensitive than ourselves, but that will only enable them to use daylight at depths 100 to 200 m deeper. Certainly below 900 m there are too few photons left for any eye to make use of them. Some bioluminescent animals retain eyes at lower depths, but their use is mainly restricted to the detection of intraspecific signals.

For good vision at moderate depth an eye must trap as much light as possible. For an instrument like a camera or an eye viewing extended fields, the illuminance (light flux per unit area) provided to the image plane is proportional to $(A/f)^2$, where A is the aperture diameter and f the focal length. A receptor of diameter d, length x and extinction coefficient k will trap an amount of light proportional to its cross section d^2, and to the fraction absorbed ($1\text{-}e^{-kx}$) (see Land, 1981a, for a full derivation). These considerations suggest a number of strategies that an eye might use to increase the amount of light available to each receptor. They include: increasing A, decreasing f, increasing d, increasing x and increasing k. Taking these in reverse order, the extinction coefficient k for vertebrate rods seems to be fixed at around 3.5% μm^{-1}; it is rather lower for the rhabdoms of arthropod eyes. Receptor length (x) needs to be great enough to trap a high proportion of the incident light, and a vertebrate rod 86 μm long will absorb 95% of the incident light. Diminishing returns operate here; doubling the length will trap only the remaining 5%, so 100 μm is probably as long as any rod needs to be. Alternatives to long receptors are multiple banks of shorter receptors, or the use of a reflecting tapetum, which can be thought of as doubling receptor length (Locket, 1977). Increasing receptor diameter d, or in vertebrate eyes increasing the size of the ganglion cell receptive fields over which rods pool their signals, will increase effective light capture, but there is a problem here because this will 'coarsen' the retinal grain, and reduce resolution. An eye's acuity depends on the angular separation of the receptors (d/f, if they are contiguous), which means that any increase in d must be accompanied by an increase in f, if resolution is not to be compromised. The aperture A can be increased, but the maximum A/f ratio has probably already been reached in a Matthiessen lens, so increasing A must lead to an enlargement of the eye as a whole. Where does this leave us? To increase A, f must increase too, and to maintain the same resolution, d must scale with f. The strategy thus has to be to increase A, f and d together, in which case the amount of light trapped by the receptors (or detected by ganglion cells) will go up as the square of the eye's linear dimension.

Thus eyes for use in dim light must be large, and if they are to have high resolution as well they must be very large. This perhaps explains the enormous size of the eyes of large squid, with a diameter of up to 40 cm; they have to catch small prey at a distance in deep water. Amongst mesopelagic fish, living in the top 1000 m, the eyes tend to be large for the size of the animal, but below 1000 m they typically become small and degenerate (Marshall, 1954, 1979). However, this is not universal, and some fish that inhabit depths where no daylight penetrates have seemingly normal eyes. Probably these eyes are used to view bioluminescence, but we know too little about life at these depths to be sure. The argument above applies to compound eyes as well, and one would expect to see a size increase with depth, certainly in the upper 1000 m. This seems to be true for the hyperiid amphipods (Land, 1989) and isopods (Nilsson & Nilsson, 1981) but it is not very obviously true for the shrimps or the euphausiids, both of which have superposition eyes that are already of a design that provides a very bright image.

In view of the very clear predictions of what should happen to eyes with increasing depth based on photometric considerations, it is perhaps disappointing that these are only partially fulfilled. Marshall (1954) quotes Hjort: "Nothing has appeared more hopeless in biological oceanography than the attempt to explain the connection between the development of the eyes and the intensity of light at different depths in the ocean."

It may be that the eyes of many epi- and mesopelagic animals are already close to the maximum limit of their sensitivity, perhaps because they are used at night as well as at depth. If we see no further adaptation to life at depth, it may simply be that no further increase in size is anatomically realistic, given that only a certain fraction of the animal's space and energy can be devoted to its eyes.

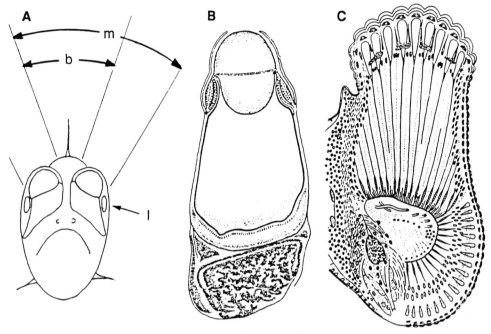

Figure 9.9. Tubular eyes of deep-sea animals. (A) The monocular (m) and binocular (b) fields of view of the fish *Scopelarchus*. (l) indicates the lens-pad, a mysterious organ that appears to channel light from the side. (B) Tubular eye of a deep-sea octopod *Amphitretus pelagicus*. (C) The 'tubular' superposition compound eye of the euphausiid *Stylocheiron suhmii* This eye is actually double, with a high-resolution part directed upwards and a low-resolution part covering the rest of the field. Compiled from Marshall (1979) after various sources.

2. Adaptations to the light distribution in the sea

From a few metres down, the view of the sky through a smooth surface is a circular patch 97° across, known as Snell's window. Beyond the edge of this window total internal reflexion occurs, and the view is of the bottom in shallow water, or the dark abyss in deep water. As one goes deeper, scattering causes the edges of the patch to blur and its width to decrease, so that by 100 m in clear water the distribution of light with altitude is nearly bell-shaped. The brightest region is vertically above, and the position of the sun in the sky no longer matters. The half-width of the light distribution is about 70°, and upwelling light from below is about 200 times dimmer than downwelling light (Jerlov, 1968). Below 100 metres the distribution scarcely changes, although the absolute levels decrease as discussed earlier. All this means that an animal earning its living by trying to sight objects - prey or detritus - against the residual daylight from the sea surface can usefully restrict its field of view to the 70° or so above it. If the downwelling light is only just adequate for the detection of objects, then vision in other directions is pointless.

There are animals in all the principal phyla that do seem to have opted for such a strategy. Amongst the mesopelagic fish there are many with vertically-pointing tubular

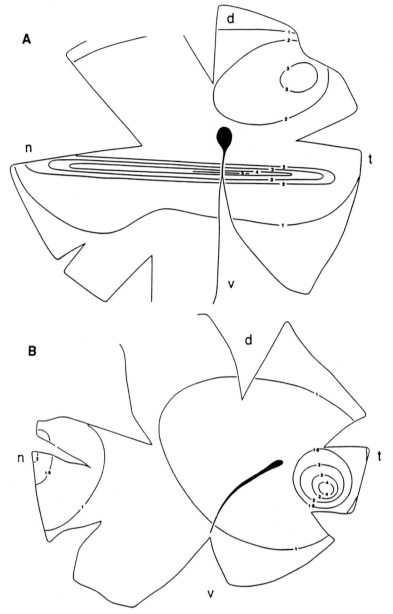

Figure 9.10. Distribution of ganglion cells in the retinae of two reef fish. (A) *Lethrinus chrysostomas* inhabits open water, and the pronounced visual streak images the underwater horizon. (B) *Cephalopholis miniatus* lives in cavities in the reef, and has a pronounced forward-pointing area of high ganglion-cell density. Numbers are ganglion cells per mm 2×10^4. d,v,n & t are the dorsal, ventral, nasal and temporal regions of the retina, of the left eye in both cases. The black area is the head of the optic nerve. Combined from figures in Collin & Pettigrew (1988*a*, *b*).

eyes, including *Opisthoproctus*, *Scopelarchus* and the hatchet fish *Argyropelecus* (Marshall, 1954, 1959; Locket, 1977). Some, like *Gigantura* and *Winteria* have forward-pointing eyes, but these fish may well hold their bodies vertically in the sea (Lythgoe, 1979). All these fish have eyes with a field of view restricted to 70° or less. Amongst the cephalopod molluscs some deep-sea octopods (*Amphitretus*) and squid (*Histioteuthis*) also have tubular eyes (Figure 9.9).

Some crustaceans with compound eyes show similar adaptations. Two groups in particular, the hyperiid amphipods with apposition eyes and the euphausiids with superposition eyes, both exhibit a gradation of forms from those with round eyes, through species with double eyes in which the upper part is emphasized (Figure 9.9C), to forms with effectively only an upward-pointing eye. This last condition is found in the deepest-living species, the euphausiid *Nematobrachion boopis* (Land *et al.*, 1979), and the hyperiid *Cystisoma* whose huge eyes have an upward field only about 10° across (Land, 1981*b*).

3. Adaptations to the complex environments of inshore waters

Although the eyes of most shallow-water fish are superficially similar to each other, their retinae contain subtle differences in the way their ganglion cells are organized that reflect particular features of the environment. It has been known for many years that mammals that live on plains tend to have 'visual streaks', elongated regions of elevated ganglion-cell density corresponding to the region around the horizon (Hughes, 1977). In this case the density variations have nothing to do with light intensity, but indicate rather the amount of neural processing that the animal finds it useful to devote to different parts of the visual field. Recently, in an elegant study of reef fishes, Collin & Pettigrew (1988*a*, *b*) found variations in ganglion-cell density that paralleled those of mammals. A wide variety of patterns was apparent. Fish that lived in holes or gullies in the reef tended to have one or more circular areas of high cell density. Those feeding above a flat bottom

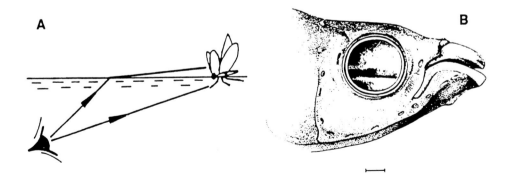

Figure 9.11. Vision at the surface. (A) shows that an object trapped in the surface film can be seen both above and below the surface, at different locations on the retina. (B) eye of *Aplocheilus lineatus* with the front removed, showing the two horizontal areae (light stripes). This fish hunts at the surface, and the two areae correspond to the two directions shown in the diagram. Scale: 1 mm. From Munk (1970).

or below the surface tended to have elongated visual streaks, above or below the eye's equator depending on the location of the region of interest (Figure 9.10). In all these cases there was an evident match between retinal anatomy and the fishes' visual priorities in the environments they inhabited.

The most impressive of all these matches was described much earlier, by Munk (1970). He found that in *Aplocheilus lineatus*, a fish that hunts prey trapped in the surface film, the retina contained not one visual streak but two, both horizontal but separated vertically by about 40° (Figure 9.11). One is directed laterally, and so views the surface from underneath, but the other is directed upwards at 47° from the vertical, and thus looks into air, just above the rim of Snell's window. The two streaks are actually looking in almost the same direction - and presumably at the same objects - but whilst one looks just below the surface, the other looks just above it.

Chapter 10

THE COLOUR SENSITIVITY AND VISION OF FISHES

J.C. PARTRIDGE

Department of Zoology, University of Bristol, Woodland Road, Bristol, BS8 1UG

Fishes are outstanding among vertebrates for the diversity of the visual environments in which they live. In the depths of the open ocean species exist in a dim, homochromatic blue world and at greater depths they live in darkness broken only by bioluminescence. By contrast, in the surface waters of the oceans other species exist under the full glare of the tropical sun. Yet others live in green coastal waters, muddy estuaries or yellow-brown rivers. The study of the vision of fishes and the 'natural laboratory' of the aquatic light environment has revealed many links between the available light and the way fish visual systems have evolved. The survey of many fish has shown that species from the same environment are often similar in their sensitivity to light of different wavelengths. More detailed examination of single species, however, has demonstrated greater complexity than such surveys suggested.

THE COLOUR OF NATURAL WATERS

Strictly speaking 'light' refers to photons that are visible to humans, with wavelengths between about 400 and 750 nm. When discussing the vision of other animals, however, it is often preferable to extend the definition of light to encompass the whole of that small part of the electromagnetic spectrum where photons have sufficient energy for detection by animals and plants, but generally too little energy to damage biological tissue (approximately 300-800 nm). In most terrestrial environments during the day this available light is more than adequate for vision at all wavelengths from the ultraviolet to the far red. In the aquatic environment, by contrast, the spectral transmission of the water may impose severe constraints on the way visual systems can function.

Light, predominantly from the sun but also from the sky, moon and stars, is modified by its passage through the atmosphere in ways which vary with latitude, time of day, temperature and the presence of cloud cover, and produce large variations in spectral irradiance at the surface of the sea. Underwater, light is modified in ways which are essentially similar to atmospheric effects, although on a more extreme scale, and much of the mathematical framework for describing the underwater light environment has its roots in atmospheric studies (Middleton, 1952; Duntley, 1963).

The passage of light through water is affected by both the scattering of light and by the spectral absorptions of: (1) water itself, (2) chlorophyll, and (3) the breakdown products

Herring, P.J., Campbell, A.K., Whitfield, M. & Maddock, L. (ed.). *Light and Life in the Sea.*
© 1990. Cambridge University Press.

of plants, known variously as Gelbstoff, yellow substances, gilvins or Dissolved Organic Matter. The relative importance of these factors varies considerably in different waters, both fresh and marine, leading to wide variation in light transmission and spectral irradiance at different depths (Figure 10.1). Pure water best transmits light in the blue region of the spectrum and has a maximum transmission close to 460 nm, a property which is unaffected by dissolved salts (Morel, 1974). As a result, the open oceans and a few oligotrophic, freshwater lakes such as Crater Lake, Oregon appear blue in colour (Jerlov, 1976; Tyler & Smith, 1970). At depth in such waters the ambient light is restricted to the blue part of the spectrum. In coastal waters, where phytoplankton and run-off from the land increase the levels of chlorophylls, Gelbstoff and particulate matter, the water colour is changed to green or yellow. The levels of these factors vary dramatically within short periods and plankton blooms and variations in run-off cause unpredictable and extreme variations in spectral transmission and hence in the light available for vision. Fresh water is usually green, yellow-green or brown but in some fresh waters, such as peat-stained lakes, the tea-coloured Gelbstoff is found in such high concentrations that the water best transmits red light. Similarly in the 'blackwater' tributaries of the Amazon it is near-infra-red wavelengths that best penetrate (Muntz, 1978).

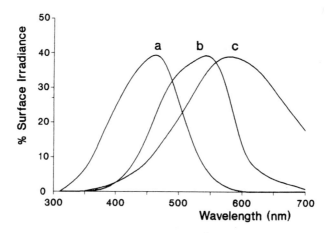

Figure 10.1. Spectral irradiance, as a percentage of surface irradiance, calculated from diffuse attenuation coefficients tabulated by Jerlov (1976) for three different marine waters which differ in chlorophyll and Gelbstoff concentrations: (a) Jerlov Type I oceanic water at 50 m depth; (b) Jerlov Type 1 coastal water at 7·5 m depth; (c) Jerlov Type 9 coastal water at 1·5 m depth. The depths for the three types of water were chosen so that smooth curves drawn through the data had equal maxima. At these optically equivalent depths all three water types restrict the available spectrum with roughly equal efficiency but with greatest transmissions in widely varying parts of the spectrum.

The variation in the transmission of light by different waters led Jerlov (1976) to propose a classification of marine waters based on downwelling underwater spectral irradiance. More recently Baker & Smith (1982) and Prieur & Sathyendranath (1981) introduced schemes which can be used to predict, with reasonable accuracy, the spectral irradiance in different waters from measurements of chlorophyll and Gelbstoff. Classifications such as these which describe or predict spectral irradiance underwater have been used in a number of studies in which broad ecological discussions have revealed correlations between ambient light and the visual adaptations of fishes (Munz & McFarland, 1977; Loew & Lythgoe, 1978). For a more detailed understanding of fish vision, however, it is the spectral radiance distribution that is of greater importance.

THE UNDERWATER RADIANCE DISTRIBUTION

Unfortunately for vision ecologists the relatively simple measurement of spectral irradiance is insufficient to describe the visual environment of an aquatic animal. Instead the measurement of spectral radiance is required, ideally describing the spectral light flux from all directions surrounding a point in the water column.

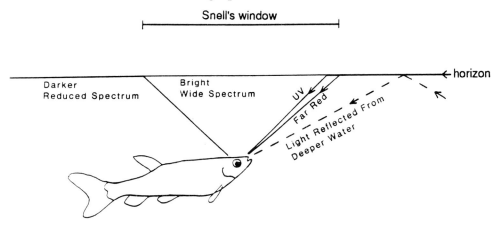

Figure 10.2. When the water surface is flat a fish will see the entire hemisphere above water condensed into a solid angle of about 97°. Light from this region (Snell's window) is bright and in shallow depths contains most of the spectrum available to terrestrial animals. Outside Snell's window light is reflected from deeper water and, particularly in open water, this light is dim and restricted in spectral range. At the edge of Snell's window light from objects near the above-water horizon is split into a 'rainbow' due to the different refractive indices of water to light of different wavelengths.

Close to the water surface the radiance distribution is particularly complex and is strongly affected by solar angle, cloud cover and especially by surface disturbance. In the rare cases where the water surface is flat an aquatic animal will see the whole above-water scene through Snell's window, a cone with a solid angle of some 97° (Jerlov, 1976; Smith, 1974). At angles greater than this the underside of the water surface acts as a mirror and the image of light from deeper water is reflected into the line of sight. For shallow-living fish the edge of Snell's window not only presents an abrupt brightness disconti-nuity but the spectral distribution of light also changes dramatically with light from within Snell's circle often containing a far broader spectrum than light reflected from deeper water. At the edge of Snell's window the light from objects close to the above-water horizon is split into a spectrum due to the slightly differing refractive indices of water to light of different wavelengths (Kaye & Laby, 1966) and as a result the spectral radiance distribution becomes particularly extreme. The angular subtense of this 'rainbow' is small, being approximately 1·4° in fresh water for the spectrum from 300 nm to 800 nm (Figure 10.2), but for small fish feeding near the surface this may be important, being equal to the angle subtended by a 1 mm zooplankter seen at a distance of about 40 mm. If viewed against the edge of Snell's window at this distance such a plankter will be seen against a highly complex 'rainbow-like' background. Interestingly, some shallow-living fishes have a retinal region of high acuity where the edge of Snell's window is normally imaged onto their retinae (Munk, 1970).

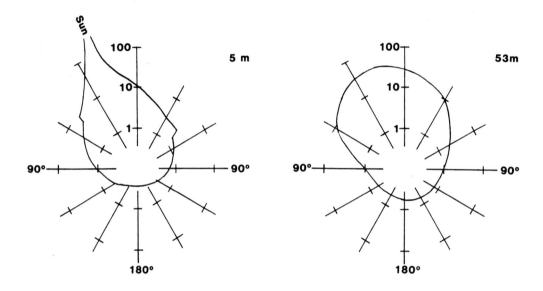

Figure 10.3. The angular distribution of light in the open ocean at two depths (data from Smith, 1974). The logarithmic scales on the radii indicate the relative radiance, normalized for horizontal light in one direction. At 5 m depth the sun is clearly visible and 'shoulders' in the radiance distribution mark the edge of Snell's window. At 53 m the directions of the sun and of Snell's window are far less apparent and the angular distribution of light is approaching its asymptotic distribution.

Very near the surface the position of the sun has a strong influence on the underwater radiance distribution, casting shadows as on land. However, as depth increases radiance rapidly approaches an asymptotic distribution that is symmetrical about the vertical axis (Figure 10.3). The symmetry of this light distribution, at all wavelengths, is one reason why mirror camouflage has been adopted by so many fishes. For mid-water species a highly reflective body surface produces an almost perfect camouflage which needs augmenting only by counter-shading to disguise a fish's three-dimensional shape (Denton & Nicol, 1966). A number of other fishes which are not exclusively pelagic have also evolved silvery flanks (*e.g.* sandeels) but in these species it may be that the reflection of bottom features has some merit as camouflage or that disguise is most important for these species in pelagic phases of their behaviour. In contrast to the open-water situation the spectral radiance distribution in shallow water and close to the bottom or other underwater surfaces is highly asymmetrical. In such habitats the ambient light varies markedly in both brightness and colour depending on direction of view (Munz & McFarland, 1977; Levine & MacNichol, 1982). The retinae of many fishes have been found to have extreme specializations of acuity corresponding to different directions of view (Collin & Pettigrew, 1988a, b) and it is quite likely that there are similar retinal adaptations to the asymmetry of spectral radiance.

To this already complex light distribution must be added another factor, the spatial and temporal variations in light distribution due to surface waves (McFarland & Loew, 1983). Surface waves focus sunlight at different depths depending on the period of the wave; surface ripples focussing at a few cm depth, longer period waves focussing at greater depths (Schenk, 1957). Wave-focussing causes illuminated objects to flicker, the frequency of flickering decreasing with increasing depth. Near the surface flicker is rapid but at 70 m depth McFarland & Loew (1983) found the predominant flicker-frequency to be approximately 1 Hz. It is not clear how important object-flicker is to fish but the photopic critical fusion frequency (the frequency at which a flickering light source is seen by an animal as unvarying) has been measured for a few species. In the shark *Mustello manazo* the critical fusion frequency was found to be 10 Hz (Kobayashi, 1962), in the skate *Raja erinicea* it was 30 Hz (Green & Seigel, 1975), in the bowfin *Amia calva* 38 Hz (Ali & Kobayshi, 1968) and in the trout *Salvelinus fonatinalis* 90 Hz (Ali & Kobayshi, 1968). If more species are surveyed it is possible that flicker fusion frequencies will tend to correlate with the depth ranges of the different species with deeper-living animals having lower critical fusion frequencies. However, there is little evidence for this suggestion at present and critical fusion frequencies may vary considerably in one species depending on light level. For example, in man flicker fusion frequencies are known to vary from 50-60 Hz in daylight to less than 10 Hz at scotopic light levels (Hecht & Verijp, 1933).

For a full understanding of their visual importance the spatial and temporal effects of wave focussing cannot be separated. However, in spatial terms objects are illuminated by complex patterns of light and dark bands and a vermiculate pattern of cell-like units (McFarland & Loew, 1983). Such patterns may have important implications both for the ways fish extract information from their visual images and for their methods of camouflage but there has been little investigation of these possibilities.

IMAGE TRANSMISSION UNDERWATER

If all aspects of the underwater light environment are considered simultaneously it would appear that it is an unlikely candidate for simple mathematical description. Nevertheless, the mathematics of radiance transfer underwater are well established (Duntley, 1962, 1963; Preisendorfer, 1964; Tyler, 1960a) and have been successfully used as a basis for the understanding of vision underwater both in man (Hemmings & Lythgoe, 1965; Lythgoe, 1968; Pilgrim *et al.*, 1989) and in fishes (Lythgoe, 1979). In the simple situation depicted in Figure 10.4, some of the light incident on fish B is reflected towards the observing fish (A) which is at a horizontal distance z. This reflected light is attenuated by scattering and absorption such that the spectral radiance (photons nm^{-1} s^{-1} steradian^{-1}) arriving at A, $R(\lambda)_z$ is:

$$R(\lambda)_z = R(\lambda)\, e^{-\alpha(\lambda)z} \qquad\qquad (10.1)$$

where $\alpha(\lambda)$ is the narrow-beam spectral attenuation coefficient. This coefficient, which is tabulated for pure water by Jerlov (1976), describes the combined effects of light absorption by water and the way light is scattered out of the path to the observer. In most underwater situations, however, vision is not limited by the reduction in light as described by Equation 10.1 but by non-image-forming 'veiling' light that is scattered into the direction of vision. This gain in veiling light is given by:

$$_bR(\lambda)_z = {}_bR(\lambda)\,(1 - e^{-\alpha(\lambda)z}) \qquad\qquad (10.2)$$

where $_bR(\lambda)$ is the background radiance. The apparent radiance $_aR(\lambda)_z$ of an object seen at distance z is then given by the combination of Equations 10.1 and 10.2 as:

$$_aR(\lambda)_z = R(\lambda)\, e^{-\alpha(\lambda)z} + {}_bR(\lambda)\,(1 - e^{-\alpha(\lambda)z}) \qquad\qquad (10.3)$$

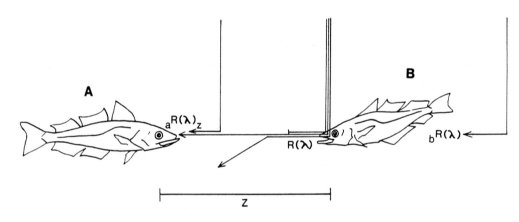

Figure 10.4. Light incident on fish B, primarily but not entirely from above, is reflected towards observing fish A which is at a distance z. Of this light, $R(\lambda)$, some is absorbed and some scattered out of the visual path. The remaining image-forming light reaching fish A is degraded by non-image-forming 'veiling' light that is scattered into the visual pathway. Veiling light reduces the contrast between light from fish B, $_aR(\lambda)_z$, and the background spacelight, $_bR(\lambda)$.

The effect of veiling light is to degrade the contrast of an image in much the same way as atmospheric fog degrades contrast in the terrestrial environment. This has important visual implications because most visual detection tasks can be reduced to a problem of contrast detection and vision underwater often involves the detection of objects of low contrast (Lythgoe, 1988). Equation 10.3 can be used to demonstrate the way in which contrast (C) between an object and the background spacelight decreases with visual range:

$$C(\lambda)_z = ({}_bR(\lambda)_z - {}_aR(\lambda)_z) / {}_bR(\lambda)_z$$

or, $$C(\lambda)_z = C(\lambda)\, e^{-\alpha(\lambda)z} \qquad\qquad (10.4)$$

VISUAL PIGMENTS

The equations presented above allow the radiance distribution, or the contrast of an object against its background to be calculated at discrete wavelength intervals for different visual ranges and values of $\alpha(\lambda)$. However, it should not be forgotten that animals do not sample light at discrete wavelengths. Instead the absorption spectra of their visual light receptors, and the visual pigments that they contain, are broad and simultaneously sample a wide part of the spectrum. Visual pigments are coloured molecules that are common to many types of light receptor in both vertebrates and invertebrates. In the vertebrate eye visual pigments are found in high concentrations in the outer segments of retinal rods and cones. These retinal photoreceptors act as the transducers of the visual system, converting the light signal from the environment into a neural signal which is transmitted to the brain. Light in the image focussed onto the retina must be absorbed by visual pigments before it can be converted into a neural signal and light that is not absorbed can play no part in vision. As a result, knowledge of the absorption spectra of visual pigments is fundamental to many facets of the study of vision.

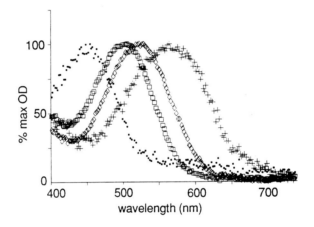

Figure 10.5. Visual pigments of the two-spot goby (*Gobiusculus flavescens*) measured by microspectrophotometry and shown as percentage maximum optical density against wavelength. The data are averages of a number of cells of four different classes: (1) □, rods, with a peak absorbance (λ_{max}) near to 501 nm, number of cells averaged, n=13; (2) ●, blue-sensitive cones ($\lambda_{max} \approx 452$ nm, n=2); (3) ◊, green-sensitive cones ($\lambda_{max} \approx 525$ nm, n=11); (4) +, orange-sensitive cones ($\lambda_{max} \approx 570$ nm, n=6).

Visual pigment absorption spectra can be measured in two ways: by the spectro-photometry of pigments extracted from retinal samples in organic solvents (Knowles & Dartnall, 1977) or by measuring the absorption spectra of single cells by microspectro-photometry (Liebman, 1972; Harosi, 1975; Partridge, 1986). Both these techniques are exacting procedures because of the light-sensitive nature of visual pigments. Neverthe-less, with purpose-built instruments used with care, both techniques enable visual pigment absorption spectra to be measured with a high degree of accuracy and the visual pigments of many species have now been measured. The absorption spectra of visual pigments are always bell-shaped, as shown in Figure 10.5, and absorb light over a reasonably wide part of the spectrum with an absorption peak at a wavelength known as the λ_{max}. At this wavelength the visual pigment, and the photoreceptor outer segment that contains it, are most sensitive to incident light. It is now known that λ_{max} values of visual pigments found in the retinae of different animals, including fishes, range from the ultraviolet to the far red; well beyond our visual spectrum at both short and long wavelengths.

VISUAL SENSITIVITY

Visual photoreceptors act not as energy measurers but as photon counters and in many visual situations the main visual problem is to obtain sufficient photons. The rate of photon flux is 'measured' by the eye as brightness but both the emission and absorption of photons are uncertain processes. As a result the detection of photons and the visual detection of objects in the environment are, at root, statistical problems (Rose, 1973). Because of this statistical problem, the amount of information in a retinal image is fundamentally limited by the number of photons that form it (Buschbaum, 1981). This limitation is particularly important for nocturnal or crepuscular animals but in humans it is significant well into levels of brightness regarded as 'daylight' (Land, 1981a).

Underwater, the environmental attenuation of light ensures that light levels are often low and that catching as many photons as possible is of paramount importance. The eyes of fishes show numerous adaptations towards the maximization of photon catch including optics designed to give a bright retinal image (Fernald, 1988) and tapeta which reflect light back through the retinal photoreceptors for a 'second chance' of being absorbed. For similar reasons a number of fishes, including crepuscular predators (Lythgoe, 1988) but more especially deep-sea fishes (Ali & Anctil, 1976), have evolved particularly large photoreceptors in which the unusually long path-length that light must travel in the outer segment, often as much as 100 µm, substantially increases the chances of photon absorption. Deep-sea fishes have also increased the density of visual pigment in the outer segment (Munk, 1966a; Partridge et al., 1989) and specific optical densities may be as high as 0·028 µm^{-1} in these animals (Partridge et al., 1989), very much higher than values between 0·010 µm^{-1} and 0·015 µm^{-1} typical of shallow-living fishes (MacNichol et al., 1973; Knowles & Dartnall, 1977; Partridge, 1986).

In any photon-limited environment photoreceptor sensitivity can be increased by matching the spectral sensitivity of the visual pigment to the ambient spectral irradiance and the degree to which this is so is shown, for the rod visual pigments of fishes, in Figure 10.6. In the deep oceans the spectrum of the downwelling light is restricted to wavelengths close to those best transmitted by water (Jerlov, 1976) and even light derived from bioluminescence is, overall, brightest in this part of the spectrum (Widder et al., 1983; Herring, 1983). As a result it is perhaps not surprising that most deep-sea fishes have visual pigments with λ_{max} values in the range 475-485 nm (Partridge et al., 1988, 1989;

Partridge, 1989). Indeed this result was predicted over 50 years ago (Clarke, 1936; Baylis *et al.*, 1936). In aquatic environments other than the deep sea, however, the match between the λ_{max} values of fish visual pigments and spectral irradiance is far less good (Figure 10.6), despite the fact that at twilight or at night the main limit to vision is photon capture. This mismatch may be explained if, as predicted on thermodynamic grounds, visual pigments with greater long-wavelength sensitivity are more susceptible to spontaneous isomerizations due to thermal excitation (Barlow, 1957*b*). Spontaneous isomerizations of rhodopsins are known to occur at extremely low rates in human and amphibian rods and yet thermal isomerization is probably the dominant factor in setting the ultimate limit to visual sensitivity (Barlow, 1988; Aho *et al.*, 1988). As a result particularly noise-prone visual pigments will be unacceptable in a photon-limited environment, even if their spectral sensitivity gives a better match to the available light than other, less noisy, pigments.

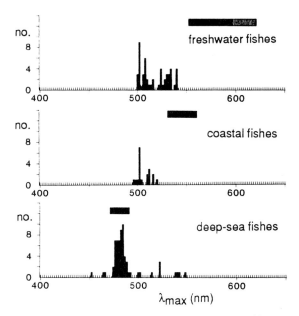

Figure 10.6. Histograms of λ_{max} values of rod visual pigments measured by microspectrophotometry of fishes from three habitats. The horizontal bars indicate the possible range of visual pigments most likely to confer maximum sensitivity in the different water types (after Lythgoe, 1988). Only the deep-sea fishes have rods best positioned to maximize photon catch. (Data mainly from Partridge *et al.*, 1988, 1989; Levine & MacNichol, 1979; Loew & Lythgoe, 1978.)

VISUAL PIGMENTS AND CONTRAST

It is quite possible that the evolutionary selection of rod visual pigments is driven largely by the need to optimize photon catch whilst employing a noise-free photoreceptor. Nevertheless, it is also likely that the visual tasks that animals must perform have an important effect on the selection of visual pigments. Most visual tasks involve the detection of changes in light levels, either with time or in space, and the larger the change the easier is the task of detection (Land, 1981a). Spatial differences in light level (contrast, as calculated in Equation 10.4) may be greater in regions of the spectrum where radiances are less than at other wavelengths. In these cases visual pigments with λ_{max} values 'offset' from the wavelengths of greatest light flux may confer an advantage in contrast-detection tasks.

In order to calculate which visual pigments will give the greatest contrast in a visual task such as that depicted in Figure 10.4 it is necessary to consider the photon catch of the photoreceptors involved. If we are interested only in ascertaining which single visual pigment will maximize contrast then we must calculate I_a and I_b, the relative rates of photon catch for photoreceptors containing the same visual pigment but receiving light from different parts of the visual field. To a first approximation these are given by:

$$I_a = \int_{300}^{800} A(\lambda).T(\lambda)._a R(\lambda)_z.d\lambda \qquad (10.5)$$

$$I_b = \int_{300}^{800} A(\lambda).T(\lambda)._b R(\lambda)_z.d\lambda \qquad (10.6)$$

Where $A(\lambda)$ is the spectral absorbance of the visual pigment and $T(\lambda)$ is a parameter which describes the spectral transmission of the intraocular tissues such as the cornea and lens.

The contrast (C') is then:

$$C' = (I_a - I_b)/I_b \qquad (10.7)$$

By such methods it is possible to demonstrate that visual contrasts are often greater when visual pigments have λ_{max} values offset from the wavelengths of maximum irradiance and several authors have discussed the advantages of 'offset' visual pigments in contrast enhancement (Lythgoe, 1979; McFarland & Munz, 1975; Munz & McFarland, 1977; Levine & MacNichol, 1979; Loew & Lythgoe, 1978). In general these authors have shown that, for grey targets, 'offset' visual pigments increase the contrast of a bright target seen against a darker background. In the situation where a dark target is seen against a lighter background, however, a visual pigment 'matching' the ambient spectrum will maximize contrast.

For natural visual tasks it is likely that targets will have spectral reflections more complex than that of a grey target and as a result the prediction of which visual pigments will maximize visual contrast is not intuitively obvious. The λ_{max} values of such pigments will vary considerably depending on the details of the visual task under consideration and target spectral reflection, visual range, spectral beam attenuation coefficients and the line of sight will all influence the selection of the 'best' visual pigment. In addition, the full importance of contrast detection in visual pigment evolution will be understood only if actual rates of photon capture by photoreceptors are calculated rather than relative rates. Without this information it is quite possible for an 'offset' visual pigment to appear to give a very high contrast but in fact to have a photon capture rate that is too low to be visually useful.

COLOUR VISION

A single class of photoreceptor is able to code only the brightness of an image, not its colour. For example a cell most sensitive to green light will produce the same output when illuminated by a dim green light or by a bright red light, the cell's response being dependent only on the number of photons it absorbs. To code for the colour of objects the retina or brain must compare the outputs of photoreceptors of different spectral sensitivities when both are 'looking at' the same part of the visual field. In this way an animal is able to extract information about the spectral reflection curves of objects independent of brightness, and these different shaped curves may be neurally coded to give the sensations of different colours.

Because of the need in colour vision to compare the outputs of different cell types the presence of different photoreceptors with more than one spectral sensitivity (*e.g.* containing different visual pigments) is not, by itself, proof that an animal has colour vision (Jacobs, 1981). Colour vision requires not only that the retina has at least two types of photoreceptor but also that the animal has the appropriate neural machinery to compare the output of the different cell types, an ability perhaps best revealed by psychophysical or behavioural experiments (Jacobs, 1981). Nevertheless, many fish appear to have evolved colour vision and in the goldfish (*Carassius auratus*) colour vision is known to be highly developed (Shefner & Levine, 1976). Indeed, Munz & McFarland (1977) postulate on environmental and behavioural grounds that multiple visual pigments, and hence the potential for colour vision, may have evolved in fishes some 400 million years before present.

A few fish, notably most deep-sea fishes, seem to have only one type of visual pigment in their retinae (Partridge *et al.*, 1988, 1989; Munz & McFarland, 1977) and in these species colour vision is unlikely. However, Denton & Locket (1989) argue that the multi-layered retinae of many deep-sea fishes could result in photoreceptors with the same visual pigment having different spectral sensitivities, thus providing a basis for colour vision. This suggestion is particularly appealing because their calculations show that such a colour vision system would give these fishes best hue discrimination in the blue to greenish-yellow part of the spectrum, which is the only part of relevance in the deep oceans.

Unlike deep-sea fishes most fish have both rods and a number of cones, commonly two or three types, in their retinae (Loew & Lythgoe, 1978; Levine & MacNichol, 1979). However, other species of fish have more numerous cones and the 14-spined stickleback (*Spinachia spinachia*), the roach (*Rutilus rutilus*), the Japanese dace (*Tribolodon hakonensis*) the shanny (*Blennius pholis*) and others are known to have four types of cone (Loew & Lythgoe, 1978; Avery *et al.*, 1983; Harosi & Hashimoto, 1983; Partridge, 1986). Some guppies (*Poecilia reticulata*) may have up to six types of cones in their retinae (Archer *et al.*, 1987; Archer & Lythgoe, in press). How these animals incorporate this number of cell types into a colour vision mechanism is a matter for speculation.

CONE PIGMENTS AND THE ENVIRONMENT

Different species of fish vary extensively both in the number of cones and in the number of visual pigments that they have in their retinae. When groups of fish from particular environments are examined, however, it is clear that there are common patterns in their visual pigment complements (Figures 10.7 & 10. 8). These patterns reflect both the transmission properties of different types of water and the behaviour of the various species. For example, surface living diurnal species, whether fresh water or tidepool, all experience a wide range of wavelengths in the ambient light with a high photon flux at most wavelengths. Like terrestrial animals (Lythgoe & Partridge, in press) these fishes all have visual pigments which extend over a reasonably wide range of the spectrum (Figures 10.7A & 10.8A). These animals all tend to have visual pigments based on vitamin A1 (rhodopsins) and share the same degree of long-wavelength sensitivity with maximum λ_{max} values somewhere near 570 nm. By contrast, diurnal freshwater fishes that live in slightly deeper water have markedly extended red-sensitivity which is related to their possession of porphyropsin pigments based on vitamin A2. These animals have the most long-wavelength-sensitive visual pigments that are known and in general they have also abandoned their short-wavelength sensitivity (Figures 10.7B & C).

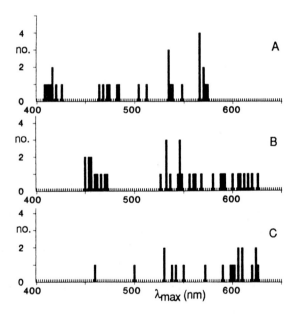

Figure 10.7. Histograms of cone visual pigments of tropical freshwater fishes from three habitats measured by Levine & MacNichol (1979). (A) Group I, surface limited or shallow-water (<2 m) dwelling diurnal species. (B) Group II, generalized midwater species. (C) Groups III & IV, crepuscular predators and benthic species. Group I fishes have violet-blue sensitive pigments not found in the other groups and some species also have ultraviolet sensitive pigments (not shown). Visual pigments in this group tend to be rhodopsins and lack the extended red sensitivity of the porphyropsins found in the other groups. Group II fishes have abandoned the violet-blue-sensitive pigments of shallow-living fish but retain blue-sensitive cones which are almost absent in species from groups III and IV.

Fishes which use vision in more photon-limited conditions have very different sets of visual pigments from those of shallow-living diurnal species and in both freshwater and marine environments deeper-living or crepuscular fishes usually possess only two cone pigments. In the case of marine species these are most sensitive in the blue and green parts of the spectrum (Figure 10.8B) but in freshwater species λ_{max} values are displaced to longer wavelengths (Figure 10.7B & C), perhaps reflecting the predominance in their surroundings of long-wavelength light.

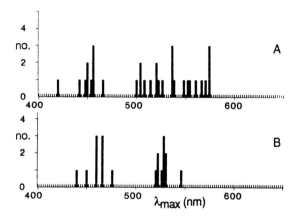

Figure 10.8. Histograms of cone visual pigments of temperate coastal marine fishes, measured by Loew & Lythgoe (1978) and Partridge (1986). (A) Shallow-living and tide-pool species. (B) species from moderate coastal depths. The shallow-living species have a range of visual pigments similar to Group I freshwater species (Figure 10.7A) although fewer have violet-blue-sensitive pigments. Deeper-living species have retained the blue-sensitive pigments but have restricted long-wavelength sensitivity.

VISION IN THE ULTRAVIOLET

Sensitivity to UV-A (320-390 nm) is well documented in both vertebrates and invertebrates and a number of fishes have retinal cones sensitive to ultraviolet light including the roach (Avery *et al.*, 1983), the Japanese dace (Harosi & Hashimoto, 1983; Harosi, 1985) and the goldfish (Hawryshyn & Beauchamp, 1985). All these fishes live in shallow fresh water and ultraviolet receptors have not yet been identified in marine fishes. However, to date the only shallow-living marine fishes that have been extensively investigated are tide-pool species and it is possible that pelagic surface-feeders such as garfish (*Belone belone*) will be found to have UV-sensitive cones.

The possession of UV vision is necessarily correlated with the possession of corneas and lenses that are transparent to ultraviolet light and short-wavelength cut-offs are tabulated for the ocular media of several species of fish by Nicol (1989) and by Douglas & McGuigan (1989). A recent study of 33 temperate marine fishes found only four that possessed lenses transparent to wavelengths shorter that 390 nm (Douglas, personal communication). However, most of the animals examined were adults and in several freshwater fishes it is known that corneal and lenticular UV transmission decreases as a fish ages (Douglas, 1989). Ultraviolet-transparent ocular media are an indication of the possibility of ultraviolet visual capacity but even if ultraviolet light can reach the retina UV vision may be lacking. For instance Hawryshyn *et al.* (1989) have shown that rainbow trout (*Salmo gairdneri*) are UV-sensitive when weighing less than 30 g but lose this UV vision at body weights greater than about 60 g, despite little change in the UV transmission of their ocular media. Similarly, in the brown trout (*Salmo trutta*) UV-sensitive cones are lost from the retina in two-year-old fish (Bowmaker & Kunz, 1987). By analogy with freshwater species it is most likely that UV-sensitivity will be found in planktivorous post-larval or juvenile marine fishes particularly those living in shallow water or in water containing low levels of Gelbstoff.

Although the importance of UV vision to marine fishes is speculative it is perhaps worth considering the potential for vision at these wavelengths. In the terrestrial environment it is known that there is visual information in the UV, particularly in reflections from leaves and flowers. This information is generally invisible to humans but is available to some insects and birds (Loew & Lythgoe, 1985) and it is likely that we are missing similar UV information from underwater objects. The UV reflections of relatively few underwater objects have been measured but Lythgoe & Shand (1983) have indicated UV reflections from the lateral stripe of the neon tetra (*Paracheirodon innesi*). Reflections in the UV have been shown by UV video to be bright (Lythgoe, personal communication) and Lythgoe & Shand (1983) have shown that the brightness of this reflection varies both with a diurnal rhythm and in response to the emotional state of the animal. Similarly, Harosi (1985) has demonstrated high-contrast UV reflection patterns on the flanks of the Japanese dace (*Tribolodon hakonensis*). It is highly likely that the examination of many marine species will reveal similar UV patterns which may be of both visual and behavioural importance.

Ultraviolet light is no more strongly absorbed by pure water than is light with wavelengths close to 600 nm (Morel, 1974) but the presence of Gelbstoff in most natural waters ensures significant attenuation at short wavelengths. Nevertheless, UV light can penetrate to considerable depths in clear seas (Smith & Baker, 1981) and McFarland (1986) has calculated that there is sufficient light for UV vision to a depth of 100 m in clear oceanic waters. Although the attenuation of UV light in other waters may be high the visual usefulness of UV wavelengths will be determined more by the effects of scattering than by its absorption. The scattering of light is highly dependent on wavelength, with short wavelengths being scattered most by both large particles (Mie scattering, proportional to $1/\lambda$) and by particles of a size similar to the wavelength of light (Rayleigh/Tyndall scattering, proportional to $1/\lambda^4$). As a result UV images are quickly degraded underwater and in most aquatic environments UV vision is probably useful only at short visual ranges. Perhaps for this reason UV-sensitive visual pigments are best known in those surface-living freshwater fishes which feed predominantly on plankton which are necessarily seen only at short ranges. In at least one species, the trout, UV receptors are lost at a stage in development when fish move to deeper water from a surface-dwelling, planktivorous, juvenile phase (Bowmaker & Kunz, 1987). However, if UV receptors are involved in the detection of plankters it is by no means clear in what capacity they operate. Bowmaker & Kunz (1987) suggest that plankters will be visible as dark, UV-absorbing objects against a UV-bright background, but Lythgoe (1988) argues that plankton reflect downwelling ultraviolet light and will be seen, in most cases, as bright objects against a relatively dark background. Whatever the UV contrast of plankters and the implications that this has for visual detection, it has also been suggested that the spectral reflections of individual plankters provide information about their nutritional value (Spinrad & Yentsch, 1987). If such cues appear in the UV the incorporation of UV receptors into colour vision mechanisms would be particularly beneficial to planktivores but at present there is little evidence about how UV-sensitive photoreceptors are integrated in fish vision.

POLARIZATION SENSITIVITY

The underwater light field is polarized due to scattering by water molecules and particulate matter and the degree of polarization is greatest at wavelengths with high attenuations (Ivanoff & Waterman, 1958). The detection of the resulting polarization patterns could provide useful cues for orientation and navigation (Waterman, 1981). Alternatively, sensitivity to polarization could provide a mechanism for the enhancement of contrasts in a manner analogous to offset visual pigments (Lythgoe & Hemmings, 1967).

A number of fishes are known to respond to polarized light (Waterman, 1975; Douglas, 1986; Hawryshyn & McFarland, 1987) and in most cases it is the cone photoreceptors that seem to be involved in polarization detection. Indeed, in the goldfish (*Carassius auratus*) the UV-, green- and red-absorbing cones are all polarization-sensitive and only the blue cones are insensitive to e-vector orientation (Hawryshyn & McFarland, 1987). The mechanism by which fish detect light polarization is not yet understood. Most fish photoreceptors, with the possible exception of the anchovies (*Anchoa mitchilli* and *A. hepsetus*; Fineran & Nicol, 1976), lack any obvious intracellular structures suitable for the detection of e-vector orientation. Consequently current hypotheses invoke the differential scattering and the oblique entry of light into outer segments to explain sensitivity to polarized light (Waterman, 1981). Nevertheless, it is likely that many fishes, both freshwater and marine will be found to be polarization-sensitive but at present there is little information about the way in which fishes use this visual input.

VARIATION IN VISUAL PIGMENTS IN INDIVIDUAL FISHES

Visual pigments consist of a protein (opsin) and a 'chromophore' which is a derivative of vitamin A, and variation in either component causes variation in the absorption spectra of the combined molecule. If the chromophore is retinal (vitamin A1 aldehyde) the resulting visual pigment is a rhodopsin; if the chromophore is 3-dehydroretinal (vitamin A2 aldehyde) the visual pigment is a porphyropsin. Other chromophores are known in invertebrates but to date these have not been identified in vertebrates (Lythgoe & Partridge, in press). Rhodopsins and porphyropsins based on the same opsin differ in both λ_{max} and in the breadth of their spectral absorbance curves, with porphyropsins being wider and generally having peak absorptions at longer wavelengths than rhodopsins (Figure 10.9). This difference is small at short wavelengths but at longer wavelengths the effect of chromophore substitution is substantial.

Chromophore substitution is well known in several species of fish and in these it is often associated with different stages in their life histories (Beatty, 1984). Some of the best-known studies of visual-pigment lability by chromophore exchange concentrated on the rod pigments of migrating fishes and in many cases obvious correlations were noted between the visual pigments and water quality. Thus migrating salmon have a predominance of rhodopsin when living in the sea but a predominance of porphyropsin

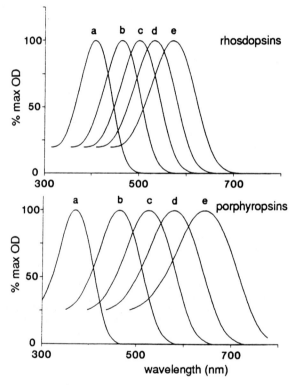

Figure 10.9. Absorption spectra of rhodopsin and porphyropsin pigments, based on visual pigment templates (see Knowles & Dartnall, 1977), shown as percentage maximum optical density. Rhodopsin pigments are narrower and generally have peak sensitivities (λ_{max}) at shorter wavelengths than their porphyropsin analogues. The λ_{max} values of the visual pigment analogues were calculated from the equation; $\lambda_{max}(A1) = 0.60 \lambda_{max}(A2) + 186$ given by Dartnall & Lythgoe (1965). (NB.This relationship is based on rhodopsin pigments with λ_{max} values in the range 492 to 562 nm and may not hold for rhodopsins outside this range.) The A1 pigments shown are the pure rhodopsins found in the guppy (Archer et al., 1987) and the A2 pigments are their theoretical porphyropsin analogues. The plotted rhodopsin:porphyropsin pairs are: (a) VP408$_1$:VP370$_2$; (b) VP464$_1$:VP463$_2$; (c) VP501$_1$:VP525$_2$; (d) VP533$_1$:VP578$_2$; (e) VP572$_1$:VP643$_2$.

when they move into fresh water to breed (Beatty, 1984). Similarly freshwater eels have porphyropsin pigments as juveniles but, in anticipation of their migration to the sea, they substitute rhodopsin pigments instead (Carlisle & Denton, 1959; Beatty, 1975). Chromophore exchange is also well known in several non-migratory fish and in these species a shift in visual pigment A1:A2 ratio is found which can usually be related to seasonal variation in underwater irradiance (Beatty, 1984). In the rudd (*Scardinius erythropthalmus*), for example, both rod and cone visual pigments shift from rhodopsin-based pigments in summer to porphyropsins in winter (Loew & Dartnall, 1976).

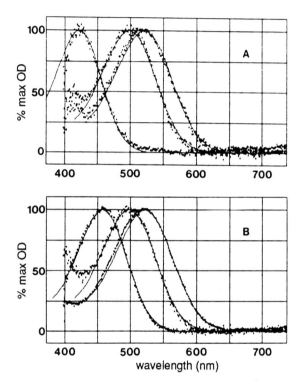

Figure 10.10. Averaged spectral absorbance curves for the two classes of cones and one type of rod found in the pollack (*Pollachius pollachius*). Data (from Shand *et al.*, 1988) are presented as percentages of maximum optical density together with the best-fitting rhodopsin templates (continuous curves). (A) 44 mm standard length (SL) juvenile. (B) 236 mm SL immature adult. The class of cone with short-wavelength sensitivity shifts in λ_{max} during development from a λ_{max} in the violet-blue part of the spectrum to a λ_{max} in the blue.

Relatively few marine fishes are known to possess visual pigments based on vitamin A2, namely various wrasse (Lythgoe, 1972) and several species of deep-sea fish (Partridge *et al.*, 1989). In these species there is the potential for visual pigment lability by chromophore substitution. However, other marine fishes are known to vary their visual pigments not by chromophore exchange but, apparently, by opsin variation.

A number of opsins have now been sequenced and it is clear that the different amino acid sequences of these proteins somehow determines the λ_{max} of the rhodopsin pigments that they form (Nathans *et al.*, 1986; Piantanida, 1988; Feiler *et al.*, 1988). Measurements of rod visual pigments indicate that some λ_{max} values are far more common than others, and that some may in fact never occur (Dartnall & Lythgoe, 1965; Jacobs & Neitz, 1985; Partridge *et al.*, 1989). The results of this 'spectral clustering' of λ_{max} measurements indicates that the λ_{max} values of the possible visual pigments are separated by intervals of some 5-7 nm. In consequence there may be a total of about 50 possible rhodopsin visual pigments and animals must 'choose', in the course of evolution, from this limited palette. It is possible that this constraint may be due to the limited number of amino acid sequences that will produce a functioning opsin. However, it is perhaps more likely that the total number of visual pigments is limited by the number of electronic resonant states that the chromophore will support (Lipetz & Cronin, 1988).

In the few animals, all mammals, whose opsin genes have been sequenced it appears that these genes are always expressed but in some fishes this may not be the case. In the eel (*Anguilla rostrata*) retinal rods contain a rhodopsin pigment with a λ_{max} of 501 nm when in coastal water but prior to migration into the Atlantic this alters to another pigment, apparently also a rhodopsin, with a λ_{max} of 482 nm which is more typical of a deep-sea fish (Beatty, 1975). Opsin shifts are also implicated in the pollack, *Pollachius pollachius* (Figure 10.10). Juveniles of this species (standard length <40 mm), which tend to live in shallow water and feed mainly on plankton, have a violet-sensitive cone (λ_{max} 420 nm) in addition to their rods (λ_{max} 498 nm) and green-sensitive cones (λ_{max} 521 nm; Shand *et al.*, 1988). However, during their development they change their diet and tend to live at greater depths (Potts, 1986) and, concomitant with these changes, the violet-blue-sensitive cone shifts in λ_{max} towards longer wavelengths. In larger fish (SL >80 mm) this cone has a λ_{max} of 460 nm in the blue region of the spectrum (Figure 10.10), the other pigments remaining unaltered from the juvenile state (Shand *et al.*, 1988). It is highly likely that the detailed investigation of the visual systems of other fish species will reveal similar changes associated with development.

CONCLUSION

It is clear that there is wide variation in fish visual pigments and that there are several ecological correlations between the visual pigment complements of fishes and the light environment. To pursue this subject further there may be merit in taking the route provided by radiance transfer mathematics and to model mathematically the function of photoreceptors in the course of certain visual tasks. However, the success of this exercise in answering the question 'why do fish have the visual pigments that they do?' may well rest in the identification and analysis of visual tasks critical to the survival of different species as well as in the successful marriage of physics, physiology and fish ethology.

Chapter 11

MESSENGERS OF TRANSDUCTION AND ADAPTATION IN VERTEBRATE PHOTORECEPTORS

H.R. MATTHEWS

Physiological Laboratory, University of Cambridge, Downing Street, Cambridge, CB2 3EG

It is generally accepted that the processes of transduction and adaptation in vertebrate photoreceptors are mediated by diffusible internal messengers. Many lines of evidence, both biochemical and physiological, indicate that cyclic GMP is the messenger of transduction in both rods and cones. Cyclic GMP has been shown directly to open the channels in the outer segment membrane which carry the dark current. Light initiates an enzymatic cascade which culminates in the hydrolysis of cyclic GMP and the closure of these channels, to yield the photoresponse.

Changes in cytoplasmic calcium concentration are now known not to be directly involved in phototransduction, but instead play a fundamental role in light adaptation. Calcium concentration is believed to decrease in the light due to a shift in the balance between influx of calcium through the outer segment conductance and efflux of calcium through a sodium-calcium exchanger. If this decrease is prevented then light adaptation is abolished, and both rods and cones saturate compressively at a relatively low intensity. Therefore calcium is the messenger of light adaptation in the photoreceptors of lower vertebrates.

INTRODUCTION

This brief review addresses the mechanism by which vertebrate photoreceptors respond to light. In particular, it describes how diffusible internal messengers are of crucial importance not only in the process of phototransduction, but also in controlling the sensitivity of the transduction mechanism. In a volume largely devoted to marine organisms, it must be stated at the outset that although our understanding of vertebrate photoreceptors is mostly based on work on the rods of amphibians, phototransduction in marine vertebrates is believed to be largely similar (see, for example, Dowling & Ripps, 1972).

THE NEED FOR INTERNAL MESSENGERS

Figure 11.1A shows a schematic diagram of a vertebrate rod. The rod is subdivided into two parts: the outer segment (os) and the inner segment (is). The outer segment is

Herring, P.J., Campbell, A.K., Whitfield, M. & Maddock, L. (ed.). *Light and Life in the Sea.*
© 1990. Cambridge University Press.

specialized for the process of phototransduction. Enclosed within it is a stack of membranous discs, which contain the photopigment rhodopsin as an integral membrane protein. On the other hand, the inner segment is responsible for electrically shaping the photoresponse and transmitting it to second order cells (reviewed by Attwell, 1986; Owen, 1987), and also contains all the necessary machinery for cellular metabolism.

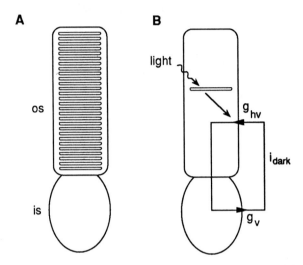

Figure 11.1. (A) schematic diagram showing the structure of a vertebrate rod. (B) diagram illustrating the electrical basis of the photoresponse, and the need for an internal transmitter for transduction.

Figure 11.1B illustrates the electrical basis of the response to light. In darkness, a steady 'dark current' (i_{dark}) flows into the outer segment (Hagins et al., 1970; Baylor et al., 1979), and out through voltage-gated conductances in the inner segment (g_v). Absorption of light by rhodopsin in the disc membrane leads to a graded decrease in the light-sensitive conductance (g_{hv}) of the outer segment membrane, and thereby to a decrease in this circulating current. As the discs are topologically separate from the outer segment membrane in which the light-induced conductance-change takes place, it is generally accepted that a diffusible messenger or 'internal transmitter' is required to provide a link between this conductance change and the light-initiated biochemical events which take place in the vicinity of the discs. Such an internal transmitter might act to decrease the outer segment conductance, and be released on illumination. Alternatively, the transmitter might be necessary in the dark to maintain the outer segment conductance, and be removed on illumination. These two possibilities are known as 'positive' and 'negative' internal transmitters respectively (discussed by Pugh & Cobbs, 1986a).

In cones, the pigment is located not in discs, but in invaginations or sacs formed from the surface membrane. Although this difference in morphology removes the strict geometrical requirement for an internal transmitter, the mechanism of transduction in cones appears to be substantially similar to that in rods, as will be described below.

RESPONSES TO FLASHES AND STEPS OF LIGHT

Figure 11.2 illustrates the well-known form of rod responses to flashes and steps of light. These photocurrent responses were recorded from a salamander rod using the suction pipette technique (Baylor *et al.*, 1979). Figure 11.2A shows responses to a series of flashes delivered in darkness. It can be seen that as the flash is made brighter, the response increases in amplitude and duration. Ultimately, for very bright flashes, the current is suppressed altogether and the response saturates, according to a simple compressive relation (Baylor *et al.*, 1979).

Figures 11.2B & 11.2C show responses to the same sequence of flashes presented during steady illumination of two different intensities. The response to the background

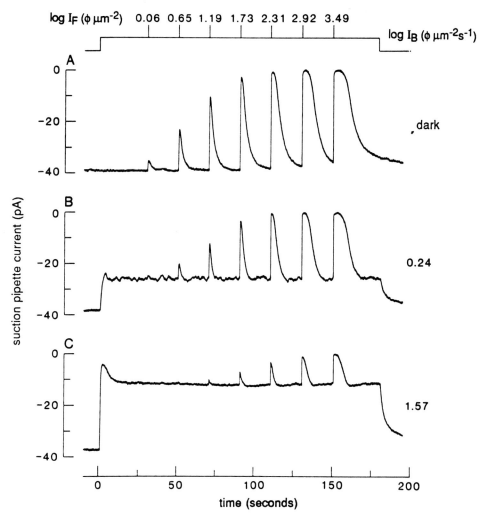

Figure 11.2. Illustrative responses of a salamander rod to flashes of light of progressively increasing intensity (A) in darkness; (B) and (C) during steady illumination. Each flash is about four times as bright as its immediate predecessor. The intensities of flashes (I_F) and steps (I_B) are noted beside the traces. (H.R. Matthews, R.L.W. Murphy, G.L. Fain & T.D. Lamb, unpublished.)

light itself settles rapidly to a new steady level. The response to any given flash superimposed on steady light is smaller and faster than in darkness: the rod has adapted to the steady light, and has become less sensitive (Fain, 1976; Baylor *et al.*, 1980). As the background intensity is increased, the sensitivity of amphibian rods decreases in inverse proportion according to Weber's law (Fain, 1976; Baylor *et al.*, 1980). The sensitivity of cone photoreceptors also changes during steady illumination in a similar Weber-law fashion (Baylor & Hodgkin, 1973, 1974).

This process of sensitivity control, which is known as light adaptation, is also believed to involve an internal messenger (Donner & Hemilä, 1978; Bastian & Fain, 1979; Lamb *et al.*, 1981), which is thought to be distinct from the internal transmitter of excitation. In rods, the messenger of adaptation is known to diffuse a short distance along the outer segment, to influence the transduction mechanism in the vicinity of discs which have not yet absorbed photons (Lamb *et al.*, 1981). So in photoreceptors there are two diffusible messengers to consider: the internal transmitter of excitation and the messenger of adaptation.

INTERNAL TRANSMITTER OF EXCITATION

The identity of the internal transmitter of excitation was long a matter of controversy (reviewed by Pugh & Cobbs, 1986*b*; Pugh, 1987). Two competing candidates had been put forward. The first, calcium ion, was proposed by Hagins (1972) to be released from the discs on illumination and to result in the closure of the outer segment conductance. Calcium was thus proposed to act as a 'positive' transmitter, and would yield the light response through an increase in its concentration. The evidence for this contention was indirect, and was largely based on the observation that an imposed increase or decrease in cytoplasmic calcium concentration (Ca^{2+}_i) was known to lead to substantial changes in the photocurrent (Yoshikami & Hagins, 1973; Brown *et al.*, 1977; Yau *et al.*, 1981; MacLeish *et al.*, 1984). The second candidate was cyclic guanosine monophosphate, or cyclic GMP. It was known on the basis of much biochemical evidence that cyclic GMP was hydrolysed on illumination (Miki *et al.*, 1975; Hubbell & Bownds, 1979; Liebman & Pugh, 1979; see review chapters in Miller, 1981) and it was proposed that this destruction led somehow to the suppression of the dark current. Cyclic GMP would thus function as a 'negative' transmitter, and cause the light response through a decrease in its concentration. The controversy over the nature and identity of the internal transmitter of excitation continued until early in 1985, when it was resolved more or less simultaneously by three major lines of evidence (reviewed by Pugh & Cobbs, 1986*b*; Pugh, 1987).

CYCLIC GMP IS THE TRANSMITTER OF EXCITATION

The first line of evidence was the demonstration by Fesenko *et al.* (1985) that cyclic GMP directly influences the conductance of the rod outer segment membrane. Their results are shown in Figure 11.3.

They used the patch-clamp technique to record from patches of outer segment membrane. A patch of membrane was excised from the rod outer segment in the inside-out configuration and was superfused with test solution. When the cytoplasmic face of the patch was exposed to cyclic GMP, the conductance of the patch increased. In contrast, varying the concentrations of other cyclic nucleotides or of calcium had no such effect. These observations provided enormously strong support for the cyclic nucleotide hypothesis.

These results have subsequently been confirmed and extended by many other groups, and much is now known about the cyclic GMP activated conductance as an entity in its own right (reviewed by Owen, 1987; Yau & Baylor, 1989). The rod cyclic GMP activated conductance exhibits electrical and ionic properties similar to those of the outer segment conductance in intact rods (Fesenko *et al.*, 1985; Matthews, 1986, 1987; Matthews &

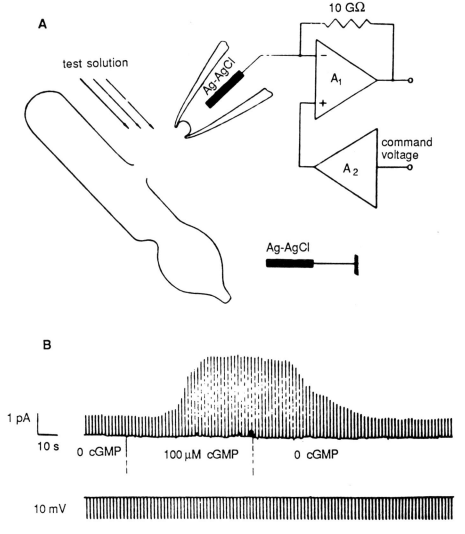

Figure 11.3. (A) method used by Fesenko *et al.* (1985) to record the cyclic GMP activated conductance in patches of rod outer segment membrane. (B) response of the patch to voltage clamp pulses during the application of cyclic GMP. From Fesenko *et al.* (1985) by permission of Macmillan Magazines Ltd.

Watanabe, 1987). Furthermore, if cyclic GMP is introduced either into an intact rod (Cobbs & Pugh, 1985; Matthews *et al.*, 1985) or a truncated rod outer segment (Yau & Nakatani, 1985*b*; Nakatani & Yau, 1988*b*) the conductance activated by cyclic GMP can be completely suppressed by light. Therefore there can be little doubt that the cyclic GMP activated conductance and the outer segment conductance are identical. Similarly, cyclic GMP has been shown to activate the cone outer segment conductance both in excised membrane patches (Haynes & Yau, 1985) and when introduced during whole-cell recording (Cobbs *et al.*, 1985).

These results from excised patches, when taken together with the substantial body of biochemical evidence that cyclic GMP is destroyed on illumination (reviewed by Stryer, 1986; Hurley, 1987; Liebman *et al.*, 1987), indicate that cyclic GMP is the 'negative' internal transmitter of excitation in both rods and cones. So the photocurrent response results from a light-induced fall in the concentration of cyclic GMP, which leads to a reduction in the cyclic GMP activated conductance.

CALCIUM CANNOT BE THE TRANSMITTER OF EXCITATION

At about the same time as this demonstration that cyclic GMP is the excitational transmitter, it became clear that calcium could not be. The calcium hypothesis required that a rise in Ca^{2+}_i be necessary for the light response (Hagins, 1972). Two approaches indicated that this is not the case.

One approach was to attempt to slow changes in Ca^{2+}_i by incorporating calcium chelators into isolated rods by means of whole-cell patch-clamp recording (Matthews *et al.*, 1985; Lamb *et al.*, 1986). Figure 11.4 illustrates the effect on rod responses of incorporating the calcium chelator BAPTA (Tsien, 1980). Following the start of whole-cell recording at time zero the dark current increased, presumably due to a reduction in Ca^{2+}_i caused by the introduction of the chelator, and the falling phase of the responses to

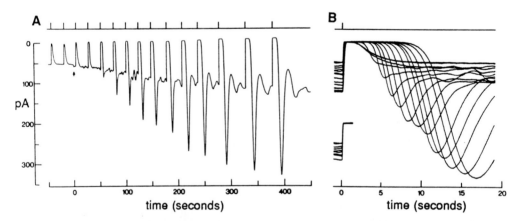

Figure 11.4. (A) suction pipette recording of the outer segment current of a salamander rod during the incorporation of calcium chelator from a patch pipette containing 10 mM BAPTA. Arrow denotes the time of patch rupture. (B) superimposed flash responses from (A) showing the progressive increase in the duration of response saturation, and the development of the overshooting recovery. From Matthews *et al.* (1985) by permission of Macmillan Magazines Ltd.

bright flashes developed a pronounced overshoot. However, the chelator had little effect on the rising phase of these responses. In other words BAPTA did not 'buffer' the light response, as would have been expected if the calcium hypothesis were correct. Indeed the responses remained in saturation for longer than in control, indicating that sensitivity had increased. The introduction of BAPTA into the rod can be shown greatly to slow changes in Ca^{2+}_i (Lamb & Matthews, 1988). Therefore the inability of BAPTA to 'buffer' the light response was inconsistent with the calcium hypothesis, and required its rejection (Lamb *et al.*, 1986).

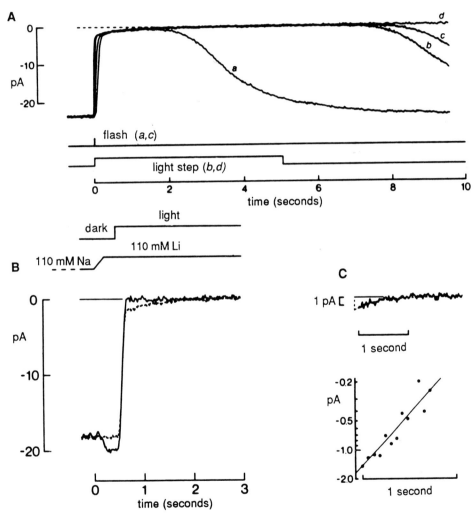

Figure 11.5. Sodium-calcium exchange currents recorded by Yau and Nakatani (1985a) during the plateau of the light response. (A) responses of a toad rod to saturating flashes and steps of light, illustrating the independence of the exchange transient on the intensity and duration of illumination. (B) abolition of the exchange transient on the replacement of external sodium by lithium. (C) approximately exponential decline of the exchange transient. From Yau & Nakatani (1985a) by permission of Macmillan Magazines Ltd.

The other line of evidence against the calcium hypothesis was provided by Yau & Nakatani (1985*a*), who proposed that calcium concentration actually falls during illumination, rather than rising. Their evidence for this idea is shown in Figure 11.5. When the dark current was completely suppressed by the onset of bright light, they observed a slowly-decaying residual component of inward current. They attributed this transient current to the extrusion of calcium by an electrogenic sodium-calcium exchange mechanism, as the transient was eliminated if external sodium was replaced by lithium, which does not support sodium-calcium exchange (Yau & Nakatani, 1984*b*; Hodgkin *et al.*, 1985). They therefore proposed that the exchanger acted to lower Ca^{2+}_i on illumination.

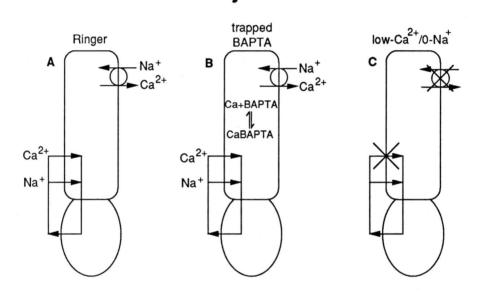

Figure 11.6. Schematic diagrams illustrating the mechanisms by which Ca^{2+}_i is controlled in vertebrate rods. (A) normal rod in Ringer; (B) following the introduction of calcium chelator; (C) during superfusion with low-Ca^{2+}/0-Na^+ solution.

Hence, Ca^{2+}_i is now believed to be controlled by a balance between calcium influx and calcium efflux. This idea is illustrated in Figure 11.6A. A proportion of the current through the outer segment conductance is known to be carried by calcium, accounting for 10-15% of the dark current (Capovilla *et al.*, 1983; Yau & Nakatani, 1984*a*; Hodgkin *et al.*, 1985; Nakatani & Yau, 1988*a*). This steady influx in darkness is opposed by the extrusion of calcium via a sodium-calcium exchanger (Yau & Nakatani, 1984*b*, 1985*a*; Hodgkin *et al.*, 1987; Lagnado *et al.*, 1988; Nakatani & Yau, 1988*a*). During the light response the circulating current decreases, and therefore the influx of calcium ions decreases also. But the exchanger continues to extrude calcium ions, and so Ca^{2+}_i falls. It was the decaying component of current corresponding to this electrogenic extrusion which Yau & Nakatani recorded. During steady illumination Ca^{2+}_i falls until a new steady state between influx and efflux is attained. Similarly, the transient suppression of circulating current during the flash response leads to a *dynamic* reduction in Ca^{2+}_i. A fall in Ca^{2+}_i during illumination was subsequently measured directly (McNaughton *et al.*, 1986; Ratto *et al.*, 1988) though this has not, as yet, been accomplished from a single cell within the physiological concentration range.

All of the evidence described above indicates that calcium cannot be the internal transmitter of excitation. However, the fact remains that calcium does have powerful effects on the transduction mechanism, a decrease in Ca^{2+}_i leading to an enormous increase in circulating current (Yau *et al.*, 1981; Hodgkin *et al.*, 1984; Lamb & Matthews, 1988). So what is the functional role of calcium in photoreceptors?

CALCIUM IS THE MESSENGER OF LIGHT ADAPTATION

Recent evidence indicates that calcium acts as the messenger of light adaptation in amphibian rods and cones. This role of calcium in adaptation has been studied by attempting to prevent light-induced changes in Ca^{2+}_i from taking place, and examining the effect on the responses to flashes and steps of light. Changes in Ca^{2+}_i were opposed in two different ways.

The first approach was to incorporate the calcium chelator BAPTA using whole-cell patch-clamp recording. In these experiments, after loading a rod with BAPTA, the patch pipette was withdrawn, thereby trapping the chelator inside the cell (Torre *et al.*, 1986; Lamb & Matthews, 1988). The aim was to increase the calcium-buffering capacity of the rod without perturbing Ca^{2+}_i in the long term. As illustrated in Figure 11.6B, an increase in buffering capacity would slow down changes in Ca^{2+}_i induced by steady light. However, the final steady level of Ca^{2+}_i should not be altered, as it depends on the balance between influx and efflux, neither of which is altered by the presence of the buffer (Torre *et al.*, 1986). Furthermore, the *dynamic* fall in Ca^{2+}_i which takes place during the flash response would also be slowed by the buffer.

The second approach, shown in Figure 11.6C, was to minimise the influx and efflux of calcium across the outer segment membrane (Matthews *et al.*, 1988; Nakatani & Yau, 1988*c*; Fain *et al.*, 1989). The influx of calcium through the outer segment conductance was minimised by greatly reducing the calcium concentration superfusing the outer segment. Calcium efflux was minimised by removing external sodium, and substituting another ion, such as guanidinium, which would pass through the outer segment conductance, but which would not support calcium extrusion by the exchanger (Nakatani & Yau, 1988*a*). Rapidly changing the solution bathing the outer segment from normal Ringer to such a low-Ca^{2+}/0-Na^+ solution provided a means of holding Ca^{2+}_i near to its original value for a period of at least 15 seconds (Fain *et al.*, 1989).

Despite the practical differences between these two methods, they give results which qualitatively are remarkably similar. The main advantage of the second approach, in which calcium influx and efflux are minimised, is that it is possible repeatedly to return to control conditions. In contrast, once chelator has been incorporated there is no straightforward way to remove it!

Figure 11.7 shows the effect of BAPTA incorporation on the response of a rod to steady light. The normalized response to dim steady illumination is shown both in control conditions, and also following the trapping of BAPTA in the cytoplasm. At the onset of the background, the responses initially rose together. The control response settled rapidly to a new steady level. In contrast, following the introduction of BAPTA, the response continued to rise for longer, reaching saturation. Subsequently the response recovered, and ultimately stabilized at the same steady level as in control.

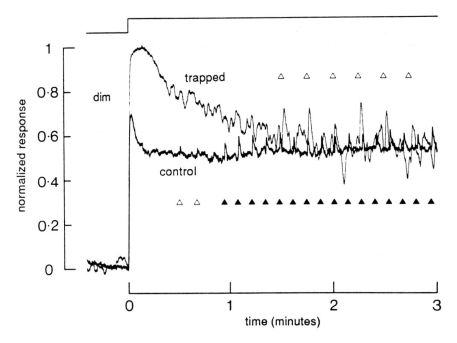

Figure 11.7. Normalized responses to dim steady light before (control) and after (trapped) the introduction of BAPTA into a salamander rod from a patch pipette containing 10 mM of the chelator. Upper trace: steady light monitor. Open and filled triangles denote dim flashes delivering 70 and 140 isomerizations respectively. From Torre *et al.* (1986).

The normal rapid relaxation of the response to background light to a new steady level is a manifestation of light adaptation (Baylor & Hodgkin, 1974). The effect of BAPTA, which slowed this relaxation, was thus to slow the development of adaptation at the onset of steady light, without altering the final level of adaptation reached. As the chelator would be expected to slow any changes in Ca^{2+}_i without altering the final steady concentration, these observations are consistent with the idea that the *static* decrease in calcium concentration which takes place in response to steady light is responsible for setting the steady operating point of the transduction mechanism (Torre *et al.*, 1986).

The effect of BAPTA incorporation on the responses to dim flashes is shown in Figure 11.8. The responses labelled D and DB were recorded in darkness preceding and following the incorporation of BAPTA. Comparison of trace DB with trace D shows that the effect of BAPTA was to delay the termination of the flash response, allowing the response to rise further along a common rising phase, thereby resulting in an increase in sensitivity. The introduction of BAPTA would be expected to oppose the change in Ca^{2+}_i

which normally results from the suppression of current during the flash response. This observation therefore suggests that this *dynamic* change in Ca^{2+}_i normally contributes to the termination of the response (Torre *et al.*, 1986).

The responses labelled L and LB were recorded in steady light before and after the introduction of BAPTA. Steady illumination resulted in substantial acceleration of the time to peak of the response and a decrease in sensitivity both in control and in the presence of BAPTA. These changes in response kinetics and sensitivity are a central part of light adaptation (Baylor & Hodgkin, 1974). It is tempting to ascribe these *static* changes in sensitivity which take place during steady light to *static* changes in calcium concentration, by analogy with the effects of the *dynamic* changes which take place during the flash response (Torre *et al.*, 1986).

These ideas can be tested further by using exposure to low-Ca^{2+}/0-Na^+ solution to hold Ca^{2+}_i near the initial level in darkness, and then examining how sensitivity varies as a function of steady intensity under these conditions (Matthews *et al.*, 1988).

Figure 11.9 summarizes the effect of superfusion with low-Ca^{2+}/0-Na^+ solution on the responses to steps and flashes of light. Figure 11.9A plots the normalized steady response to background light against steady intensity. In Ringer (filled circles) the amplitude of the

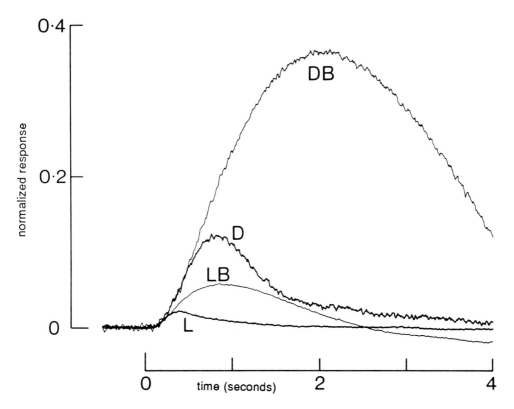

Figure 11.8. Normalized rod responses to dim test flashes in darkness and in steady light, both before and after the introduction of BAPTA. Trace D, dark adapted control; trace L, light adapted control; traces DB and LB are dark and light adapted responses respectively following the incorporation of BAPTA. From Torre *et al.* (1986).

response increased gradually as a function of intensity. This shallow response-intensity relation is a direct consequence of light adaptation (Baylor & Hodgkin, 1974). However, when the steady light was presented during exposure to low-Ca^{2+}/0-Na$^+$ solution the

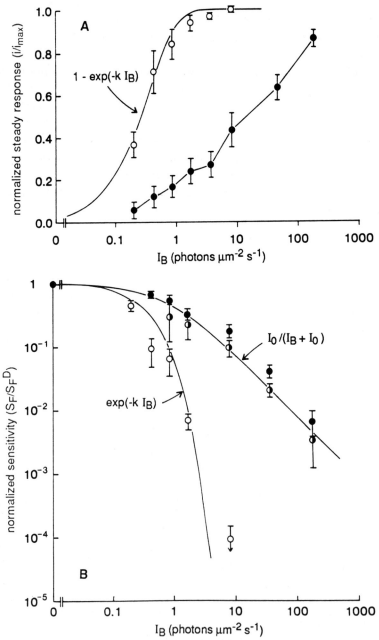

Figure 11.9. Normalized rod steady response (A) and incremental sensitivity (B) as functions of background intensity, both in control (filled circles) and in low-Ca^{2+}/0-Na$^+$ solution. Open symbols: steady light was presented during exposure to low-Ca^{2+}/0-Na$^+$ solution. Half-filled symbols: exposure to low-Ca^{2+}/0-Na$^+$ solution after equilibration to steady light. From Matthews *et al.* (1988) by permission of Macmillan Magazines Ltd.

response increased much more steeply with intensity (open circles). These points could be fitted by the exponential saturation equation, which has previously been used to fit the response-intensity relation at early times before adaptation has taken place (Lamb *et al.*, 1981).

Figure 11.9B plots normalized sensitivity as a function of steady intensity for the same cells. Data for the normalized sensitivity in control Ringer (filled circles) follow classic Weber-law adaptation, given by the solid curve (Baylor & Hodgkin, 1974; Fain, 1976; Baylor *et al.*, 1980). The points shown as the open circles were obtained by changing to low-Ca^{2+}/0-Na^+ solution and then presenting the background. The normalized sensitivity in low-Ca^{2+}/0-Na^+ falls much more steeply with increasing background intensity than was the case in Ringer. The data are well fitted by compressive saturation of the form expected from the nonlinear response-intensity relation obtained above in Figure 11.9A (Matthews *et al.*, 1988). So when changes in Ca^{2+}_i are prevented, adaptation is abolished and the rod simply saturates at an embarrassingly low intensity. Further analysis of the responses to steps and flashes of light shows that under these conditions the rod sums incident photons with an invariant integration time (Matthews *et al.*, 1988; Nakatani & Yau, 1988*c*; Fain *et al.*, 1989).

When, however, the background was presented before the change to low-Ca^{2+}/0-Na^+ solution, the points given by the half-filled symbols were obtained. These fall close to the Weber curve, and indicate that if the rod is allowed to attain its normal steady response to the background, and thereby the appropriate light-induced Ca^{2+}_i, the switch to low-Ca^{2+}/0-Na^+ does not greatly affect the pre-existing degree of adaptation.

These and other results indicate that calcium is the messenger of light adaptation in the rods of lower vertebrates (Torre *et al.*, 1986; Matthews *et al.*, 1988; Nakatani & Yau, 1988*c*; Fain *et al.*, 1989). Hence calcium acts as a 'negative' messenger during light adaptation, as it is the light-induced fall in calcium concentration which leads to this decrease in sensitivity. Similar experiments show that calcium also plays a comparable role in salamander cones (Matthews *et al.*, 1988; Nakatani & Yau, 1988*c*; Matthews *et al.*, 1990). Calcium must exert these effects indirectly, by modulating the biochemical mechanism which controls the concentration of cyclic GMP, the internal transmitter of transduction, rather than through any direct action on the outer segment conductance.

EFFECTS OF CALCIUM ON THE TRANSDUCTION MECHANISM

The transduction mechanism is illustrated schematically in Figure 11.10 (reviewed by Stryer, 1986; Hurley, 1987; Liebman *et al.*, 1987). The transduction process starts with the absorption of a photon by the photopigment rhodopsin. Photoisomerized rhodopsin (Rh*) then interacts catalytically with a GTP-binding protein (T*), also known as transducin, causing it to bind GTP in exchange for GDP. The alpha subunit of transducin then activates a phosphodiesterase (PDE*), causing it to hydrolyse cyclic GMP (cGMP). The resulting fall in cyclic GMP concentration leads to the closure of channels in the outer segment membrane, to yield the photoresponse.

Calcium could modulate this process in two distinct ways. First, calcium is known to inhibit guanylate cyclase (Lolley & Racz, 1982; Pepe *et al.*, 1986; Koch & Stryer, 1988)

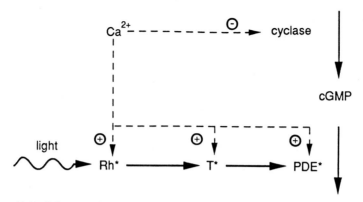

Figure 11.10. Schematic diagram illustrating the sequence of steps believed to lead to the light response. Cyclic GMP (cGMP) is produced by guanylate cyclase. The light-initiated cascade of reactions which takes place in the vicinity of the disc membrane is illustrated below. Rh* represents photoisomerized rhodopsin; T*, activated transducin; PDE*, activated phosphodiesterase; cGMP, cyclic GMP. Dotted lines show possible sites of action for calcium.

which is responsible for producing cyclic GMP. Therefore the light-induced fall in Ca^{2+}_i would lead to an increased rate of cyclic GMP production, resulting in speeded recovery of the light response and decreased sensitivity. This possibility is supported by physiological evidence that illumination produces a delayed increase in cyclase velocity (Hodgkin & Nunn, 1988). Second, calcium may act to prolong the activation of one or more of the intermediate stages in the enzymatic cascade leading to cyclic GMP hydrolysis (Torre *et al.*, 1986; Fain *et al.*, 1989). Therefore when Ca^{2+}_i falls in the light, the transduction mechanism would recover sooner than in darkness. There is some physiological evidence that elevation of Ca^{2+}_i above the normal level in darkness can increase the lifetime of one or more of these early stages in the excitation cascade (Hodgkin *et al.*, 1986; Torre *et al.*, 1986). However, biochemical evidence for such an effect of calcium in the physiological concentration range is divided (Robinson *et al.*, 1980; Kawamura & Bownds, 1981; Barkdoll *et al.*, 1989). So it remains to be determined whether light adaptation is caused by effects of calcium on the cyclase alone, or whether effects of calcium on the early stages in the transduction cascade are also involved.

In summary, two internal messengers are important in vertebrate rods and cones. First, cyclic GMP, as the internal transmitter of transduction, is responsible for the electrical response to light. Second, calcium acts as the internal messenger for light adaptation, and controls the sensitivity of the photoreceptor. Calcium completes a powerful negative feedback loop signalling the magnitude of the circulating current to the biochemical processes which control cyclic GMP concentration, and thereby the circulating current. While excitation by light is now largely understood, the mechanisms by which calcium affects these processes controlling cyclic GMP metabolism to cause adaptation are much less certain.

I wish to thank Dr T.D. Lamb for helpful comments on the manuscript. Supported by St John's College, Cambridge.

Chapter 12

PHOTORECEPTION IN SQUID

HELEN R. SAIBIL

Department of Crystallography, University of London Birkbeck College, Malet Street, London, WC1E 7HX

Light absorbed by the photoreceptor membranes of the retina is transformed by a series of biochemical steps into changes in receptor potential. The visual pigment, rhodopsin, and associated enzymes which relay the visual signal belong to a widespread family of receptor-enzyme complexes which transmit sensory, hormonal and neurotransmitter signals into cells. In the invertebrate visual system, rhodopsin activates enzymes which increase cytoplasmic calcium. The highly specialised and structurally ordered photoreceptors of the squid retina provide a favourable system for studying the mechanisms of these enzyme pathways.

INTRODUCTION

The camera eyes of cephalopods capture incident photons and convert them into amplified depolarisations of membrane potential. The light is absorbed by rhodopsin, an abundant membrane protein present in highly ordered arrays of photoreceptor membrane. In the higher invertebrates, these are formed by cylindrical projections of the cell membrane, or microvilli. Photoexcited rhodopsin activates a cascade of enzymes, resulting in changes of the concentrations of small, soluble substances in the photoreceptor cytoplasm (Stryer, 1986). From studies of hormone action, these small molecules are known as second messengers. They regulate the conductances of ion channels, setting the receptor potential which is eventually transmitted via the optic nerves to the brain. The series of steps by which the photoreceptor cells transduce the visual signal are also used in many pathways of transmembrane signalling in eukaryotic cells.

This review will cover the biochemical and molecular aspects of photoreception in the squid eye. Matthews (Chapter 11) reviews the events occurring in the rods of vertebrates. For more complete reviews of the structure and function of the cephalopod eye, see Messenger (1981), Land (1981a) and Saibil (1990).

Herring, P.J., Campbell, A.K., Whitfield, M. & Maddock, L. (ed.). *Light and Life in the Sea.*
© 1990. Cambridge University Press.

COMPONENTS OF THE SIGNALLING PATHWAYS IN PHOTORECEPTORS

Rhodopsin

The first membrane receptor protein to be isolated, because of its great abundance, rhodopsin has the form indicated in Figure 12.1 (Dratz & Hargrave, 1983; Findlay & Pappin, 1986). To date more than 20 hormone and neurotransmitter receptor sequences have been found to have homologies with rhodopsin. The 11-cis retinal chromophore is buried within a ring of seven transmembrane α-helices, and *cis-trans* photoisomerisation of this group initiates the visual response. The sequence of *Octopus* rhodopsin has been determined by cDNA cloning (Ovchinnikov *et al.*, 1988). Although *Octopus* rhodopsin is homologous to the vertebrate and *Drosophila* pigments (Applebury & Hargrave, 1986), it contains large insertions in the cytoplasmic region, which are also present in squid rhodopsin (M. Hoon & J.B.C. Findlay, unpublished). By a mechanism still unknown, the conformation change is propagated to the polypeptide loops projecting from the cytoplasmic surface, and all the subsequent actions of rhodopsin are thought to occur in this region. The photoisomerised state is called metarhodopsin, which is a stable form in invertebrates. Only one spectral type of rhodopsin is thought to be present in the squid retina, with the exception of the bioluminescent species, *Watasenia scintillans*, in which three visual pigments of differing wavelength maximum have been found. These are proposed to serve for detection of luminescence in the very low light level environment of the deep sea (Matsui *et al.*, 1988a).

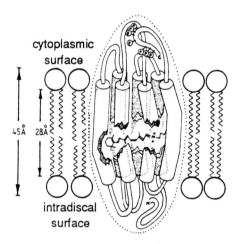

Figure 12.1. Structure of bovine rhodopsin in the photoreceptor membrane, based on amino acid sequence and physical and chemical characterisation. The intradiscal surface is topologically equivalent to extracellular space. In the case of other receptors in the rhodopsin family, the chromophore group is replaced by a ligand binding site, which must be accessible from the extracellular medium. Excitation of the receptor protein results in a change of conformation in the cytoplasmic loops, and the G-protein is activated by this region. Reproduced from Dratz & Hargrave (1983) by permission of Elsevier Publications.

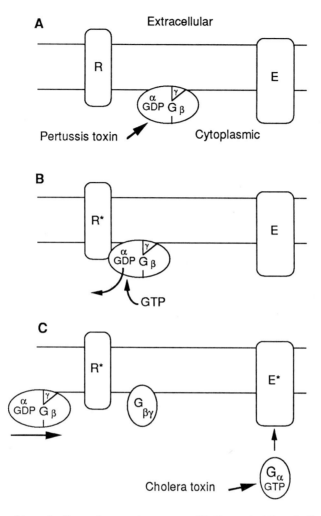

Figure 12.2. Signal transduction pathway using receptor (R), G-protein (G) and effector enzyme (E). The extracellular surface is shown facing upwards in this diagram. The guanine nucleotide binding site is on the α-subunit of the trimeric G-protein (A). After receptor-catalysed nucleotide exchange (B), this subunit separates from the βγ subunits and activates the effector enzyme (C). The βγ subunit may also serve as a modulator for certain effector enzymes. After dissociation of stimulated G-protein from the receptor, other G-proteins which come into contact with the receptor may be activated. The G-protein is restored to its unstimulated, trimeric state after hydrolysis of GTP by the α-subunit. The receptor is inactivated by phosphorylation. Sites of action of bacterial toxins are indicated.

GTP-binding proteins

Rhodopsin signals and amplifies photoexcitation by the activation of many GTP-binding proteins (G-proteins) on the cytoplasmic surface of the membrane by catalysis of guanine nucleotide exchange (Figure 12.2) (Stryer & Bourne, 1986). The nucleotide binding site is on the α-subunit of this trimeric protein. Variants of this subunit confer different receptor and target enzyme specificity. Once exchange is completed, the α-subunit is released from the βγ part, and is free to activate the effector enzyme, another membrane surface bound protein, regulating the second messenger concentration. The βγ unit may activate other target sites. This shuttle mechanism assumes free diffusion of the proteins on the membrane surface. Metarhodopsin is also free to activate more G-proteins. A kinase subsequently phosphorylates the carboxy terminal region of metarhodopsin, and this is thought to promote binding of a soluble 48 KD protein and stop further activation of G-proteins. GTP hydrolysing activity in the α-subunit returns it to the GDP-bound state, which triggers reassociation of the trimer and inactivation. Ways of detecting G-proteins are by receptor-activated GTP analogue binding or GTPase activity, and by covalent modification with cholera and pertussis toxins which are specific for different subtypes of the protein.

Phosphodiesterases and ion channels

In vertebrate rod photoreceptors, the second messenger substance is cyclic GMP (cGMP), and its effector enzyme is cGMP phosphodiesterase. Cyclic GMP opens the light-sensitive sodium channel in rods, and its light-induced hydrolysis produces the hyperpolarising response of these cells. In the invertebrates light causes depolarisation, and the mechanisms are only partially known. One effector enzyme is a different type of phosphodiesterase, specific for the phosphorylated form of a membrane phospholipid, phosphatidylinositol bisphosphate (PIP_2). This enzyme is also known as phospholipase C (PLC). Hydrolysis of PIP_2 by PLC releases two messenger substances, the water-soluble headgroup, inositol trisphosphate (IP_3) and the lipid diacylglycerol. Although it is important in cell activation and growth control, no effect of diacylglycerol has been reported in photoreceptors. But the calcium-mobilising agent IP_3 mediates the light-induced calcium increase in invertebrate photoreceptors (Payne et al., 1988). It is not yet clear whether calcium alone controls the conductance increase which gives the depolarising light response. Phospholipase C plays an important role in many cell types, but this recently discovered system is not as well understood as cyclic nucleotide regulation. Squid photoreceptors are particularly useful for studying IP_3-calcium regulation because of the high concentrations of signalling enzymes and structural organisation of these photoreceptors. However, physiological experiments have mostly been done on other species because cephalopods are difficult to maintain in the laboratory.

We will now turn to the structure of squid retinal photoreceptors and examine what is known about the organisation of their signalling system.

CELLULAR ORGANISATION OF SQUID RETINAL PHOTORECEPTORS

Retina

The retinula cells extend through the whole of the squid retina and comprise the bulk of its tissue. The photoreceptor microvilli are hexagonally packed in the rhabdom structure characteristic of higher invertebrates. Figure 12.3 shows the arrangement of retinula cells in the retina, and the organization of microvilli in the rhabdoms (Zonana, 1961; Cohen, 1973; Saibil, 1982). The retinula cells are divided into the photoreceptive outer segments (also called distal or rhabdomeric segments) facing the lens, and inner segments, with a layer of pigment granules in the middle. There is also a population of glia whose cell bodies are located just below the pigment granule layer. They send fine processes through the photoreceptor layer to the amorphous limiting membrane at the distal surface of the retina. The retinula cell somas are filled with lamellar bodies and mitochondria. Further down in the retina are the nuclei and plexiform layer with receptor collaterals and centrifugal fibre synapses. At the base of the inner segments, the visual cells narrow into the sensory axons which terminate in the optic lobe (Young, 1962).

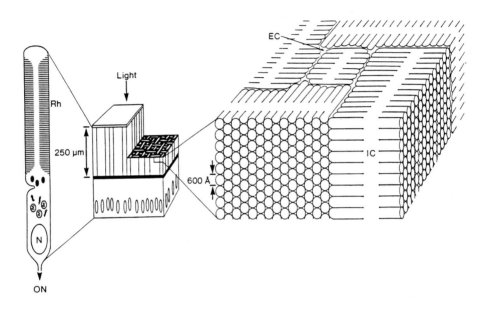

Figure 12.3. Arrangement of retinula cells in the retina and of microvilli in the rhabdoms, showing the ribbon-like retinula cells and the orthogonal orientation of microvilli in neighbouring cells. Rh, rhabdom; ON, optic nerve; EC, extracellular compartment; IC, intracellular compartment. From Saibil & Hewat (1987) by copyright permission of the Rockefeller Unversity Press.

Because the retinula cells are very long (500 µm) and there is no pigment epithelium in this non-inverted retina, intracellular transport is a major requirement of the receptors, for granules of screening pigment and in membrane turnover. The outer segment cytoplasm is full of longitudinally oriented microtubules and filamentous material (Cohen, 1973; Saibil, 1990).

Photoreceptor potentials and ionic mechanisms

Hagins (1965) showed that the site of illumination of the photoreceptor outer segments is the source of the photoreceptor potential. Pinto & Brown (1977) recorded intracellular potentials and found that the receptor potential is largely generated by increased Na^+

influx, and that light stimulation is followed by an increase in cytoplasmic free Ca^{2+}, which mediates light adaptation. The response waveforms of cephalopod retinae are very similar to those of turtle cones, apart from a change in sign (Duncan & Pynsent, 1979).

Photoreceptor microvilli

The receptor mosaic forms a continuous sheet of closely packed, hexagonal arrays of microvilli in alternating orientations (Figure 12.4). The microvillar membranes are tightly linked together, and cytoskeletal filaments are attached to the membranes by side-arms. An SDS polyacrylamide gel of the isolated microvillar proteins and their separation into detergent-soluble (membrane) and insoluble (cytoskeletal) fractions is shown in Figure 12.5. Very similar results are seen with cuttlefish and octopus photoreceptors. The identified proteins in this preparation are rhodopsin (Hagins, 1973; Paulsen et al., 1983), actin (Saibil, 1982) and the β-subunit of GTP-binding protein (Tsuda et al., 1986).

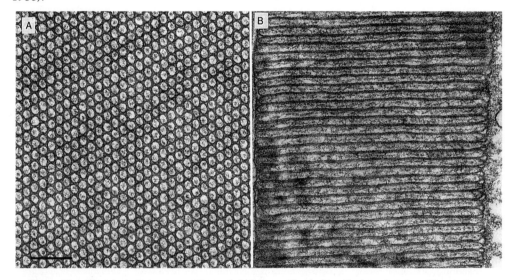

Figure 12.4. Electron microscope sections of rhabdoms, illustrating the hexagonal packing in two orthogonal views of the microvilli. Adjacent membrane bilayers appear fused, and the central filament and side-arms are irregularly preserved. The sections were prepared by the procedures described in Saibil & Hewat (1987) and counterstained with uranyl acetate and lead citrate. Calibration bar, 250 nm.

Membrane structure

Squid retina is unique in having a macroscopic array of photoreceptor microvilli in a paracrystalline lattice of 600 Å unit cells (Worthington et al., 1976; Saibil, 1982; Saibil & Hewat, 1987). Details of the membrane structure have been obtained from X-ray diffraction and electron microscopy studies (Figure 12.6). The transmembrane links create ordered domains which may be involved in the alignment of rhodopsin molecules, to produce the dichroic absorption implied by the high polarisation sensitivity of the cephalopod retina (Tomita, 1968).

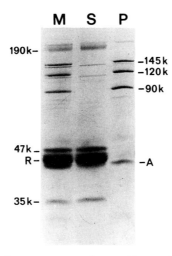

Figure 12.5. SDS gels of squid photoreceptor membranes. M, proteins of *Alloteuthis* photoreceptor microvilli; S, fraction soluble in octyl glucoside; P, fraction insoluble in octyl glucoside; R, rhodopsin; A, actin. The 35 kD band is Gβ. 10% polyacrylamide gels. Reproduced from Saibil & Hewat (1987) by copyright permission of the Rockefeller University Press.

Cytoskeleton

The microvilli are supported by a filamentous core (Figure 12.4) containing actin and several other major proteins (Figure 12.5). Genetic evidence from *Drosophila* suggests that the side-arms binding actin to the membranes contain a form of myosin head (myosin I, Montell & Rubin, 1988). The cytoskeleton in rhabdomeric photoreceptors is difficult to fix for electron microscopy and its disordered appearance causes it to blur out in averaged images. However, Tsukita *et al.* (1988) have obtained clearer images of the cytoskeletal filaments using rapid freezing methods. They compared dark- and light-

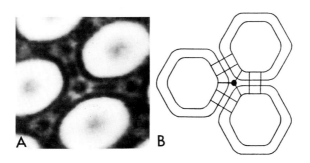

Figure 12.6. Structure of the microvillar membranes. (A) was obtained by averaging images of about 200 microvilli, using a correlation averaging method to correct for lattice disorder, and (B) diagram of the two-fold and three-fold membrane junctions in squid photoreceptor microvilli derived from X-ray diffraction and electron microscope studies. The methods are described in Saibil & Hewat (1987). Reproduced from Saibil & White (1989) by copyright permission of Plenum Press.

adapted retinae, and found that the cytoskeleton structure is disrupted in light-stimulated microvilli. This may take place on a time scale compatible with an involvement of
the cytoskeleton in phototransduction, possibly as a consequence of light-induced
changes in cytoplasmic messenger concentrations.

SIGNALLING ENZYMES ACTIVATED BY SQUID RHODOPSIN

In squid photoreceptor outer segments, rhodopsin is the unique receptor and comprises the major fraction of the membrane protein. In the invertebrate system, there is
evidence that GTP-binding protein activation of phosphatidylinositol bisphosphate
(PIP_2) phospholipase C is an important pathway (Payne, 1986; Tsuda, 1987).

GTP-binding proteins

In the isolated microvillar membranes, GTPase activity is stimulated three- to four-
fold by light (Calhoon et al., 1980; Vandenberg & Montal, 1984a; Saibil & Michel-Villaz,
1984). Light also stimulates guanine nucleotide binding (Vandenberg & Montal, 1984a).
The two bacterial toxins which have been used to identify α-subunits of GTP-binding
protein, cholera and pertussis toxins, both catalyse light-dependent labelling of proteins
in cephalopod photoreceptors. Vandenberg & Montal (1984a) found a 45 kD cholera
toxin substrate in squid membranes. In Octopus membranes, Tsuda et al. (1986) found a
40 kD pertussis toxin substrate. These two distinct GTP-binding protein α-subunits can
be observed in squid photoreceptor membranes (Baverstock et al., 1989). Pretreatment
with either toxin inhibits about half the light-activated GTPase activity, and pretreatment
with both toxins abolishes GTPase activity of the microvillar membrane preparation.
Both of these G-proteins are strongly inhibited by elevation of free calcium ions to
micromolar concentrations.

Phospholipase C

Vandenberg & Montal (1984b) showed that radiolabelled ATP added to squid photoreceptors was incorporated by phosphorylation into rhodopsin and inositol phospholipids, and that PIP and PIP_2 decreased in the light. By incorporating ^{3}H-inositol into the
retina, Szuts et al. (1986) found a rapid light-stimulated production of IP_3. Brown et al.
(1987) also found light-induced IP_3 release, and reduction of PIP_2. These effects were not
blocked by treatment with pertussis toxin, suggesting that the 40 kD Gα is not the
activator of PLC. Adding exogenously labelled PIP_2 to purified membranes, Baer & Saibil
(1988) found light-stimulated hydrolysis which required GTP, implicating a G-protein
in the activation of PLC. The corresponding physiological work has mostly been done on
Limulus and fly photoreceptors. Injected IP_3 has been shown to depolarize and adapt the
receptors, via the release of calcium from smooth endoplasmic reticulum adjacent to the
microvilli in these species (Payne et al., 1988). Calcium also mediates a negative feedback
effect on the response. Evidence for the functional importance of this pathway comes

from correlation of GTPase with depolarisation (Blumenfeld *et al.*, 1985) and from the discovery that a mutation which abolishes the photoresponse in Drosophila (*norpA*) codes for a PLC (Bloomquist *et al.*, 1988).

Cyclic GMP regulation

Weaker evidence that cGMP acts as a messenger in invertebrate phototransduction comes from biochemical studies of the squid retina and intracellular recordings of *Limulus* ventral photoreceptor. A light-stimulated increase in cGMP is seen in isolated photoreceptor membranes and occurs rapidly in the intact squid retina (Saibil, 1984; Johnson *et al.*, 1986). Johnson *et al.* found some cGMP-sensitive regions in *Limulus* photoreceptors. These 'hot spots' responded to cGMP injection in a manner that mimicked the response to light. However, cGMP-regulating enzymes have not been characterized in the invertebrate system.

CONCLUSIONS

Squid photoreceptors provide an excellent system for biochemical studies of receptor-activated GTP-binding proteins and PIP_2 phospholipase C. They are likely to exhibit interactions between parallel or diverging pathways, as a single receptor activates at least two GTP-binding proteins. An advantage over studies of the hormone system is the rapid and precise activation by light. However, there is a lack of physiological studies on the squid retina, because the animals are delicate. Structurally, the photoreceptor microvilli comprise an extensive and highly ordered array of membrane-cytoskeleton connections, and these may also be involved in the biochemical events of photoreception.

I am grateful to the staff of the Marine Biological Association, Plymouth and to Dr S. von Boletzky of the Laboratoire Arago, Banyuls-sur-mer, France, for the excellent supply of cephalopods, essential for work in this field.

Chapter 13

HERRING BEHAVIOUR IN THE LIGHT AND DARK

J.H.S. BLAXTER AND R.S. BATTY

Dunstaffnage Marine Laboratory, PO Box 3, Oban, Argyll, PA34 4AD

Herring have well-developed eyes and a complex acoustico-lateralis system. They require light to particle-feed, to school and to avoid stationary nets. Video techniques, using infra-red light invisible to the fish, show how they behave in low light intensities and in darkness. The schools disperse and activity drops but the fish retain an ability to filter-feed; they can also avoid vibrating obstacles.

Using larval herring of different age it is possible to observe behaviour when the sensory systems are at different stages of development. Using artificial sound and light, as well as predatory fish, as stimuli, it can be shown that dramatic increases in responsiveness occur when rods recruit to the retina and when the otic bulla and lateral line become functional.

INTRODUCTION

The Atlantic herring (*Clupea harengus* L.), a very abundant member of the clupeid family of teleost fish, is a schooling species, preying on zooplankton and performing extensive horizontal feeding and spawning migrations on a seasonal basis. It lays its eggs on the sea bed, a most unusual feature in marine fish, and it exploits both pelagic and deeper environments by making diel vertical migrations - towards the sea surface at dusk and towards the sea bed at dawn.

Observations now show that the herring's life style has many, sometimes unexpected, features. Although vision was at one time thought to be the dominant sense (see Blaxter & Hunter, 1982) recent unravelling of the complexities of the inner ear and its associated head lateral line belie this. New techniques, involving video systems with the use of infra-red lighting, at wavelengths invisible to the fish, allow observations of fish behaviour to be made in dim light or in darkness, when vision will be less effective or not functioning.

This account first introduces the sensory systems of the herring and reviews earlier work on behaviour; it then deals with the differences in behaviour between light and dark conditions as determined from recent investigations using these new techniques.

SENSORY SYSTEMS

The herring eye has a duplex retina in the juvenile and adult stages (Blaxter & Jones, 1967). The larvae, however, have a pure-cone retina for the first few weeks of life and the

Herring, P.J., Campbell, A.K., Whitfield, M. & Maddock, L. (ed.). *Light and Life in the Sea.*
© 1990. Cambridge University Press.

rods are progressively recruited once the larvae reach a length of about 25 mm. The juveniles and adults have a specialised ventro-posterior region of the retina - the *area temporalis* - where a high density of narrow single cones gives good acuity along the upper-fore visual axis. The duplex retina undergoes photo-mechanical or retinomotor movements of the melanic masking pigment and visual cells during light or dark adaptation (Figure 13.1), a phenomenon also found in other teleosts with duplex retinae. These movements are first observed in herring larvae at a body length of about 30 mm during early rod recruitment and they take place over a light intensity of 1·0-0·1 metre candles (1 mc ≡ 0·5 µW cm^{-2}).

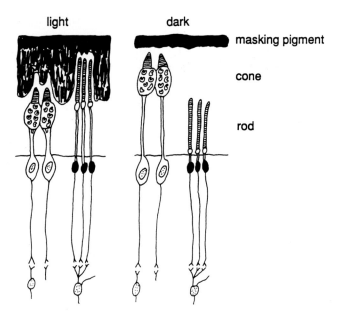

Figure 13.1. Diagram of a light-adapted and dark-adapted herring retina showing retinomotor movements of the masking pigment and visual cells (adapted from Blaxter & Jones, 1967).

The herring ear is unusual in having the utricular part of the inner ear best developed. Close by the utriculi are paired otic bullae (Denton & Blaxter, 1976), bony spheres partly filled with gas and connected to the gas in the swimbladder via fine ducts (Figure 13.2). The bullae become gas-filled after some weeks of larval life at a body length of 22-26 mm. A bulla membrane separates the gas from liquid perilymph in the dorsal part of the bulla. Movements of the bulla membrane, resulting from changes in hydrostatic or acoustic pressure, are transmitted to the perilymph outside the bulla (and around the utriculus) through a fenestra in the dorsal bulla wall. The utricular macula is stimulated by these movements, which essentially provide a local amplification device. In addition the lateral line system, which is confined to the head, is coupled to these perilymph movements around the inner ear by a compliant lateral membrane in the wall of the skull beside the bullae. The lateral line system consists of an elaborate series of canals radiating from the recesses in which these lateral membranes lie.

The presence of otic bullae and the coupling between the bullae and the head lateral line is a unique feature of clupeoids. The well developed eyes and elaborate ears and lateral lines of the herring suggest the species is adapted for complex behaviour by day with the ability to maintain some key functions by night.

The structure of the chemo-sensory system is less well known, although Dempsey (1978) described nasal pits of herring larvae that are present at hatching and develop subsequently into nasal organs.

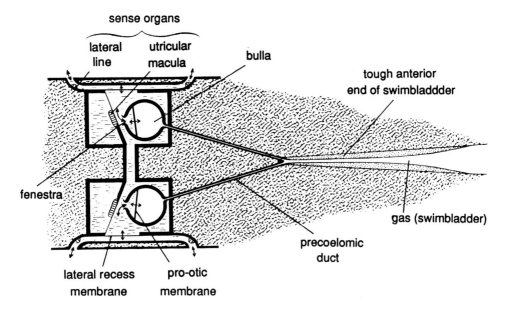

Figure 13.2. Diagram of the relationships between the otic bullae, ear, lateral line and swimbladder in the herring as viewed from the top of the head (from Denton & Blaxter, 1976).

FEEDING

Particulate feeding, the seizing of food items by predatory snapping of the jaws, is visually mediated in both larval and adult herring (Blaxter, 1964, 1968; Batty, 1987). The rate of feeding falls as the light intensity declines (Figure 13.3) with a threshold (defined as a 50% response) at about 0·1 mc. Some feeding may still occur at 0·01 mc depending on the size of food and its contrast with the background (Batty & Blaxter, unpublished). Where light is limiting herring first select large targets such as adult *Calanus* in preference to small targets such as juvenile *Calanus* and other zooplankton.

Recent studies show that adult herring are also facultative filter feeders and can switch to a feeding mode in which particles of a suitable size and density are filtered by the gill rakers (Gibson & Ezzi, 1985). Gill rakers start to develop in larvae of about 16 mm body length (Gibson, 1988) but their development is not complete until after metamorphosis at a body length of about 70 mm. A filter-feeding option is not therefore available to the larvae. Filter-feeding takes place in both light and complete darkness (Figure 13.4),

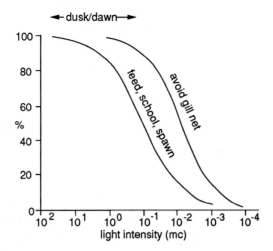

Figure 13.3. The relationships between light intensity (in mc) and behaviour of herring expressed as % of the behaviour level in bright light. The threshold is taken as the light intensity at which the behaviour level (numbers feeding, schooling etc.) falls to 50% of that in bright light. Data reworked from a variety of sources.

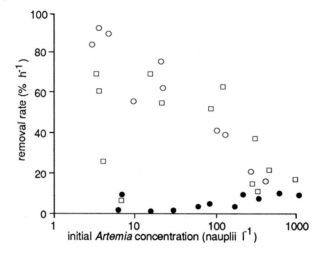

Figure 13.4. Removal rate of *Artemia* nauplii by juvenile herring expressed as % loss per hour of the initial concentration ● in the dark, ○ in the light (redrawn from Batty *et al.*, 1986), □ in the light (redrawn from Gibson & Ezzi, 1985).

although in darkness the removal rate of food is much reduced since the herring swim less strongly (Table 13.1).

The 'ground truth' for confirming such laboratory observations in the sea is inadequate. De Silva (1973) found that 65% of the stomachs of juvenile herring off the Scottish west coast were empty by night and 0% were empty by day. The extent to which herring filter-feed in the sea and the stimulus situation for triggering filter-feeding is, however, unknown.

Table 13.1. *Swimming speed of juvenile herring in body lengths per second related to schooling and filter-feeding in the light and dark (from Batty et al., 1986; Blaxter & Batty, 1987)*

	light	dark	test
schooling	0.97		
not schooling		0.60	} p < 0.001
filter feeding	2.17 } ns	0.70	} p < 0.001
not filter feeding	2.24	0.42	

SCHOOLING

The parallel orientation (polarization) of individuals in a herring school breaks down as the light intensity falls, the inter-fish distance increases and the school disperses (Figure 13.3). The light intensity threshold for schooling varies from 0·5-0·003 mc in experimental conditions depending on the number of fish and the size of the tank (Blaxter & Parrish, 1965). There are good observations from the North Sea that confirm these thresholds. Blaxter & Parrish (1965) counted the number of adult herring schools at different light intensities during the summer fishery using echo-sounding techniques (Figure 13.5). The schools dispersed at dusk at light intensities (measured at the depth of the schools) between 0·5-0·01 mc. They re-formed at dawn over a somewhat higher light intensity range *i.e.* 0·1-10 mc. In a rigorous study Welsby *et al.* (1964) used high resolution sector-scanning sonar to detect individual fish in the River Forth, Scotland. They found that juvenile herring were very dispersed at night at light intensities (measured under-water) below the schooling threshold determined in tank experiments.

Other evidence from observations at sea is more equivocal. Craig & Priestley (1960) took flash photographs of herring on a spawning ground off the Scottish west coast and found some degree of parallel orientation during the night. Harden-Jones (1962) reported that herring in the southern North Sea remained in tight groups at night. In neither investigation was the light intensity measured and in the latter report it was not clear whether the individuals in the groups were polarized.

VERTICAL MIGRATION

In behaviour closely linked to schooling, and probably to feeding, adult herring move towards the sea surface at dusk and towards the sea bed at dawn. Such diel vertical migration has not been studied experimentally, but observations in the North Sea (Postuma, 1957; Blaxter & Parrish, 1965) show that the dusk upward movement is triggered by the light intensity on the sea bed falling to 10-0·1 mc and the dawn downward movement is completed over the same range of intensity on the sea bed. During the vertical migration the herring 'follow' a light preferendum of 5-10 mc (Figure 13.6).

The diel vertical migration of larval herring has been widely reported with contradictory findings (see Blaxter & Hunter, 1982; Heath *et al.*, 1988). Usually changes in the level

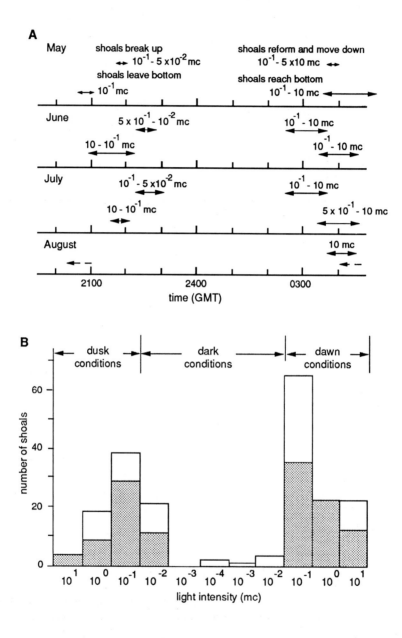

Figure 13.5. Adult herring behaviour in the North Sea fishery. (A) Diagrammatic representation of vertical migration and schooling in the dusk/night/dawn period showing the range of light intensity (measured at the depth of the fish) over which various behaviours occur. (B) Histogram showing the number of schools observed at different light intensities (measured at the top of the school). Note that 'shoal' and 'school' are used synonymously. The events are interpreted from echo-sounding records that show the extent of aggregation or dispersion of groups of fish but do not indicate the 'structure' (degree of polarization) of the group. From Blaxter & Parrish (1965) with permission of the International Council for the Exploration of the Sea, Copenhagen.

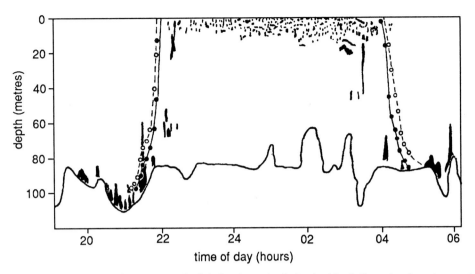

Figure 13.6. Vertical migration of adult herring schools in the North Sea related to time and 5 -●—●- and 10 lux (mc) -○ - -○- isolumes as interpreted from echo-sounding records. Redrawn from Postuma (1957).

of the 'centre-of-density' of the population have been observed depending on the time of day, often with varying degrees of dispersion. The analysis is complicated by size-dependent amplitudes of movement, size- and light-dependent escape from sampling gear, wind-induced turbulence and temperature discontinuities. Selivertsov (1974) first caught Atlanto-Scandian herring larvae at the sea surface at dusk as the light fell to 1·0 mc. Stephenson & Power (1988), however, found a semi-diel vertical migration of herring larvae off Nova Scotia with larvae nearest the sea bed at 0400-0700 h and 1900-2200 h in late October - early November.

In laboratory-simulated vertical migrations Blaxter (1973) found larvae moving to the surface of a vertical tube when the surface light level fell to 0·1-0·01 mc. Extra-retinal sensitivity may be involved since Wales (1975) showed that eyeless herring larvae also moved to the surface of vertical tubes in response to light falling to 0·001 mc. The pineal gland may be implicated in this response.

GENERAL ACTIVITY

As the light intensity is reduced in tank experiments herring schools disperse and the activity, as measured by swimming speed, falls. However, the presence of appropriate food particles for filter-feeding increases swimming speed (Batty et al., 1986, table 1). Lowered activity of herring at night has been observed from Russian submarines (Radakov, 1960) when individuals were seen moving slowly at different angles of tilt to the horizontal. Tilt angles have been measured in herring by day and night because they affect the acoustic target strength of each fish, so requiring different calibration factors in acoustic biomass estimations carried out by day or night. Foote & Ona (1985) report two experiments in which the tilt angles of herring were greater at night, *i.e.* the fish were angled upwards from the horizontal plane. Herring are negatively buoyant and at night less active fish presumably require a higher tilt angle to get sufficient dynamic lift at lower swimming speeds to hold their depth.

Larval herring also change their activity in tanks depending on the light conditions (Batty, 1987). The proportion swimming falls as light intensity decreases and is minimal in complete darkness. Activity also depends on whether food is present. In darkness or below the feeding light intensity threshold, food and other substances such as amino acids cause an increase in activity (Dempsey, 1978; Batty, 1987). Above the light intensity threshold for feeding the swimming speed increases if food is present, except in the older larvae (Figure 13.7).

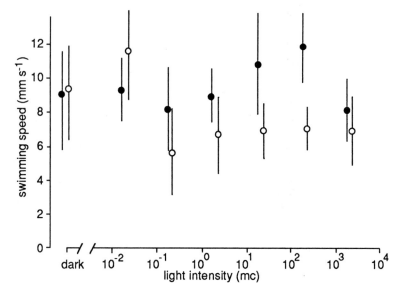

Figure 13.7. Swimming speed of herring larvae 12 mm long in the presence ● and absence ○ of food (*Artemia* nauplii) at different light intensities. Redrawn from Batty (1987).

RESPONSE TO NETS AND OBSTACLES

Traditionally adult herring have been caught by stationary gill nets, usually during the late dusk-early night period. Herring avoid such nets by day, using vision, with the light intensity threshold, determined from tank experiments, ranging from 0·03 to <0·001 mc depending on the contrast between the net and background (Blaxter & Parrish, 1965). Herring swim into monofilament nylon nets by day as a result of their low visibility. In tank experiments using parts of model trawls (nets dragged along the sea bed and producing a much wider variety of stimuli) avoidance responses ceased over a wide range of light intensities from 0·7-0·005 mc depending on the conspicuousness of the part of the trawl under test (Blaxter & Parrish, 1966). Indeed, herring even failed at night to respond to a row of moving lights underwater, suggesting a drop in general awareness.

A number of attempts have been made to test the responsiveness of adult herring to various obstacles in both light and dark conditions in order to analyse the relative importance of the eye and the ear in determining behaviour. Herring usually swim into transparent sheets of polythene placed across their swimming path in tanks in both light and complete darkness (Blaxter & Batty, 1985). In darkness individuals will occasionally turn aside and avoid such obstacles, reacting at a distance of the order of one centimetre. Since the obstacle is not producing acoustic stimuli these herring may be detecting acoustic reflections of their own swimming movements or other sound sources. If herring are presented with a 'curtain' of rising air bubbles across their swimming path, giving maximal visual and acoustic stimulation, they sometimes swim, in the light, through small gaps in the barrier where the bubbles are most separated. In darkness they consistently turn away from the barrier at a distance of 5-15 cm. It appears to be more difficult for them to identify gaps using acoustic stimuli alone.

RESPONSES TO ARTIFICIAL VISUAL AND ACOUSTIC STIMULI

Herring larvae in tanks respond to flashes of light or looming artificial visual stimuli by making a C-start or fast startle response - a very fast flexion of the body occupying less than 20 ms. With looming stimuli the larvae usually turn away from the direction of stimulation (Batty, 1989) and habituation occurs. These responses only appear late in larval development at the same time as rods recruit to the retina.

Juvenile herring in the sea are attracted by continuous lights underwater at night (Blaxter & Parrish, 1958) with no evidence of habituation. If the lights are moved vertically the fish follow them, so creating an artifical vertical migration. Such phototaxis seems to have no adaptive value and Verheijen (1956) described it as a mass disorientation similar to that of migrating sea birds attracted to a lighthouse.

Tank experiments have been done on the responses of adult herring to both transient and continuous artificial sound sources in the light and in complete darkness. Herring of all ages respond to transient sounds by making fast-startle responses as long as a gas-filled bulla is present (Blaxter *et al.*, 1981). Although the proportion of fish responding is slightly less in darkness the ability to turn away from the direction of the sound is equally good in both conditions (Table 13.2). Transient sounds are likely to be given by predators making a fast attack and it is significant that the ability to make directional escape responses from sources of sound stimuli is independent of the light conditions, *i.e.* is solely dependent on the acoustico-lateralis system. Herring, however, rapidly habituate to transient sounds. Presumably very active muscular responses cannot be repeated too

Table 13.2. *Response rates of juvenile herring to a transient 100 Hz sound and the proportion escaping away from the source of sound (from Blaxter & Batty, 1987)*

	light	dark	test
response rate	31%	27%	not significant
directionality	86%	85%	not significant

frequently for metabolic reasons and are reserved for emergencies. Such emergencies could occur during attack by a predator or during rapid movements by neighbouring fish in the school that might cause a collision.

Continuous sounds of about 20 Hz, provided by a vibrating rod in the path of adult herring swimming round a tank, cause slower avoidance responses in both light and complete darkness (Blaxter & Batty, 1985). As the fish circle the tank, about 2 m in diameter, and approach the sound source they are presented with a 'looming' sound stimulus and turn from it at a distance of 3-15 cm. The turn takes 80-200 ms (*i.e.* it is not a startle response) and there is no habituation. Since continuous sounds of this sort are not unlike the stimulus generated by a fish tail, although somewhat higher in frequency, they are probably involved in signalling inter-fish distance in schools or non-polarized aggregations. Hence the lack of habituation is an adaptation to allow continuous monitoring of the position of neighbouring individuals in the school or to avoid collisions in darkness.

RESPONSE TO PREDATORS

Both invertebrate and fish predators have been used in tanks to test the role of different herring larval sensory systems in predator evasion (Bailey & Batty, 1983; Fuiman, 1989; Fuiman & Blaxter, unpublished). The younger larvae are very unresponsive to predatory *Aurelia aurita* medusae and the predation rate by *Aurelia* is the same in both light and dark conditions. The younger larvae are also very unresponsive to fish predators. Responses to potential fish predators depend on the developmental stage of the larvae and whether an encounter is threatening (an attack) or non-threatening. Older herring larvae become very responsive to threatening encounters at a stage when the otic bulla fills with gas and rods recruit to the retina. The role of vision is to reduce responsiveness in non-threatening encounters so avoiding metabolically-wasteful and unnecessary movements. In threatening encounters the response distance is very short. Herring larvae escape from attacks by dodging aside using a C-start (fast-startle response), when the predator is very close, so close that it is too late for the predator to 're-programme' its attack. Short response distances are therefore advantageous.

Table 13.3. *Summary of herring behaviour by day and night*

day	dusk/dawn	night
Remain deep	Move up vertically at dusk	Remain near surface
School with high activity	at 10-0·1 mc	Dispersed with low activity
Particle and filter-feed	Move down at dawn at 0·1-10 mc	Filter-feed in high particle
Spawn	Light/dark adaptation	concentrations
Avoid gill nets	threshold of 0·1 mc	Enmeshed by gill nets
unless monofilament		Avoid bubble barriers
Swim through bubble barriers		Not herded by trawls
Herded by trawls		Avoid transient and
Avoid transient and		continuous sound sources
continuous sound sources		

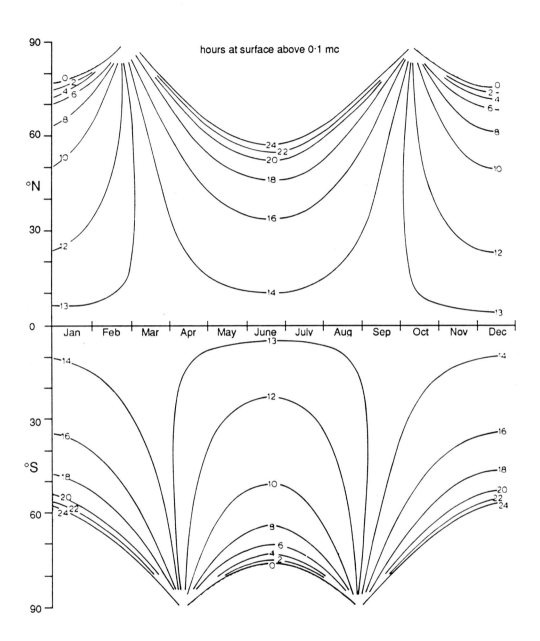

Figure 13.8. Hours per day at the sea surface at which the light intensity remains above 0·1 mc, taken as an average threshold for light-dependent behaviour to cease. Note the very large seasonal variations in 'available' light at the higher latitudes. Adapted from Blaxter (1970).

CONCLUSIONS

A summary of the changes in behaviour of herring that occur between daylight, dusk/ dawn, and night conditions are shown in Table 13.3. Herring retain a substantial repertoire of behaviour by night despite a general tendency to reduce activity. Thus energy is conserved in the dark, the herring becoming appropriately re-activated to filter-feed or evade predation if required.

In making assessments of the role of vision in the sea, it is important to measure light intensities at the depth of the fish. In moonlight or during short summer nights light intensities may well be above the threshold for light-dependent behaviour. The spread of thresholds for light-dependent behaviour can be explained by the range of visual tasks involved, the variations in contrast between visually perceived objects and their background, and such factors as motivation. There is good agreement between the visual thresholds as determined from behavioural experiments, and the light intensity changes that trigger the retinomotor responses. Thus, light-dependent behaviour seems largely to be mediated by cones, high acuity responses being mediated via the pure-cone *area temporalis*. Nevertheless the peripheral retina, with a lower density of cones and many rods, must play a part in scanning a much wider field, so improving the general awareness of the fish.

If a light intensity threshold of 0·1 mc is taken as an average value it is possible to calculate the hours per 24-hour day at which the light at the sea surface is above threshold (Figure 13.8). It can be seen that major seasonal differences occur at high latitudes in the extent to which light-dependent behaviour can occur. This must have important implications for herring distribution, both horizontal and vertical, feeding and predation and is a factor that needs to be incorporated into models of survival or food budgets of species like herring.

The sensory systems of herring should not be treated in isolation. The mechanoreceptors can operate throughout the 24-hour day. Behaviour like schooling seems to be mediated by combinations of stimuli - attraction of neighbours by vision and repulsion by mechanoreception giving a dynamic balance that determines inter-fish distance. As the light intensity wanes and the visual attraction weakens the school tends to disperse. Responsiveness to predators, at least in the late larval stages, seems largely under the control of the acoustico-lateralis system with vision modulating response distance but also inhibiting 'wasteful' responses if nearby predators are not threatening. The study of early life history stages provides great insight into the role of sensory systems as they progressively develop and elaborate.

Chapter 14

CREPUSCULAR BEHAVIOUR OF MARINE FISHES

GEOFFREY W. POTTS

Marine Biological Association, The Laboratory, Citadel Hill, Plymouth, PL1 2PB

It is likely that all marine fish living in the photic zone will show behaviours and physiological adaptations that are controlled and regulated by the day/night cycle. Most work from tropical and subtropical seas indicates that at dusk a change-over period occurs when diurnal species seek shelter and nocturnal species emerge to night-time feeding areas. The reverse occurs in the dawn twilight period. These crepuscular activities often involve vertical and horizontal migrations over considerable distances and are controlled by ambient light levels. There is a sequence of cover-seeking and emergence which appears to be related to the relative vulnerability of species still active during this time when many predators are foraging. Other aspects of crepuscular behaviour are discussed.

INTRODUCTION

Fish that live in shallow coastal waters, or within the sea's photic zone exhibit patterns of behaviour that are regulated by ambient light levels, and in particular the light/dark cycle (*e.g.* Woodhead, 1966; Hobson, 1968, 1972, 1973, 1974, 1986; Winslade, 1974; Lythgoe, 1979; Helfman, 1986). This behaviour is controlled by physiological mechanisms, mainly vision and the associated adaptations shown by the fishes' eyes (Lythgoe, 1979; Nicol, 1989; Denton, Chapter 8; Land, Chapter 9; Partridge, Chapter 10). Activity patterns during twilight periods are described as crepuscular and are represented in most species of fish as transitional behaviours. Thus species which are primarily active during the day begin to seek shelter at dusk, while nocturnal species emerge from their daytime resting places prior to beginning nocturnal foraging activities. A small number of predatory species are predominantly crepuscular, being most active during the twilight period when they pursue and feed upon diurnal and nocturnal species. Hobson (1968), classifies fishes as either diurnal, nocturnal or crepuscular depending upon when their primary feeding times occur.

A detailed review of the literature shows that field work has mostly been carried out in tropical and subtropical seas where clear waters and stable climatic conditions are ideal for prolonged observations (*e.g.* Hobson, 1965, 1974; Starck & Davis, 1966; Collette & Talbot, 1972). An increasing body of information indicates that temperate species also show crepuscular behaviours, although less well defined than is shown by tropical species. Tropical seas, as sites for studying crepuscular behaviour of fishes, have the

Herring, P.J., Campbell, A.K., Whitfield, M. & Maddock, L. (ed.). *Light and Life in the Sea.*
© 1990. Cambridge University Press.

further advantage of a much reduced tidal range, therefore reducing the confusing and sometimes overriding effect of tides over light cycles (Gibson, 1986). While recognising the importance of understanding the rhythmic activities of fishes, especially the endogenous and exogenous factors controlling the diel rhythms, these will not be considered in depth here and Thorpe's (1978) book should be consulted for further information.

TWILIGHT AND VISION

Before describing the crepuscular behaviour of marine fishes it is useful briefly to consider what is meant by twilight and its implications on the visual systems and behaviour of marine fish. There are several definitions of twilight based upon the cycle of the sun below the celestial horizon. Thus civil twilight occurs when the sun is at 6°, nautical 12° and atronomical at 18° below the celestial horizon. For zoogeographic distribution of behavioural differences it is important to recognise that the twilight period at different latitudes is of different length, being shorter in the tropics where it is estimated to occupy only 5% of the 24-hour cycle. During this period it is estimated that the light may drop from 100 lux to 0·01 lux in a period of 30 minutes. While the neural and pupillary mechanisms of fish are adapted to cope with this rate of change few eyes can accommodate the large light differences between day and night and generally become less efficient at twilight (Lythgoe, 1979; Blaxter & Batty, Chapter 13).

DIEL ACTIVITY PATTERNS

Hobson (1968, 1972) in a series of extensive underwater observations categorised fish as being diurnal, nocturnal or crepuscular. Those active during the day represented 50 to 70% of the species, those at night 20 to 30%, and crepuscular species constituted 5 to 10% (Emery 1978; Helfman, 1978, 1986). Helfman continues his review by giving details of the behavioural, ecological and evolutionary factors that relate to the diel patterns of behaviour, drawing examples from temperate as well as tropical environments. Diurnally active species divide most behaviours between feeding and a quiescent state of vigilance in which the individual fish are ready to respond to and avoid predators (Hobson, 1968). Feeding can assume many different forms and with vision the primary sensory modality it is not surprising that many diurnal species are piscivores with a number of invertebrate specialists and benthic substrate pickers. Families characteristically diurnal include the Balistidae, Acanthuridae, Ammodytidae, Scaridae, Labridae, Pomacanthidae, Anthiidae and Gobiidae with individual representatives of many other families. At twilight, diurnal species are visually at a disadvantage and will seek shelter, moving to safe crevices for the night period where they remain until dawn when the light level is adequate for them to resume normal feeding behaviour. Many show burrowing behaviour at night (Winslade, 1974; Meyer et al., 1979; Hobson, 1986) and parrot fishes have been shown to secrete a mucous cocoon in which they remain during the hours of darkness (Winn, 1955; Winn & Bardach, 1959). Most herbivorous fishes are diurnally active.

Nocturnal species are mainly generalised invertebrate and crustacean predators relying on non-visual sensory systems (chemical or tactile senses) for detecting prey, unless, like the Anomalopidae, they use their own luminescence. While most will feed opportunistically during the day they typically form non-feeding schools which reside in a state of predator vigilance. Other nocturnal species, noticeably the Holocentridae and Anomalopidae, stay in crevices during the day and emerge to disperse to nocturnal feeding grounds after dark. Families characteristically nocturnal also include the Apogonidae, Anguillidae , Haemulidae, Lutjanidae and the luminescent Leiognathidae.

Species active at twilight comprise two main groups; those nocturnal or diurnal species in a state of transition between day and night, and those that are primarily crepuscular. The boundary between nocturnal, diurnal and crepuscular is not always clearly defined and the tendency of predatory fish to show some degree of opportunism means that there is a degree of overlap. The crepuscular species are often generalised predators that may at times be represented by nocturnally active species moving from the diurnal schools to nocturnal feeding grounds. Those species that are truly crepuscular are all piscivores exploiting the disorientation of other fish (Potts, 1983) during the period of light transition and are often characterised by large eyes better adapted to twilight levels than other species (Major, 1977). The carangids, lutjanids, serranids and several species of gadoid in temperate seas are examples of predators commonly active at dawn and dusk.

SEQUENCE OF CHANGE-OVER

During twilight periods Hobson (1972) found that there was a clearly defined sequence of activity shown by diurnal and nocturnal species and that the timing of their sequences relative to sunset and sunrise was remarkably precise. At dusk he described a series of horizontal and vertical migrations of fish in which midwater plankton pickers such as anthiids and pomacentrids (Davies & Birdsong, 1973) descend to the reef while other families such as scarids and acanthurids undergo horizontal migrations along set pathways (Dubin & Baker, 1982). These migrations by diurnal fish are to nocturnal retiring places where they are afforded some protection from nocturnally active predators.

During the dawn and dusk twilight periods fish undergo a period of transition between their daytime and night-time activities. The times at which they begin this transfer depends upon the species involved and the relative light level. Detailed fieldwork by Hobson (1972) and Colette & Talbot (1972) indicate that the timing of activities is very precise for individual fish and families of fish. Most fish are able to anticipate the change-over period and begin to behave in preparation for the transition. Thus wrasse begin to move into crevices five minutes before sunset and by 10-15 minutes after sunset most species are already in hiding. Many other families show similar precision. During twilight it is not only the benthic and epibenthic species that seek shelter, but also midwater plankton picking species which slowly descend to the reef where they remain in crevices until dawn.

This sequence of change-over is not just confined to short movements over the reef and in some cases may involve movements over hundreds of metres to offshore feeding grounds. Often the onset of such migrations is preceded by restlessness within the school, individuals starting and returning to the group until finally enough start moving together to entrain a mass migration of the group to the new area. In the Haemulidae these migrations are completed and dispersal achieved between 5 and 20 minutes after sunset. The time of onset of these migrations is controlled by light and can be extremely precise, the fish leaving within a minute or two of the same time each evening. A remarkable study over a 3-year period indicated that the time of onset of migration shows greater precision with increase of age of the fish (Ogden & Erhlich, 1977; Gladfelter, 1979; McFarland et al., 1979). Work by Helfman, Meyer & McFarland (1982) and Helfman & Schultz (1984), showed that the timing and route of this crepuscular migration is learned and that transplanted individuals soon synchronise with the school to which they are added. Potts (1980, 1981, 1983) discovered that prey species would show an ordered retreat to the shelter of the reef in the event of predatory activity in the area. Species would leave their normal daytime resting place and move close to the reef and eventually take shelter in crevices or caves. It was at times possible to identify the predator by the sequence of retreat and it is significant that the sequence during the twilight transitional period closely matches that seen during diurnal predation. It strongly suggests that the sequence of crepuscular change-over has evolved primarily as an antipredation behaviour with certain, most vulnerable, species taking shelter first and others in order of decreasing vulnerability following.

The importance of shelter-seeking is emphasised by the fact that many species not only return to the same crevice night after night (Lavett Smith & Tyler, 1972) but also defend it against intruders (Robertson & Sheldon, 1979). The resulting confusion of such activity could be exploited by predators and the ordered sequence during the change-over was put forward by Domm & Domm (1973) as a mechanism to avoid or minimise such confusion. Helfman (1986) discusses this and suggests that more observations are needed to confirm the hypothesis.

About 10-20 minutes after sunset most diurnal species have gained the shelter of the reef and there follows what Hobson (1972) called the 'quiet period' when very few fish are in evidence. It is at this time that crepuscular predators become most active and in particular the serranids, carangids and lutjanids. These fish typically live in close proximity to the reef and emerge at twilight to capture prey silhouetted against the surface water. This type of hunting is characteristic of many reef fishes and has also been recorded among inshore temperate gadoids (Potts, 1986; Edwards, personal communication).

The 'quiet period' coincides with the period when light levels are changing most rapidly and is, therefore, most clearly defined and extensive in tropical reefs, lasting approximately 20 minutes. By contrast the period between shelter-seeking by diurnal species and the emergence of nocturnal species is less clearly defined in temperate seas and the 'quiet period' is not always easy to identify. Most of the literature refers to the change-over period of the coral reef fishes and the cover-seeking activities in crevices.

However, many resort to burrowing in fine particulate substances and sandeels or sandlances may be cited as examples (Winslade, 1974; Potts, 1986; Hobson, 1986). The Ammodytidae are some of the most important species in the NE Atlantic, not only for their fisheries potential, but also because they provide a vital food source for many commercially important fish species as well as sea birds, seals and whales. Their burrowing and emergence patterns are important factors in rendering them available to the many species that depend upon them for food.

Between 20 and 30 minutes after the 'disappearance' of diurnal species the emergence of nocturnal fishes begins and like the cover-seeking also shows a set sequence. This occurs within half an hour of sunset and can involve extensive vertical and horizontal migration. One of the few field studies on juvenile grunts (Haemulidae) in the Caribbean showed that at night they fed on invertebrates among offshore eel-grass beds and migrated to these at dusk, returning to daytime retiring places at dawn. It was concluded that these horizontal migrations, which were often several hundred metres in extent, were controlled by changing light levels during the twilight period (McFarland et al., 1979).

At dawn the reverse sequence is seen with the nocturnal species returning to diurnal retiring places and the re-emergence of diurnal species to their daytime feeding grounds. Once more a peak of predatory activity is seen during this transitional period when large-eyed crepuscular predators become active.

PREDATION

To be most effective, predators need to maximise the frequency with which they encounter prey and exploit situations when the prey are vulnerable. The twilight period offers advantages to many large-eyed predators over their prey (Hobson, 1972; Colette & Talbot, 1972; Ehrlich et al., 1977). Munz & McFarland (1973) and McFarland & Munz (1976) have shown a relationship between the behaviour of predatory fish and vision. In many the eyes are visually better adapted to twilight levels of illumination and, therefore, better able to exploit prey species and those seeking shelter or re-emerging during this transitional period (see details above). Predators most active in the twilight period are for the most part inconspicuous in colouring such as snappers (lutjanids) or jacks (carangids) and are primarily benthic or epibenthic (serranids). Their prey consists mainly of fish descending to the shelter of the reef which are vulnerable while still silhouetted against the water surface. It is interesting to note that some species resort to different hunting tactics under different conditions and the Pacific Salmon has been shown to have two different hunting modes depending upon the light levels. Thus at twilight levels the fish are scotopic and resort to silhouette feeding while at higher light levels they attack horizontally (Ali, 1959, 1976). Other species show these hunting methods: the skipjack tuna attacks upwards at fish silhouetted against surface waters while the dolphin fish (Coryphaena) in bright surface waters strike laterally. The distinctive change-over period, typical of the tropics, is not always evident in temperate marine conditions (Ebeling & Bray, 1976; Hobson et al., 1981). The reasons were identified as the protracted twilight period, high turbidity and the number of species found in temperate seas that feed both at night and day without clear diurnal rhythms.

Recent work by Edwards and Potts has shown that even in the ill-defined twilight period of the temperate seas there is an increase in twilight predation with some species. Pollack were shown to increase the number of attacks after the onset of the twilight period. The two-spot goby, *Gobiusculus flavescens* and the predator, *Pollachius pollachius* both feed actively during the day between bouts of nesting activity, but at dusk the gobies reduce feeding and increase in vigilance before seeking shelter among the crevices of kelp holdfasts. At dawn the gobies show an increase in feeding behaviour and at the same time are subjected to increased predation by pollack.

From the extensive literature reviewed by Hobson (1968, 1972, 1974, 1981) and Helfman (1986) it is evident that many of the behaviours observed in fish during the twilight period are either exploitation by predators of the transitional behaviours of fish seeking shelter, or emerging, or the antipredator tactics of prey species (see Figure 14.1).

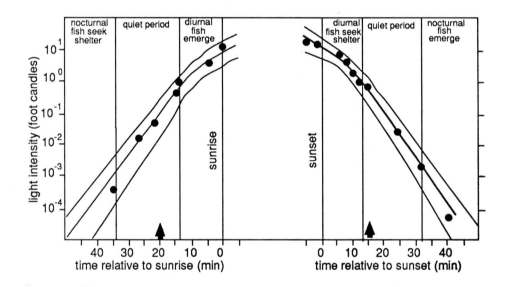

Figure 14.1 Diagram summarising the patterns of crepuscular behaviour shown by a range of marine fishes. The solid circles mark the emergence and cover-seeking behaviours of different species in relation to ambient light levels. The arrows show the time for peak activity for crepuscular predators (modified and simplified after Hobson, 1972).

REPRODUCTIVE BEHAVIOUR AT DUSK

Lobel (1978) records that many normally diurnal species show an increase in breeding activity at dusk and this may be particularly common in species with planktonic eggs. It has been suggested that gametes released in the evening and just prior to nightfall are less vulnerable to planktivores than those released during the day (Hobson & Chess, 1978, 1986) and have the opportunity to disperse at night (Johannes, 1978). Since many fish that spawn in this way swim away from the shelter of the reef it may be suggested that adult fish are increasing their own risk of predation in order to minimise the risk to their gametes (Helfman, 1986).

It may be seen from the above examples that twilight is a critical period in the lives of many species of marine fish. It is a period when piscivorous predators are most active in exploitation of prey as they seek cover or emerge from the shelter of reef systems. To counteract this predator-pressure vulnerable species have evolved behavioural mechanisms that improve their chances of survival.

It is a classical evolutionary 'arms race' where predators are most active during a period when prey are vulnerable and when the behaviour of small and vulnerable species is evolving to minimise the risk of encountering a predator. Only relatively recently have we begun to understand the underlying physiological mechanisms controlling these reactions (Shand et al., 1988), but much more research is needed.

Chapter 15

LIGHT AND THE CAPTURE OF MARINE ANIMALS

P. L. PASCOE

Plymouth Marine Laboratory, Citadel Hill, Plymouth, PL1 2PB

A wide variety of light-producing mechanisms and light-induced behavioural responses are utilized by marine animals both for prey attraction and predator avoidance. These responses have been exploited by man since prehistoric times with fishing techniques using light ranging in scale from simple to sophisticated and covering the marine environment from shore to deep ocean. The diverse applications of light in the capture of marine animals are reviewed. Experimental studies on the effect of artificial light on the efficiency and selectivity of various types of fishing gear are described and discussed.

INTRODUCTION

The main concern of this review is the applications of both natural and artificial light in the capture of marine animals. Most other capture methods depend on the visual sensitivity, movement and distribution of the animals concerned, which in turn are often governed by the behavioural and physiological responses to light in their environment. Aspects of this subject have been reviewed elsewhere (Clarke & Denton, 1962; Woodhead, 1966; Blaxter, 1970, 1976; Segal, 1970) and are be covered by other authors in this volume.

Most species react to light, usually by means of photokinetic or phototactic responses. Photokinesis is a response in overall activity *i.e.* the velocity or rate of change of movement related to change in the intensity of light, regardless of its direction. The activity of animals responding in this way usually increases with increasing light intensity. Phototaxis is a locomotory response relating to the direction of a light stimulus, being either positive or negative *i.e.* towards or away from the light source. The behaviour of animals with respect to light in the sea, particularly artificial light, has been examined by several authors, often with different interpretations (Scharfe, 1953; Blaxter & Parrish, 1958; Verheijen, 1958; Woodhead, 1966; Blaxter & Currie, 1967).

The majority of light-fishing techniques involve the attraction of animals towards light. Several theories to explain this attraction have been proposed; these include positive phototaxis, preference for optimum light intensity, investigatory reflex, feeding, schooling and disorientation. These are discussed by Protasov (1968) and Ben Yami

Herring, P.J., Campbell, A.K., Whitfield, M. & Maddock, L. (ed.). *Light and Life in the Sea.*
© 1990. Cambridge University Press.

(1976), but obviously different explanations are applicable to different species and conditions. However, the responses of many species of animals to light can often be predicted and are exploited by animals themselves and particularly by man. Many animals, particularly in the dimly-lit mesopelagic depths and below, utilise bioluminescent light-producing mechanisms for prey attraction and predator avoidance. Examples of these mechanisms are detailed elsewhere (see Herring, 1977, 1978, Chapter 16; Herring & Morin, 1978; Buck, 1978).

Man's use of light in fishing probably dates back to prehistoric times, with techniques ranging from simple to sophisticated and covering the marine environment from shore to deep ocean. These diverse applications of light in the capture of marine animals, including some recent experimental studies, are discussed below.

THE USE OF LIGHT BY MAN

Bioluminescence

Some of the bioluminescent mechanisms that have evolved in animals are adapted and used by man to attract and capture other species. Photophores in the skin and mantle of squid are used as luminescent bait in the commercial long-line fishery in the deep waters off Madeira. A special technique is employed in slicing the tissue to expose the active photophores. The black scabbard fish, *Aphanopus carbo*, is the predominant species caught in this way. Similar examples exist with luminous fishing lures being produced from the light organs of the flash-light fishes, *Photoblepharon* and *Anomalops*, in Indonesia, or by infecting bait with luminous bacteria. This latter technique for use with long-lines, traps and trawls was investigated experimentally by Makiguchi et al. (1980) and will be dealt with in a later section.

Another potentially important, but indirect utilisation of bioluminescence in fishing is the detection and assessment of large shoals of fish, crustaceans or squid by the luminescence produced by the dinoflagellates, or other planktonic organisms, they disturb. The use of image intensifiers carried by aircraft for this purpose has been investigated by Roithmayr & Wittmann (1972), Cram (1973) and Cram & Hampton (1976).

Man-made light

The use of artificial light in fishing covers a vast span of time and sophistication, from the simple techniques of early man to the technically elaborate, commercial fisheries of the present day. Its use has spread to many areas of the marine environment, with each area and most target species presenting unique problems to be overcome in the method of lighting and the design and deployment of the gear involved. Ben Yami (1976) presents a good review of the applications in use up to a decade ago, so this chapter will be restricted to the principal types and techniques of fishing with artificial light including recent additions or modifications. Other reviews on this subject include those of Kurc (1969), Sidelnicov (1981) and Flores (1982).

CHEMILUMINESCENCE

The chemical reactions producing bioluminescence within animals can be artificially produced and modified by man. Chemiluminescent lures of the 'light stick' type are therefore used in much the same way as the bioluminescent 'bait' methods mentioned above. The majority of applications involve small tubes of chemiluminescent material placed on or near hooks and are used mainly in recreational or sport fishing. They could potentially be scaled up for use with nets and pumps if appropriate. The advantages of chemiluminescence over natural systems are that totally man-made luminescence is usually more controllable, can be stored, easily transported and made available virtually anywhere in large quantities. The possible disadvantages are the expense of production and the fact that for many target species chemiluminescent lures may not be effective.

FIRE AND LAMPS

The earliest use of artificial light for fishing would have been with fires and primitive shore-based methods as mentioned below. Fishing with fire has obvious limitations which led to the natural progression from fire to lamps. Their gradual sophistication has produced an enormous growth in the practice of light-fishing. Lamps can be powered in a variety of ways (oil, gas, electricity), the output is usually controllable and variable and they can often be easily modified for use under water, thus allowing tremendous diversity in their application. Man is now capable of introducing artificial light into virtually any area of the marine environment and continues to exploit and study its effect on marine animals. The techniques of fishing with fire and lamps can be classified in numerous ways but perhaps the most appropriate in the present context is to consider the gear or method of capture with which they are employed.

Primitive methods

Many of the techniques included here may date back to prehistoric times and generally involve the capture of relatively small numbers of animals with basic, non-mechanised gear used from the shore. Some of the simplest methods involve the use of open fires on beaches or hand-held burning torches in conjunction with spears, harpoons, bows and arrows, dip-nets, beach-seines or simply a group of people to drive the fish onto the shore (see Scharfe, 1953; Ben Yami, 1976).

In some areas simple methods of light fishing are employed to capture 'flying' and jumping marine species from small boats (von Brandt, 1984). The Yami tribe on the island of Botel Tobago (SE of Taiwan), for example, attract flying fish to their boats with torches and catch them in the air with long-handled scoop nets. This method has been superseded by more efficient techniques and is now performed only ceremonially. Another interesting but presumably inefficient method is used for shrimps in Kerala, southern India, and involves light attraction and a chain dragging the bottom to frighten them into jumping from the water into a pair of canoes tied together at an acute angle and tilted towards each other.

Such techniques were practised in small primitive communities throughout the world and, although natural fire has now been replaced by other souces of artificial light, essentially the same methods are still practised in many areas today. The main effect of light in these methods is to attract the target species to a restricted area near the light, but with the use of bright torches a dazzling, stunning or immobilising effect may also be obtained. A recent example is the use of underwater lamps and spear-guns by scuba divers.

Offshore - boats and nets

The desire for greater productivity from light fishing led to the increase in the size and mechanisation of shore-based nets and the use of boats to transport the lighting equipment to offshore areas and to deploy larger and more efficient nets. There are many examples of off-shore fisheries utilising lights in conjunction with nets and other gear (see Ben Yami, 1976; von Brandt, 1984). Their level of sophistication and productivity varies enormously; some of the main types are dealt with below.

1. *Lift nets*. In principle, a sheet of netting is stretched and suspended by poles or a similar framework, dipped or submerged below the water, often with a light suspended above it, then lifted when animals have aggregated in the illuminated zone. If small, these nets can be operated by hand, but there is usually a need for a lever or pulley system to cope with the required lifting force *e.g.* the stationary 'Chinese' lift nets used in south-west India. The use of the lift-net technique from boats in conjunction with overhead lights has been developed in numerous ways by Russian, Japanese, Scandinavian and Philippine fisheries. Methods include single vessels using more than one net each, single nets suspended under catamarans, trimarans or even large nets suspended between two or four boats (von Brandt, 1984; Wickham, 1970). The target species for this type of fishery is usually the smaller, pelagic shoaling species such as sardines, saury or sprats. However, this type of net is used successfully for some squid species, particularly in California and the Philippines (Kato & Hardwick, 1976; Hernando & Flores, 1981).

2. *Stationary nets and traps*. Species that will follow a slowly moving lamp can be attracted and lured into set nets such as the so-called Chinese torch net or gill nets. There are several variations of the torch net, which is mainly used in the sardine fisheries of S.E. Asia, but the principle is to lure shoals of fish into a dust-pan shaped net with a moving light and then lift the bottom of the net thus trapping the shoal. The nets can be anchored, or held open in a current by one or two boats and the light, which was originally a burning torch, is now replaced by lamps used above and below the water surface.

Sardine-like species can be caught in much the same way with gill nets. They are attracted to and will follow a lamp which is then carried by boat in a zig-zag route over a gill net placed at the desired depth. Similarly, the catches of gill nets set in deep midwater or near the bottom might be increased by placing underwater lamps or luminescent lures on or near the net.

Light can also be used to lure or frighten fish and crustaceans into traps of various designs and sizes. There seem to be few commercial applications for lights with smaller traps; in fact, the efficiency of trapping certain crustaceans appears to decrease with light

(Makiguchi *et al.*, 1980), but there are useful applications for larval fish and zooplankton (Doherty, 1987).

3. *Surrounding nets.* Surrounding nets of the lampara or purse seine type appear to be used with light attraction in all regions of the world with an exploitable stock of pelagic shoaling fish. Many variations of net and technique exist, often used with several boats, one as the main catcher vessel and other small boats or rafts to carry and manoeuvre the lights, and thus the fish shoals, into position (Wickham, 1971; Ben Yami, 1976; von Brandt, 1984). Lamps are used on, above and below the water surface with this type of net and shoaling species caught include, sardinella, sardine, sprat, scad, mackerel, herring (particularly juveniles) and even cod and salmon. Squid are also caught by this method in many areas of the world both as the primary species and an important by-catch (Kato & Hardwick, 1976). Purse seining in particular can be devastatingly efficient as catches of over 30 tonnes per set are not infrequent. Over-fishing is a serious problem with this technique, as high numbers of immature fish are often taken. Consequently restrictions on the light output and mesh sizes have been made by the Japanese and Norwegian fisheries and, at one time, the technique was completely banned in Canada.

Pumps

The now classic method of fishing with underwater lamps and powerful pumps is that of the Caspian Sea, kilka (*Clupeonella* spp.) fishery (Nikonorov, 1959, 1968). A prerequisite for this type of fishery is that the target species should exhibit the 'kilka type' phototactic response to underwater light *i.e.* it should be attracted strongly and quickly to the light and remain in very close proximity. The gear required includes a long fishing hose with powerful underwater lamps fitted at the funnel shaped end, winches and cranes to handle the hose and the facilities to cope with the water and fish pumped to the ship. Due to the damage caused by impellers, airlift pumps are usually employed which require powerful air compressors. The lighting regime can be complex, involving changes in intensity, colour and depth of the light during fishing, and is often combined with the use of electric currents to enhance the attraction or stun the fish prior to capture. The use of different coloured lights is particularly interesting with this technique. Red light, as it penetrates least, was found to increase the catches of some species as it lured them closer to the suction nozzle. The technique is therefore sophisticated and expensive to set up which contributes to its limited use. Russian fisheries are the main users of this method of light fishing and have experimented with species other than kilka, *e.g.* saury, scad, anchovy and sardinella, with some success (Nikonorov, 1968). It was also found to have potential with the squid, *Loligo opalescens*, off California (see Ben Yami 1976).

Jigging and handlining

Handlining for squid, mackerel and tuna using light attraction was first developed by the Japanese and although handlines are still employed today, the mechanisation of this technique, particularly of squid jigging, has produced the biggest commercial application of fishing with light in the world. Fishing for mackerel and tuna involves attraction by both light and bait; moderately powered lamps (100-500 W) attract the fish towards the surface and chumming with bait helps retain them in the fishing area. The gear is a relatively simple pole and line arrangement with a hook baited with strips of fish flesh, and is now often mechanised (see Ben Yami, 1976; von Brandt, 1984).

Squid jigging is a major fishery for many countries of the world (Japan, USSR, Korea, Taiwan, China, Poland) as its level of sophistication and productivity has increased along with high market values of squid. Large vessels, of up to 300 t displacement, equipped with up to 200 Kw of lamps (50+ x 2-4 Kw), 50 or more automated jigging machines and a crew of about 20 men, can catch around 20 tonnes per day and have a storage capacity of about 300 tonnes. With market prices of $1000-2500 per tonne for squid the deployment of these factory ships to remote fishing areas for weeks at a time is still a profitable exercise. The Antarctic, for example, is fished by fleets from several countries including those of northern Europe (Patterson, 1987; Rodhouse, 1988). Species exploited by this method are mainly pelagic shoaling squid found over the continental shelves, *e.g.* ommastrephids and loliginids. Details of the history and scale of the Japanese fishery for the common squid, *Todarodes pacificus* (Ommastrephidae), is given by Murata (1989). In Japan between 1950 and 1970, annual catches of this species alone totalled more than 600,000 tonnes in some years, representing 70-80% of the total domestic catch of cephalopods during this period. The majority (around 90%) of this catch was taken by the squid jigging fishery. However, catches have been decreasing in the last two decades; evidence suggests that over-fishing has depleted the stocks. At present, world catches of cephalopods are in the region of 2 million tonnes per year, much of it being taken by jigging, and apparently considerable potential exists for further expansion of squid fisheries in particular.

Squid jigging is used all year round with maximum catches during moonless nights and peak tides. The lights used are normally incandescent lamps arranged overhead in one or two rows along the centre of the ship, white or blue light being found most effective. As the response of most squid is to aggregate around the boundary between light and shadow the positioning of the lights is such that the shadow line formed by the ship's side is below the jigging machines. Other lighting methods, such as underwater lamps, are sometimes employed to increase the initial attraction effect, and the use of jigs that actually incorporate small battery-powered lamps proves useful when the shoals are more dispersed. Jigs or rippers themselves come in a variety of designs, the modern type usually consisting of a fusiform, brightly coloured, plastic body with a single or double crown of barbless hooks below it. Many jigs (20 to 30) can be used on each vertical line which after lowering is automatically moved up and down in a 'jigging' motion. Squid engulf these lures with their arms as they would their prey and become impaled by the numerous hooks. The lines are then wound up over the large reels, and once inverted, the squid normally fall off the hooks, and in the case of factory ships are carried away automatically by way of chutes and conveyor belts to be cleaned and packed in ice in the hold.

Fishing continues throughout most of the night, the early hours of the morning often being the most productive. The technique itself is therefore quite simple but combined with sophisticated fish-finding methods, the scaling up of the fishing effort and the computerised control of mechanised jigging machines, it has developed into one of the most efficient systems employing light attraction. Further details on the methodology and development of squid jigging are given by Ben Yami (1976), Hamabe *et al.* (1982) and Flores (1982).

Trawling

The development of light-fishing techniques for commercial trawls has been very limited. It has been shown that natural light and vision are important factors in the use of this type of gear and suitable target species do respond to artificial light. Presumably the technical and safety problems involved in using high power electrical cable between a generator and the trawl has discouraged such developments. However, the recent use of battery-powered lamps on trawls suggests that many of the difficulties in developing this technique may now be overcome. Research studies in this field are described below.

RESEARCH AND EXPERIMENTAL STUDIES

Virtually all methods of light-fishing in use today will have been developed from or modified by experimental studies of some kind. The necessity for experimentation and scientific data increased with the desire to extend or develop techniques to cover more target species or different habitats and to make the techniques more efficient and possibly selective. Virtually all methods of capture are thought to rely to some extent on light and the responses of animals to it. Therefore any research into the relationships between light, both natural and artificial, the marine environment and the vision and behaviour of animals has potential relevance to the subject of capture. Over the last 30-40 years there has been extensive research in these fields, much of which will be reviewed elsewhere in this volume. Some recent investigations into the phototactic responses (positive and negative) of some species include: the use of sequentially operated lamp strings and moving lights to control and lead fish over distances up to 1 km (Wickham, 1973), the avoidance response to strobe lights and air bubble barriers (Patrick *et al.*, 1985; Sager *et al.*, 1987), and even the training of fish to move to a flashing light for a food reward (Stewart, 1981). The major fishing nations of Japan, USSR, the USA and Canada, have been responsible for most of the applied fisheries research, including the use of artificial light (see Ben Yami, 1976). However, workers at the marine laboratory of the Department of Agriculture and Fisheries for Scotland at Aberdeen have carried out extensive research into the responses of fish to trawling gear, and the role of light and vision in the behaviour and physiology of various commercially exploited species (Blaxter, 1970, 1975, 1976; Blaxter & Currie, 1967; Blaxter & Parrish, 1958, 1964, 1965; Parrish, 1969; Wardle, 1983, 1986, 1987; Glass & Wardle, 1989). The results of many of these studies show that light at a sufficiently high intensity is responsible and necessary for the 'normal' behaviour patterns of clupeids and other species and for the ordered behaviour of fish schools in the path of a towed trawl or moving net.

A study of the avoidance of towed nets by the euphausiid crustacean *Nematoscelis megalops* by Wiebe *et al.* (1982), presents evidence that this species is able to detect a midwater net and initiate an avoidance response. The detection of the net is thought to be either by its contrast with the background or by the bioluminescence produced by animals caught or disturbed by it. A suggestion is made that a reduction of these signals may be achieved by temporarily 'blinding' the animals in front of the net using intermittent flashes of strong artificial light.

Japanese workers, as mentioned previously, have experimented with a novel method of light fishing, involving the use of a luminous bacterium, *Photobacterium phosphoreum*, to infect 'bait' used for long-lines, traps or trawls (Makiguchi *et al.*, 1980). The effect of this light source on the catches varied with the species, habitat and gear involved. In trawls for example, the number of squid and prawns appeared to increase with the luminous bacteria whilst that of fish decreased. However, the number of comparisons made in some of these experiments was insufficient to show that the differences were statistically significant.

The behaviour of animals towards towed fishing gear, their reaction to light and noise and the low capture efficiency of many research nets, particularly in midwater, stimu-lated the initiation of research at the Marine Biological Association, Plymouth, into the effect of artificial light on the capture efficiency of research trawls. A simple, low-voltage, battery-powered underwater lighting system was developed with a depth capability in excess of 2000 m and was used on rectangular midwater trawls (RMTs) and bottom trawls in five geographical areas. The main animal groups of interest in these studies are those of the larger nekton with the capability of perceiving light and making phototactic responses at a sufficient speed to move them into or away from the path of the net, *i.e.* fish, crustaceans and cephalopods. In comparing the catching performance of different fishing gear, all the variable factors which may influence the catch, apart from that being studied, have to be considered and, if possible, eliminated when making valid quantita-tive comparisons. Our ability to accomplish this with small research trawls is dependent to a large extent on the sophistication of the gear and its monitoring systems.

Midwater trawls

Details of the lighting system and its use on RMTs with mouths of 8, 10 and 50 m² in two areas of the north-east Atlantic are given by Clarke & Pascoe (1985). The RMT8 (8 m² mouth) was used at a depth of 800 m in Biscay, the RMT10 at 800 m off Madeira and the RMT50 at 300 m and 1500 m off Madeira. Comparisons between the catches with the light on and those with the light off for these trawls are given in Table 15.1. Further analyses of the total fish volumes caught by each trawl according to time of day are shown in Figure 15.1. The overall results at 800 m show that the use of a light on the RMTs increases the total volume, and probably the number, of fish caught compared with trawls with the light off. The size of the fish also shows a similar increase. However, the catch of crustaceans, particularly decapods, decreases with use of the light.

At 300 m and 1500 m with a larger trawl (RMT 50), the catches of crustaceans again appear to be significantly less with the light, whilst those of the fish do not show the

Table 15.1. *Comparison between trawl catches with lights on and with lights off*

		A. RMT 8 Biscay 800m					B. RMT 10 Madeira 800m				
		light on		light off		%	light on		light off		%
		mean	SD	mean	SD	difference	mean	SD	mean	SD	difference
fish											
total	vol.	620	206	350	124	+77 *	334	153	187	77	+79 *
largest	vol.	50	24	37	18	+35	75	79	34	21	+121 *
10 largest	vol.	232	95	170	93	+36	136	113	102	69	+33
Crustaceans (decapods)	vol.	(62)	(17)	(101)	(48)	-63	48	14	88	35	-83 *
Cephalopods	vol.	23	25	23	37	0	60	86	46	80	+33
	no.	2	1.5	2	1.6	0	7	3	4	2	+75 *
		C. RMT 50 Madeira 300m					D. RMT 50 Madeira 1500m				
fish											
total	vol.	594	255	652	308	-10	1618	331	1322	429	+22
largest	vol.	53	38	64	77	-21	108	23	131	36	-21
10 largest	vol.	169	87	204	105	-21	562	59	562	123	0
Crustaceans	vol.	239	63	645	170	-170 *	399	104	638	220	-60
Cephalopods	vol.	101	38	122	191	-21	215	144	95	67	+126
	no.	21	7	14	5	+50 *	22	3	19	7	+16

% difference is the difference between the mean values expressed as a % of the lower figure.
Number of samples (on/off) were 15/15, 12/12, 9/9 and 5/5 for A, B, C and D respectively.
* $P<0.05$ in t-tests on the paired comparative samples.
Volumes are in ml.

significant increase found at 800 m. The numbers of cephalopods caught in these nets were often too low to show statistically significant differences but, in general, the numbers tended to increase with use of light. Detailed analyses of the catches from these studies have been made to the species level by Swinney *et al.* (1986, and in preparation) for the fishes and Hargreaves & Clarke (in preparation) for the crustaceans. It was found that individual species appeared to react in different ways to the presence of the light on trawls. Figure 15.2 shows graphs of some fish species (numbers or weights) caught in comparative pairs of hauls taken at 800 m in Biscay. These all show an increase in catches with the light on, thus contributing to the overall result given in Table 15.1. The weight of *Stomias boa* is shown to increase with the light on during the day but decrease with the light on during the night. This effect is due to differences in the size of individuals in each case as the same is not shown in the analysis of numbers. Examples of the effect of lights on individual fish species at different depths off Madeira are shown in Figure 15.3. Although few species inhabit the complete range of depths studied, the results of some species (*e.g. Chauliodus danae*) do follow the overall results for total fish catches (see Figure 15.1), as at 800 m and 1500 m the catch increases with the light, whilst at 300 m the catch decreases with the light.

Further work of this nature has been carried out with a RMT 50 off Madeira, varying the light intensity used by equipping the underwater light with 20 W, 70 W and 150 W lamps. A comparative series of four hauls, one with the light off and one with each of the three different lamps, was repeated 15 times at various times of day. A summary of the results obtained is given in Table 15.2. This large RMT used in the fairly rich waters off

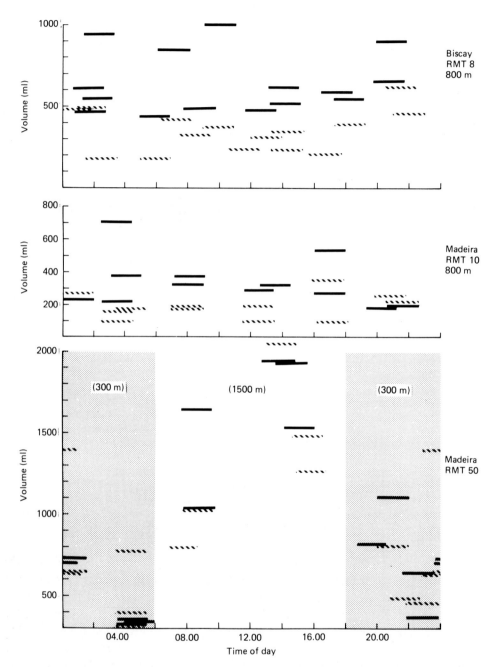

Figure 15.1. Total fish volumes caught by each trawl according to the time of day. Trawls with lights are shown by solid bars, trawls without lights by broken bars. Each bar represents the duration of the horizontal part of a trawl. Shaded area indicates the approximate time of darkness off Madeira. Depths of trawls are indicated in metres. (From Clarke & Pascoe, 1985.)

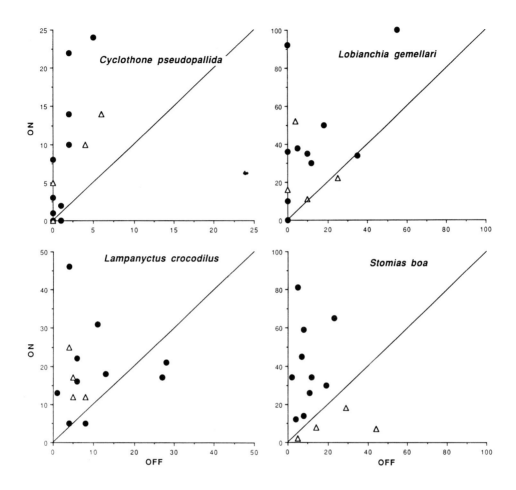

Figure 15.2. The number (left) and weight (right) of four species of fish caught in 14 sets of hauls paired for similar times of day with an RMT 8 in Biscay, plotted as light-on values against light-off values. If the light had no effect on the catches, points would be expected to lie along the lines drawn. ● represent hauls by day and, △ hauls by night.(Redrawn from Clarke & Pascoe, 1985.)

Madeira provided excellent catches of the major taxa, even the cephalopods were caught in sufficient numbers to show significant differences in the hauls. As was found in the earlier work, the use of a light tends to increase the volumes of total fish catches, and those of the ten largest fish, but decreases the volumes of the crustaceans. The results for cephalopods show a significant increase in the number of individuals caught when lights of any of the three intensities were used and there was also an increase shown in the number of species caught. Although there is some indication that the effect of the light, be it positive or negative, is enhanced with increasing light intensity, the results of t-tests run on the samples at different intensities do not show that this is statistically significant. To confirm this, further samples would be required, perhaps with more precise comparative trawling and an increase in the range of light intensities.

More recently, collaboration with scientists at the Institute of Oceanographic Sciences, Deacon Laboratory in Surrey, has enabled the work to be extended to another geographical area and to become much more sophisticated in the control of variables during comparative hauls. The net used was a modification of the RMT 1+8 multi-net (Roe & Shale, 1979), and essentially consists of three RMT 8 nets which can be opened and closed in sequence by means of acoustic signals from the ship (see Figure 15.4). The lights and battery pack are mounted on a bar above the nets and can also be switched acoustically. The net monitor fitted allows fine control of depth during the tow, and the flow meter gives a recording of the effective speed of the net and hence the volume of water sampled can be calculated (Roe *et al.*, 1980). A total of 30 hauls, paired as closely as possible at the same time of day, were made within a depth range of 775 m - 825 m off west Africa (20°N 20°W). Construction of the net used and preliminary analysis of the results are given by

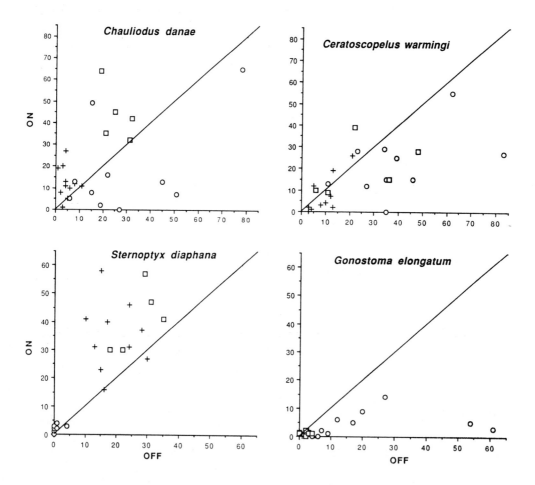

Figure 15.3. The numbers of four species of fish caught in paired RMT hauls at three different depths off Madeira, plotted as light-on values against light-off values. Lines of equal numbers are drawn as in Figure 15.1. + represent paired hauls with a RMT 10 at 800 m, ○ a RMT 50 at 300 m and, □ a RMT 50 at 1500 m.

Pascoe & Aldred (in preparation). Catches were treated initially in a similar way, with the fishes being divided into *Cyclothone* spp. (predominantly *C. livida*) and 'other fish', and the number of individuals as well as the volumes of each category were recorded. A summary of the results is given in Table 15.3. The effect of the light on catches appears very similar to that of previous work at this depth in other regions. All the categories of fish show an increase when the light is used, most being statistically significant. The volume of crustaceans once again decreases in hauls with the light on. The number of cephalopods caught in this series was insufficient for valid comparisons of the paired hauls, but total numbers were increased slightly with use of the light (Table 15.3). Further analyses of these samples to the species level are to be made.

Interpretation of these results is not a simple matter, as many factors may contribute to differences in the catches obtained. It is shown that overall differences in total catches of the major taxa are often attributed to a small proportion of the species involved. Without direct observation or photographic evidence of the behaviour of animals in the path of the trawls any interpretation of the results must remain speculative. The theories put forward for positive phototaxis in marine animals usually relate to a stationary or

Table 15.2. *Comparison between trawl catches with lights of different intensities for RMT 50 hauls at 800m near Madeira*

light/watts		off		20		70		150		all lights		
nom. lumens		0		450		1900		5000		-		
no. of hauls		15		12		15		15		42		
		mean	SD	mean	SD	mean	SD	mean	SD	mean	SD	% diff
total fish	vol.	758	276	1219*	618	1053*	336	1280**	420	1182**	460	+56
10 largest fish	vol.	350	175	453	251	351	159	664	1052	492*	650	+41
Crustaceans	vol.	253	123	149**	41	186	42	167*	62	169**	51	-50
Cephalopods	no.	12.9	4.9	23.0**	9.1	24.9**	13.4	30.6**	11.0	26.4**	11.6	+105

* P<0·05, ** P<0·01 in t-tests on paired comparative samples (lights on *vs* lights off).
% diff, compares all samples with lights on with those with lights off, calculated as in Table 1.
Volumes in ml.

Table 15.3. *Comparison between trawl catches with light on and catches with light off for a multiple RMT8 at 800 m at 20°N 20°W*

		light on		light off		% difference
		mean	SD	mean	SD	
total fish	vol.	145.3	43.7	89.5	32.5	+62.3**
10 largest fish	vol.	55.2	30.8	40.8	25.6	+35.3
Cyclothone spp.	no.	267.7	99.0	166.8	49.7	+60.5**
'other' fish	no.	14.6	3.8	10.2	2.2	+43.7**
Crustaceans	vol.	20.4	5.6	23.9	6.0	-16.8
Cephalopods	no.	0.77	0.34	0.61	0.37	+26.4

Figures are standardised as volume (ml) or number of animals/haul/10,000 m^3 of water sampled.
% difference derived as in Table 1.
** P<0·01 in t-test.
On/off: n = 15/15.

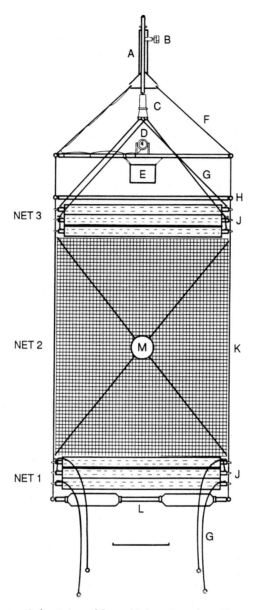

Figure 15.4. Diagrammatic front view of the multiple rectangular midwater trawl (8 m² mouth) fitted with underwater light (RMT 8 ML). It is shown hung in the vertical position with net 2 in the open position, net 1 having opened and closed, and net 3 yet to open. The scale line represents 1 m. (A) Net monitor. (B) Flow meter. (C) Release gear. (D) Light. (E) Battery pack. (F) Towing bridle. (G) Opening and closing bridles. (H) Fixed bar. (J) Sliding net bars. (K) Side wire. (L) Weight bar (291 kg in air). (M) Cod-end bucket.

slowly moving light and some therefore are probably not relevant to the 'unnatural' stimulus of a moving light combined with the other noise and disturbance created by a midwater trawl. Therefore, one can only speculate that perhaps some species are initially attracted to the light source and then at closer range the unnaturally bright illumination dazzles or disorientates them in the mouth of the net effectively preventing an escape response. Those animals which appear to decrease in the catches made with the light on presumably exhibit a negative phototactic response at some distance from the net and therefore move away from the path of the trawl.

Bottom trawls

The same lighting system has also been used on small bottom trawls in the shallow waters (60 m) off Plymouth and at a depth of 1000 m in the Rockall Trough off the west coast of Scotland (56°N 9°W). Results in shallow water (Clarke *et al.*, 1986) show that few species appeared to be affected by use of the lights. In total, seven species show significant differences in catches with and without lights. Three species (*Merlangius merlangus*, *Trisopterus minutus* and *Trachurus trachurus*) showed an increase, mainly in weight, using the lights, but only during the night. Four species (*Eutrigla gurnardus*, *Micromesistius poutassou*, *Merluccius merluccius*, and *Limanda limanda*) showed a decrease in the catches with the lights on. At this shallow depth the effect of the light appeared greater at night, presumably due to the ambient light levels effectively decreasing its contrast during the day. Development of light fishing with this gear seems limited, but possible areas worthy of investigation are the selectivity of the gear and the use of lights to create barriers to herd or guide fish towards a trawl.

Comparative trawling with lights on bottom trawls in deeper water (1000 m) produced several interesting differences in the catches. Preliminary results showing comparisons of the most numerous fish families, crustaceans and cephalopods are given in Table 15.4. Large increases in the numbers of deep-sea sharks (Scyliorhinidae and Squalidae), alepocephalids and notocanthids are obtained when using the light, and, in contrast to the midwater studies, the decapod crustaceans also show an increase in number.

Interpretation of the results must once again remain speculative. A bottom trawl with bridles and boards creating noise and possibly visual signals in the benthic environment,

Table 15.4. *Comparison between number of animals caught with lights on and those caught with lights off in bottom trawls at 1000 m in the Rockall Trough*

	light on		light off		% difference
	mean	SD	mean	SD	
Scyliorhinidae + Squalidae	18.3	18.1	3.6	3.2	+402*
Chimaeridae	53.0	29.1	42.3	29.8	+25
Alepocephalidae	71.6	41.4	18.5	16.4	+287*
Synaphobranchidae	154.1	133.5	170.3	169.6	-10
Notocanthidae	50.5	41.0	8.7	7.9	+480*
Macrouridae	276.5	143.4	275.0	165.3	0
Moridae	51.5	23.4	53.0	32.5	0
Decapod Crustacea	33.3	28.5	9.2	6.6	+262*
Cephalopoda	2.8	2.8	1.9	2.7	+48

* $P<0.05$ in t-tests on paired comparative samples.
On/off: n = 11/11

presents many different interrelating factors which may influence the catch size and composition. Many of the animals living on or near the bottom at 1000 m appear visually well adapted, often with huge eyes, so ambient light, mainly of biolumunescent origin, presumably plays an important role in much of their behaviour. Several factors may contribute to an increase in the catch when using lights:

1) Positive phototactic responses to the light may bring more animals into the path of the trawl.

2) The phototactic responses of some species may in turn attract their predators.

3) Dazzling or sensory blanking effects may reduce the animals ability to make avoidance responses.

Several of the species which show an increase when using the lights *e.g. Centroscymnus coelolepis, Deania calcea* and *Alepocephalus bairdii* are all large fish, so a substantial increase in numbers represents considerable differences in biomass caught by the trawls. These species have little, if any commercial value at present, but the results do suggest that light-fishing at these depths may have a potential other than that of purely scientific interest.

Studies on the visual physiology of some of the species caught during this work (see other sections in this volume) may help to interpret the effects of light on trawls. Little is known of the behaviour of deep-sea animals from direct observation; indirect evidence, such as that provided by experimental trawls with lights, or by inference from the visual anatomy or bioluminescence, is a first step towards understanding the importance of light in the interactions of the deep-sea fauna and their reactions to fishing gear.

Chapter 16

BIOLUMINESCENT COMMUNICATION IN THE SEA

PETER J. HERRING

Institute of Oceanographic Sciences Deacon Laboratory, Wormley, Godalming, Surrey, GU8 5UB

Communication in the broad sense includes a very wide range of interactions between species. Bioluminescent communication in the sea involves the emission of a light signal, its transmission through the seawater, and its visual reception by the target. The sender and receiver may be the same individual (for reflected luminescence) though in the most highly developed systems there are two participants, both of whom seek to maximize the efficient transfer of the information encoded in the signals. Features of the bioluminescence of different groups of marine organisms are considered in this context, and are related, where possible, to associated behavioural evidence for the selective value of the bioluminescent signals.

INTRODUCTION

Optical communication involves the emission of a light signal, its transmission through a medium and the visual reception of the signal. Sender and receiver may be the same individual, but in the most highly developed systems at least two participants are involved, both of whom seek to maximize the efficient transfer of information - or disinformation (Marler, 1968). In well-lit environments the light signals are produced by reflection of daylight or moonlight and the signals modulated by reflective patches or pigments. Important elements of most such displays are the movements, postures and orientation of one or both participants, which may become highly ritualised. The intensity of the signal is limited by the degree to which ambient light can be reflected and the contrast is enhanced by the juxtaposition of reflective and non-reflective areas. Repetitive signals are usually produced by body movements rather than by intrinsic changes in the reflector surfaces themselves. The very rapid colour changes of cephalopods, and of a few fish species, are exceptions to this rule.

The evolution of bioluminescent signals has frequently occurred in environments where the ambient light is too dim to be an effective source of reflected light. The deep oceans comprise the most extensive and ancient of these and it is here that bioluminescent signals are most widely employed. Unfortunately, very little is known about the signals themselves, though the anatomical basis for signal generation and modification is more easily identified (*i.e.* photocytes and photophores). It is ironic that the subtleties of bioluminescent communication are much better understood in the relatively few terrestrial species, particularly fireflies, than in the much more numerous and varied

Herring, P.J., Campbell, A.K., Whitfield, M. & Maddock, L. (ed.). *Light and Life in the Sea.*
© 1990. Cambridge University Press.

marine ones. It is salutary to recognize that although it may be experimentally possible to discriminate between certain bioluminescent signals (*e.g.* Nealson *et al.*, 1986) this does not necessarily indicate that the same criteria are, or even could be, employed by the organisms themselves.

The classification of light signals and interactions has been extended by Lloyd (1977, 1983), following Otte (1974), into twenty four categories of phenomena. These are appropriate to firefly communication but are too specific to be useful in the marine environment, given the present fragmentary knowledge of the signals themselves, let alone of their consequences. The objective of this Chapter is therefore to emphasize the main *observations* of luminous behaviour in marine species and to consider the communicative role of luminescence in this context. More detailed discussion (from different perspectives) can be found in Buck (1978), Morin (1983), Young (1983) and Herring (1986).

BACTERIA

The luminescent species of the three main genera of luminous marine bacteria (*Vibrio, Photobacterium* and *Alteromonas*) are not responsive to stimulation but all glow steadily once induced, though at relatively low individual intensities (10^4-10^5 q s^{-1} cell^{-1}). Seliger (1975, 1987) has argued that the primitive luciferases evolved in order to utilize oxygen directly as an electron receptor in reducing atmospheres (but see McCapra, Chapter 17). If the flavin cofactor involved in the oxidation of aldehydes to fatty acids were $FMNH_2$ the system would be luminescent. If this bacterial 'proto-bioluminescence' were bright enough to elicit responses from motile organisms *new* selective pressures would apply and the light emission would then have a signalling (= communicative) value. Luminous bacteria growing on detritus or faecal pellets, and perhaps derived from the gut flora of larger animals, may attract predators to their light and therefore be more likely to be transferred to a new nutritious environment - the predator's gut - than if they are non-luminous (Ruby & Morin, 1979; Morin, 1983; Andrews *et al.*, 1984). The process of luminescent induction ensures that luminescence occurs only at high densities, when there will probably be sufficient individual cells for their light output to be perceived. It is known that a weak light is attractive to many marine organisms (*e.g.* Nicol, 1959) and illuminated prey may be more attractive to predators than dark prey (Moeller *et al.*, 1972). The vicarious communicative role of symbiotic bacterial luminescence, expressed in those fish and squid with bacterial light organs, could be regarded as the ultimate evolutionary success of the light signal in ensuring the survival of the bacteria. The connection of most bacterial light organs to some part of the gut lumen is an indication of their probable evolution.

DINOFLAGELLATES

Many dinoflagellates produce single short duration flashes (~100 ms) of high intensity (10^8-10^{10} q s^{-1}). The luminescent ability in most species is inhibited by illumination and exhibits a circadian rhythm of responsiveness to stimulation. There is now an impressive

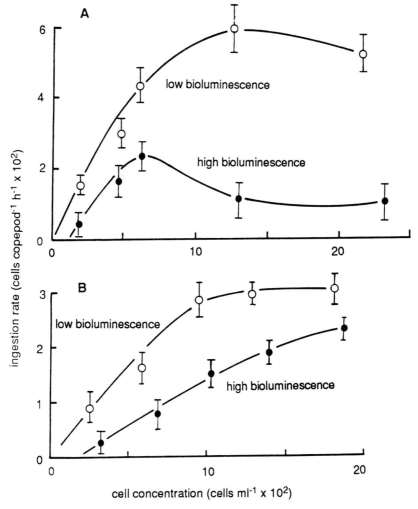

Figure 16.1. The effects of high and low bioluminescence capacity in dinoflagellate populations on their ingestion rate by two species of copepod. (A) *Calanus pacificus* feeding on *Gonyaulax catenella*. (B) *Acartia clausi* feeding on *G. acatenella*. From Esaias & Curl (1972), in which the experimental details can be found.

array of experimental evidence that a dinoflagellate flash, at least at the higher end of the intensity scale, has an effect on the behaviour of such typical predators as copepod crustaceans, and reduces the grazing pressure. Copepods feeding on dinoflagellates have a lower ingestion rate in luminous cultures than in non-luminous ones (Figure 16.1) (Esaias & Curl, 1972; White, 1979), with flash stimulated high-speed swimming bursts in the luminous cultures interrupting feeding activities (Buskey *et al.*, 1983). Direct observations of interactions between copepods and luminous dinoflagellates have confirmed that it is the feeding activities, not the disturbance produced by swimming, that initiates both most of the flashes and the copepods' consequent 'startle' response (Buskey *et al.*, 1985). Although changes in shear, acceleration and pressure can induce dinoflagellate luminescence (Anderson *et al.*, 1988) copepod swimming by itself does not produce stimuli of sufficient magnitude. Simulated dinoflagellate flashes produce a similar behaviour, indicating that it is the flash itself, rather than any other dinoflagellate attribute (*e.g.* toxin), that initiates the response in the copepod (Figure 16.2); the action spectrum for the behavioural response of two copepods has a maximum close to that of the spectral emission of the dinoflagellate (Buskey & Swift, 1983, 1985). In itself this does not identify a selective advantage for the individual dinoflagellate, but the presumption must be that in many such encounters the individual escapes being eaten. While the justification, in selective terms, for such high intensity flashes (*e.g.* of *Pyrocystis*, *Noctiluca* and *Gonyaulax*) is attractive, there are many other dinoflagellates with much lower luminescence intensity, the efficacy of whose flashes in deterring a predator has yet to be demonstrated. Although it is probable that copepods provide a major component of the grazing pressure on dinoflagellates there is no information concerning the responses (if any) of other dinoflagellate consumers.

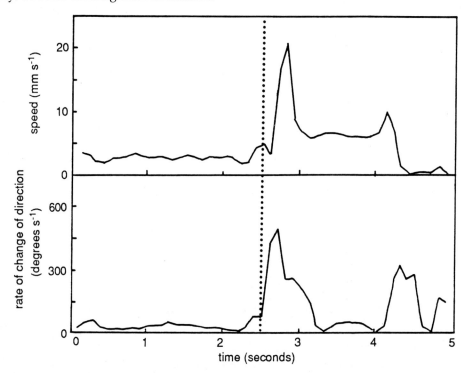

Figure 16.2. The effect of a simulated dinoflagellate flash (dotted line) on the rate of change of direction and swimming speed of a single copepod *Calanus finmarchicus*. From Buskey & Swift (1985).

A second communicative element of dinoflagellate luminescent signals is that which constitutes the 'burglar alarm' effect, by which a potential predator is highlighted by the luminescence and thus becomes more vulnerable to its own predators (Burkenroad, 1943). This is a consequence often attributed to dinoflagellate flashes, for the illumination of the outlines of fish or dolphins in water containing large numbers of luminous dinoflagellates is an easily observed phenomenon (Porter & Porter, 1979; Buck, 1978;

Morin, 1983). Indeed Hobson *et al.* (1981) suggest that the attack style of some crepuscular and nocturnal predators is adapted to minimize the stimulation of luminous dinoflagellates in the vicinity. It has also been suggested (Koslow, 1979) that the timing of the diel vertical migrations of some copepods, which arrive at the surface for feeding *before* sunset, is a strategy to avoid visible interactions with luminous dinoflagellates and therefore to decrease nocturnal predation.

CNIDARIANS AND CTENOPHORES

Despite the variation and complexity of the bioluminescent responses of cnidarians and ctenophores the communicative role of the signals is almost completely unknown. Indeed Morin (1974) concluded that "In no single case is there evidence for the role that bioluminescence may play in the interactions of these coelenterates with their predators, competitors, prey or among themselves". The simplest observed luminous responses are the local unpropagated flashes of some hydromedusae (*e.g. Aequorea*) (Davenport & Nicol, 1955) while in anthozoans, particularly sea pens, single flashes may propagate across the surface of the colony as a wave of light (Nicol, 1958*b*; Bilbaut, 1975); multiple flashes may also occur at each site and in certain situations multiple waves of light may result from a single stimulus (Nicol, 1955; Buck, 1973). The waves are propagated in some cases by epithelial conduction, *e.g.* in the siphonophore *Hippopodius* (Bassot *et al.*, 1978), but in others the propagation is via the neural network. Uncoupled multiple responses can also be observed in some scyphozoans (*e.g. Pelagia, Atolla*) but the most elaborate responses are probably those of hydropolyps, siphonophores and ctenophores, in which bursts of flashing occur at each site (80-100 ms flash duration in *Obelia*) and the flashing propagates as one or more waves of luminescence over the individual or colony (Chang, 1954; Morin & Cooke, 1971*a, b*; Widder *et al.*, 1989). In ctenophores the photocytes are localized in the comb plates (Freeman & Reynolds, 1973); siphonophores are more typical of the Cnidaria (Morin, 1974) in that some have widely distributed photocytes (*Hippopodius*) whereas in others (*e.g. Nanomia*) they are localized (Freeman, 1987). In addition some hydrozoans, scyphozoans and ctenophores can emit luminous slime or particulate material (Widder *et al.*, 1989).

There is no evidence in these animals that any luminescent signal is initiated by, or directed at, another individual of the same species (but *cf.* Labas, 1977), despite the fact that some luminous species have ocelli. Morin (1976) has analysed the physical, behavioural and directional characteristics of pennatulid bioluminescent displays. He concluded that the luminous signals are probably directed at visually orientating mobile predators, such as demersal fishes, and may additionally provide a warning of nematocyst defensive capabilities. A similar view was taken by Kastendiek (1976), working with *Renilla*, although he concluded that the luminescence did not affect invertebrate predators. Direct proof of this had been provided by Bertsch (1968) who noted that when the nudibranch *Armina* fed on *Renilla* it was not affected by the luminous waves it stimulated.

The very limited number of other observations on the communicative value of coelenterate luminescence also support its primarily defensive value (*e.g.* Widder *et al.*,

1989). The deep-sea hydromedusan *Colobonema sericeum* Vanhöffen has been repeatedly observed from submersibles to produce waves of light when touched and when more vigorous attempts were made to capture it the animal shed all or parts of its tentacles as a luminous mass while the darkened bell swam off (Robison & Widder, unpublished). The use of light signals in an apparently equivalent decoy role is a feature of some siphonophores (Mackie *et al.*, 1987) and of certain ctenophores which will eject luminous material (*e.g. Mertensia, Eurhamphea, Ocyropsis*) (Hamner *et al.*, 1975; G.R. Harbison, personal communication). The multiple flashes of ctenophores and siphonophores are among the brightest of luminous signals (Widder *et al.*, 1989) and their value as a defensive tactic is tacitly assumed, but has never been demonstrated. Relatively few predators of planktonic coelenterates (*i.e.* ctenophores, siphonophores, medusae) have yet been identified, though some ctenophores (*e.g. Beroe*) feed extensively on others. Observation of whether such interactions initiate luminescent signals, and whether the signals produce any response, would be most interesting.

The multiple flashes of many cnidarians and ctenophores raise the possibility of information being encoded in the flash frequency or spatial pattern. There is no indication that this is so, and the assumption must be that repetitive flashing reinforces the deterrent effect on a predator (providing it has the appropriate visual acuity and flicker fusion frequency to discriminate the individual flashes and their sites). However, the multiple responses may be more a consequence of the organization of the neural or epithelial conductive pathways than directly selected for their repetition rates or spatial patterns.

ANNELIDS

Many marine annelids are luminous (Herring, 1978), including a littoral oligochaete (*Pontodrilus*). The expression of the bioluminescence is as varied as the taxonomic diversity of the emitting species, but its communicative role is little known and there are very few reports of natural, as opposed to experimentally induced, luminescence. Many luminous annelids (*e.g.* cirratulids, terebellids and *Chaetopterus*) are tube dwellers and produce a luminous slime as well as being able to flash rapidly. *Chaetopterus* can eject the luminous mucus from its tube and the light probably has a defensive signal value, though Morin (1983) considers it may also have a role in reducing the intrusion into the tube of undesirable photophobic inquilines. The physiological responses occurring on experimental stimulation of *Chaetopterus* (Nicol, 1952, 1954) and similar annelids demonstrate the expected relation between stimulus strength and bioluminescent response (until fatigue sets in). Although multiple flashes can be produced by a single stimulus, more complex flash patterns do not occur.

The bioluminescent signal of the polynoids (scale worms) is more complex. On initial stimulation there is a rapid single flash in the scale, which may propagate down the body. On more intense stimulation the scales are shed, each one flashing repetitively for tens of seconds. The flash frequency of 1-5 s^{-1} shows no experimental consistency and changes gradually (Nicol, 1953; Bilbaut & Bassot, 1977) and presumably has no informational content, though enhancing its power as a diversionary object.

Syllid worms demonstrate a bioluminescent emission which clearly transcends the simple communicative role of the signals noted so far and which involves a basic luminescent dialogue between males and females. Sexual swarming occurs in species of *Odontosyllis* at particular periods of the moon. In Bermuda the females appear first, swimming in tight circles and emitting a greenish luminous secretion. Males are attracted to them and circle round, producing short intermittent flashes, and the gametes are shed in a luminous cloud. The females have a brighter light than the males, and the males orient visually to them (they can also be attracted to a hand-held light). The communicative role of the males' luminescence is not clear; it may provide the appropriate trigger for the shedding of eggs or merely maintain the sexes in the same vicinity while other stimuli initiate the shedding of gametes (Markert *et al.*, 1961). The behaviour of *O. phosphorea* in California is different; the males appear first and are joined later by the females, both sexes flashing as well as producing a steady luminescence over the body and a luminous secretion (Tsuji & Hill, 1983). The maximum visual sensitivity of *Odontosyllis* is close to the bioluminescence emission maximum (510 nm) and different from that of two non-luminous polychaetes (Wilkens & Wolken, 1981).

CRUSTACEA

There are bioluminescent species of ostracods, copepods, amphipods, euphausiids, mysids and decapods (Herring, 1985). The expression of the luminescence varies markedly between species and the most complex behavioural patterns of bioluminescence reported in any marine organism are those identified in certain ostracods of the genus *Vargula*. *In situ* observations of these animals have given a remarkable insight into the parameters that can be employed in the displays, which approach the intricacies of those observed in lampyrid fireflies (Morin, 1986; Cohen & Morin, 1986, 1989; Lloyd, 1983).

The simplest luminous response of these ostracods is the production of a cloud of persistent blue luminescence when attacked by a predator. This signal probably distracts or confuses an attacker and the escape of the ostracod has been observed in several encounters (Morin, 1983). A much more complex display pattern usually occurs for a limited period after twilight and involves males only (Morin, 1986). Trains of long-lasting (several seconds) luminescent pulses are secreted into the water by a rapidly swimming male, who is usually accompanied by a swarm of satellite, non-signalling, males. The time interval between individual pulses of light (and hence the distance, in a male moving at constant speed) usually reduces along the display path, as does the duration of each pulse. Long, widely separated pulses gradually change into short pulses bunched close together. Two different groups of *Vargula* species emphasize different elements of the display: one emphasizes the first part of the display with its shortening pulse trains, the other the later part, with its more even pulse trains. One species, *V. graminicola*, is in a third category in which *all* the males in a cluster produce very short pulses (280 ms) in entrained synchrony with each other. This species' display path is vertically upward, but others may have upward, downward, horizontal, oblique, zig-zag or L-shaped pathways (Figure 16.3). Different species include within their displays a number of environmental, temporal, spatial, social and intensity variables, providing each with a unique suite of characteristics. Females are presumed to be attracted by the appropriate specific displays but do not respond with any luminescent signal of their own. A luminous pulse, or train of pulses, in one display may stimulate another male a few metres away and initiate a second, synchronous, display by leader-follower entrainment. This will produce a wave front of entrained synchrony in evenly spaced populations. The long duration of each pulse, relative to the interval, results in an apparent synchrony of the signals over a part of the display path.

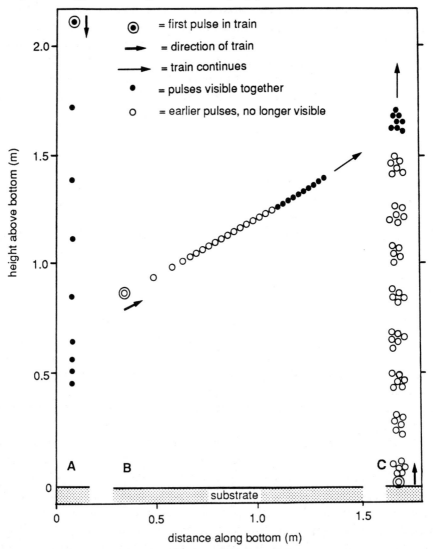

Figure 16.3. The three main types of luminescent displays by Caribbean species of the ostracod genus *Vargula*. (A) *V. shulmanae* showing a progressively shortening downward train. (B) *V. contragula* showing an evenly spaced oblique train. (C) *V. graminicola* showing a group pulsing upward train. From Morin (1986).

It is possible that the suite of display characteristics is a more accurate indication of specific identity than is the combination of more usually employed morphological characters. *Vargula kuna*, for example, has two display types which differ in a number of characters, the most obvious being that one is upward and one downward. The two groups of individuals are not morphologically distinguishable and may represent two sibling or cryptic species (Morin & Cohen, 1988; Cohen & Morin, 1989).

The initiation of one male's display by the luminescence of another provides the behavioural basis for the observation that *Vargula* species respond to a bright artificial light by luminescing, though not in any identifiable display pattern (Tsuji *et al.*, 1970). Similar luminescent responses to illumination have been observed in Indian Ocean populations of *Cypridina*, particularly *C. dentata*, though there is no indication as to which sex was involved. During copulation both sexes of this species are reported to luminesce brightly (Daniel & Jothinayagam, 1979). There have been no reports of displays in this oceanic species comparable to those in the shallow-water Caribbean species of *Vargula*.

There are indications of a correlation between eye size and luminescence in *Vargula* species. Males have larger eyes than females in most species and there may also be a link between eye size and the complexity of the display (J.G. Morin, personal communication).

The widespread oceanic genus *Conchoecia* contains many species with specific patterns of luminous glands (Angel, 1968) and with the capability of producing either a luminous secretion or intraglandular luminescence. There are no data on the type of signals produced *in situ* by these animals.

Luminous species of copepod are to be found in at least seven families (Herring, 1988*b*). Most of these have luminous glands which secrete material into the surrounding seawater. The few available observations on the interactions of live animals indicate that the light signal is primarily a defensive one, induced by a predator (David & Conover, 1961). Three potential predators, however, were unaffected by simulated copepod luminous flashes (Buskey & Swift, 1985). The glandular secretions do not form a discrete cloud, like those of *Vargula*, but are more particulate in appearance and are produced with simultaneous rapid escape movements. Observations of both free-swimming and tethered animals have shown that the luminescence may sometimes remain associated with the aperture of the glands; the sudden appearance of the luminescence and the rapid movement of the animal may contribute to the confusion of a predator. In the smallest luminous copepod, *Oncaea conifera* (Giesbrecht) (<1·5 mm), the luminescence is wholly intracellular and live animals will flash repeatedly during a short burst of swimming activity, producing a spatially separated train of flashes. Copepod luminescence pulses have decay times of up to several tens of seconds (*e.g. Gaussia*, Barnes & Case, 1972) but are generally of the same order (1-5 s) as those of ostracods. Copepods have not been reported to respond directly to light stimuli. Experimental studies of the responses of copepods to their own simulated luminescence (600 ms flashes) have shown that coastal and oceanic species both exhibit 'startle' reactions; this has led to the suggestion that copepod luminescence may have a 'burglar alarm' value (Buskey & Swift, 1985; Buskey *et al.*, 1987).

Bioluminescent signals from amphipods have not yet been observed under *in situ* conditions. However, the luminescence is usually intracellular and often located at the extremities of the limbs and/or body (Herring, 1981). When they are held, species of the genus *Scina* adopt a particular rigid posture, with the extremities extended, and flash brightly from the distal tips of the elongate limbs. Some species eject a coloured fluid from the mouth at the same time. This luminescent signal appears to be a defensive response and the secretion may perhaps be distasteful. The luminescence of other amphipods can be considered as a similar response, though the peripheral limb luminescence of *Megalanceola* could be interpreted as mimicking the radial symmetry of a luminous medusa (Herring, 1981).

Euphausiid crustaceans, with the exception of *Bentheuphausia* and some deep-living species of *Thysanopoda*, all have ventrally located light organs on the eyes, thorax and abdomen, with some generic differences in light organ numbers (Herring & Locket, 1978). Despite their abundance, and the observations that have been made on captive animals, little is known of the ecological value of the luminescence. It is widely presumed that the light acts as a counter-illumination system, eliminating the animal's silhouette against downwelling daylight, though the evidence on which this is based is still circumstantial. Light organs are absent or reduced in deep-living species, the spectral and angular distribution of the luminescence is appropriate for such a function and both the mobile body light organs, and the eyes with their rigidly coupled light organs, can be rotated (in response to changes in the angle of ambient illumination) to maintain their ventral directionality during changes in the animals' orientation (Herring & Locket, 1978; Land, 1980; Grinnell *et al.*, 1988). Euphausiids will respond to another individual's luminescence and additional communicative functions have been suggested in schooling behaviour (Mauchline, 1960; Boden *et al.*, 1961) and sexual interactions. The latter role is based on sexual and seasonal differences in the bioluminescent response to electronic flash stimulation (Tett, 1972) and to the sexual dimorphism of the photophores in the genera *Nematoscelis* and *Nematobrachion*.

The only luminous genus of mysids, *Gnathophausia*, produces a cloud of light (λ_{max} 485 nm) into the water from glands on the maxilla (Widder *et al.*, 1983). This is clearly a defensive reaction and is easily stimulated by handling. The spectral sensitivity of the animal is maximal at 510 nm, has a broad bandwidth (Frank & Case, 1988*b*) and shows no close correlation with the emission characteristics, in keeping with the latter's interspecific defensive role. Many decapods emit a similar, but bluer (λ_{max} 460 nm), secretion from the mouth with the same defensive signal value (Herring, 1983; Herring & Roe, 1988). Beebe (1935) described how "a rocket-like burst of fluid was emitted with such violence that the psychological effect was that of a sudden explosion". Species of *Stylopandalus*, *Thalassocaris*, *Chlorotocoides* and *Sergestes* have ventrally directed light organs derived from hepatopancreas tubules. The signal from these organs clearly has a counter-illumination value. It has been shown for *Sergestes similis* that the intensity of luminescence follows that of ambient overhead light, that the different light organs respond in synchrony and, as in euphausiids, that the light organs are rotated to compensate for changes in body orientation (Warner *et al.*, 1979; Latz & Case, 1982).

Epidermal light organs are present in some penaeid shrimps and in the meso- and bathypelagic genera *Sergia, Gennadas, Oplophorus, Janicella* and *Systellaspis*. They are not movable, but are directed ventrally, and their light has the appropriate angular and spectral distribution (notably different from that of the secretion) to support their use as a counter-illumination signal. Their anatomical distribution has similarities with that of euphausiid light organs and they are smaller, or absent, in deep bathypelagic species. Their distribution has a slight sexual dimorphism in some species of *Sergia*. There are no observations on their *in situ* employment but in a recent analysis of spectral sensitivity those genera with cuticular photophores (*Oplophorus, Systellaspis, Janicella*) had peak e.r.g. sensitivities at 400 nm and 500 nm, whereas those with only secretory luminescence (*Notostomus, Acanthephyra*) had a sensitivity maximum at 500 nm only (Frank & Case, 1988*a*). The authors suggest that this might allow a spectral discrimination between light organ luminescence and secretory luminescence.

MOLLUSCA

The bivalves *Pholas* and *Rocellaria* secrete clouds of luminous mucus into the water when disturbed in their burrows, and the response, like that of the polychaete *Chaetopterus*, is presumed to have a deterrent value against predators that may attack the exposed siphons. The luminescence of the three genera of luminous nudibranchs and one marine pulmonate is also presumed to be a defensive signal based on mechanical stimulation (Herring, 1988*a*).

The variety of light organ structure and distribution in cephalopods is second to none, but unfortunately it is not matched by observations of luminous interactions in living animals. Several species of loliginid and sepiolid squids have bacterial light organs in the mantle, the contents of which can be vigorously squirted out by *Heteroteuthis*. Luminescence can also be internal; the intensity can be modified (Haneda, 1956) and may act as a counter-illumination system (Young, 1977). Luminescence is reported to be produced during copulatory behaviour of one species of *Loligo* (Hamabe & Shimizu, 1957). The match of ambient light by the ventral photophores of *Abralia* and *Abraliopsis* (over a 15,000-fold intensity range, Figure 16.4) provides powerful evidence for counter-illumination (Young & Roper, 1976, 1977; Young *et al.*, 1980). This is supported by the finding that spectral changes induced by temperature correlate with the requirements for a spectral match of daylight at their daytime depth and of moonlight during their nocturnal migration closer to the surface (Young & Mencher, 1980). Apart from ventral light organs such as these, present in so many mesopelagic squid, there are only anecdotal accounts of other light organ signals, mainly of short flashes from specific organs in response to mechanical or electrical stimuli (Young *et al.*, 1982; Herring, 1988*a*).

Some cephalopods have a complex sexual dimorphism of their light organs (Herring, 1988*a*). Indeed the males of *Lycoteuthis* were, until recently, regarded as a separate genus (*Oregoniateuthis*), partly on the basis of the light organ patterns (Toll, 1983). Sub-adult and mature females may develop additional arm-tip light organs or, in some octopods, a unique circumoral light organ. Males of *Ctenopteryx* develop an additional posterior light organ. There is no evidence how these organs are employed during *in situ* communication. Despite the spectral variations in the bioluminescent emission maxima of cephalopods (Herring, 1988*a*) there is no evidence of any correlated visual adaptations other than the recent suggestions of Matsui *et al.* (1988*a*) for *Watasenia*. The immense morphological and distributional variety of cephalopod photophores reflects a plethora of communicative opportunities. It is reasonable to assume that the behavioural potential matches the anatomical diversity.

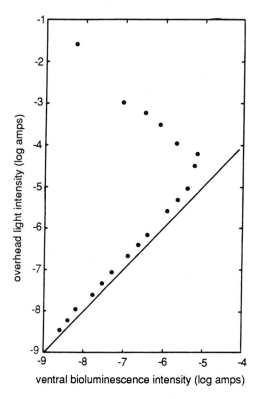

Figure 16.4. Ventral bioluminescence of the squid *Abraliopsis* sp. exposed to different overhead light intensities. The solid line indicates the expected values if the match is perfect. At the higher overhead intensities the animal can no longer match the background with its bioluminescence. From Young *et al.* (1980).

ECHINODERMS

There are many luminous echinoderms (Herring, 1974) but only for ophiuroids have *in situ* observations been possible and rapid flashing been reported. In view of the absence of image-forming photoreceptors in the echinoderms it must be presumed that the light signals are directed at potential predators, either as a startle response or an aposematic warning of distastefulness.

Many ophiuroids emit short (60-70 ms) flashes at frequencies of 8-10 s⁻¹ and successive waves of light may propagate along each arm in response to a single stimulus (Brehm, 1977). It has been possible to document the luminous responses of two species of

Ophiopsila to encounters with other species in the community, and to carry out some experimental manipulations (Grober, 1988*a, b*; Basch,1988). Despite the fact that these ophiuroids occur in dense populations the response of one individual does not initiate that of another. All the observations indicate that the flashes are directed at other species during either deliberate attacks or accidental collisions. Fishes and large crustaceans trigger graded responses, from a weak glow to multiple flashes. As a result predator behaviour is changed (Figure 16.5), with increased changes of direction. Predators in experimental conditions which feed readily on *Ophiopsila* by day, do not do so at night, when luminescence would be visible, but do attack non-luminescent ophiuroids (and anaesthetized *Ophiopsila*) under these conditions. In dark encounters blinded predators (portunid crabs) damaged more *Ophiopsila* than did visually intact specimens. There is some evidence that *O. riisei* is unpalatable and that the flash is an aposematic signal whose relation to unpalatability can be learned by a crab without necessarily involving the death of the prey. The dense aggregations of ophiuroids may reinforce the deterrent effect and even provide a refuge from predators for smaller animals which do not cause, or are not responsive to, defensive flashes.

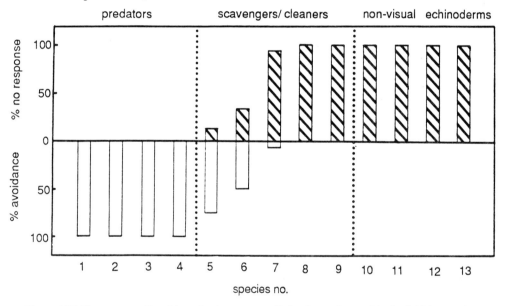

Figure 16.5. The percent of 'avoidance' or 'no response' behaviours observed in the field for a variety of coral reef animals in response to luminescent flashing by the ophiuroid *Ophiopsila riisei*. Species nos. 1, 6, 7 are crabs, 2, 5, 8, 9 are shrimps, 3 is an octopus, 4 a fish, 10 an asteroid, 11 an ophiuroid, 12 and 13 are echinoids. From Grober (1988*b*).

The luminescence response may persist in an autotomized arm, providing the predator with a decoy signal and the potential prey with time to escape. The distal propagation of luminous waves also leads predators to the less vital arm tips. Morin (1983) also noted the potentially stunning effect of an *Ophiopsila* flash on a nereid worm and the 'burglar alarm' effect of increased predation on young fish after they had initiated flashing by the ophiuroids.

TUNICATA

Luminescence in pelagic tunicates is probably limited to the colonial *Pyrosoma* and to certain larvaceans. A *Pyrosoma* zooid responds to a stimulus with a slow flash which in turn may stimulate adjacent individuals (or even colonies) resulting in the propagation of a wave of light over the colony. The long duration (>10 s) of a response often results in the whole colony glowing brightly. The onset of luminescence is preceded by the arrest of cilia on the gill bars. This could be advantageous in a defensive situation if the ciliary arrest also results in sinking and in the consequent movement of the colonies out of an unfavourable environment (Mackie & Bone, 1978).

Larvacean luminescence is very different. Certain species produce granular inclusions in the house rudiment. These inclusions luminesce, and remain potent even when the house has been abandoned. The presence of the inclusions is correlated with the presence of oral glands but there is no evidence that any other tissue is luminous. The communicative role is not clear. The inclusions may act as a distraction to a predator while the animal leaves the house. The discarded houses subsequently provide the equivalent of a planktonic minefield (Galt, 1978; Galt & Sykes, 1983; Galt *et al.*, 1985).

FISHES

The light organs of fishes are at least as diverse as those of cephalopods. Their contribution to the channels of communication in the sea will be even greater as a consequence of the greater abundance of luminous fishes at all depths. Observations (and interpretations) of luminous behaviour are tantalisingly meagre. Many of the shallower species employ luminous bacteria as their light sources, maintained in specific organs which are often associated with the gut lumen. The most detailed observations have been made on the anomalopids and leiognathids, which exhibit some common features (Morin *et al.*, 1975; Herring & Morin, 1978; McFall-Ngai & Dunlap, 1983). The anomalopid genera *Photoblepharon* and *Anomalops* have been observed by divers to emerge from dark caves or crannies to forage over the reef when the moon is down. The fishes have a large bacterial light organ under each eye which can be rotated or occluded to turn off the light (described by Morin *et al.*, (1975) as a 'blink'). The fishes move as male/female pairs, or as larger groups over the reef and the luminescence is used to locate planktonic prey, to maintain the school, to attract prey to the luminescent area produced by the school and as a defensive pattern of 'blink-and-run' behaviour when under threat. In addition the male/female pairs away from the main schools showed distinct patterns of luminescent 'blinking', indicating that sexual communication was an important element of their behavioural repertoire, as well as territorial defence. Morin (1981) suggests that the use of luminescence enables these fishes to utilize smaller planktonic prey than other nocturnal fishes. The efficacy of the luminescence in locating plankton is supported by the fact that animals which have lost their luminescence (in aquaria) have great difficulty locating prey, until provided with equivalent levels of artificial light (McCosker, 1977).

The behavioural versatility of the anomalopids in the employment of their single pair of light organs is approached by the leiognathids (McFall-Ngai & Dunlap, 1983). These fishes have a single circum-oesphageal light organ whose output is controlled directly by a muscular shutter and indirectly by other tissues, including the swimbladder, ventral musculature and melanophores. The intensity of the ventral luminescence can be modulated in response to overhead illumination, thereby providing a counter-illumination camouflage (Hastings, 1971; McFall-Ngai & Morin, unpublished). A flash from the opercular region is produced in response to handling and free-swimmimg specimens of *Gazza minuta* can also produce a pair of collimated beams of light, projected antero-ventrally from the posterior margin of the opercular cavity. Such displays may aid in prey capture or in schooling communication between individuals (*cf.* anomalopids). These fishes also produce a short bright flash over the lower part of the body in response to a moving object in their path, perhaps as an anti-predator signal. The parallels between the luminous behaviour of leiognathids, anomalopids and monocentrids (all of which employ luminous bacteria) have been considered by McFall-Ngai & Dunlap (1983). Marked sexual dimorphism of the light organs occurs in several species of leiognathid (Haneda & Tsuji, 1976; McFall-Ngai & Dunlap, 1984) but no correlated luminescent behaviour patterns have been reported.

Fragmentary though the observations of luminous behaviour are in these particular fishes, they nevertheless exceed those available for any others. Most oceanic fishes do not employ bacterial light sources, but have their own intrinsic systems. The anglerfishes provide one of the few exceptions to this rule. In their case the bacterially illuminated esca of the females has always been assumed to be a lure - but there is no direct evidence for this. Observations on the use of the non-luminous 'lure' in shallow water ceratioids (Pietsch & Grobacker, 1978) provides compelling circumstantial evidence, but whether the subtleties of escal architecture encode any specific information as well as a generalist 'bait' mimicry is unresolved. Anglerfish can flash the escal light source, and it may have a greater communicative repertoire than initially anticipated. Its sexual dimorphism has encouraged the view that it may provide a communicative beacon for the male, but evidence is lacking.

Experimental studies of captive specimens of lanternfish and hatchetfish, in particular, have provided very strong evidence for the counter-illumination function of ventral luminescence. Live lanternfish modify the intensity of their emission (in a matter of seconds) in response to changes in ambient light intensity (Case *et al.*, 1977; Young *et al.*, 1980). The responses are mediated through a feedback system involving both the eyes and the pineal organ. The spectral distribution of the light of myctophids and hatchetfish is an accurate match of that of downwelling daylight in the sea and the angular distribution of the luminescence of hatchetfish (and others) also matches that of light in the sea (Denton *et al.*, 1972, 1985) (Figure 16.6). Although the selective value of counter-illumination has not yet been demonstrated *in situ* (Buck, 1978) the experimental evidence is now overwhelming that this is at least one of the communicative roles of luminescence in mesopelagic fishes (Herring, 1982). The versatility of the single bacterial light organs of the anomalopids and others should caution against the assumption that counter-illumination is the *only* function of ventral light organs.

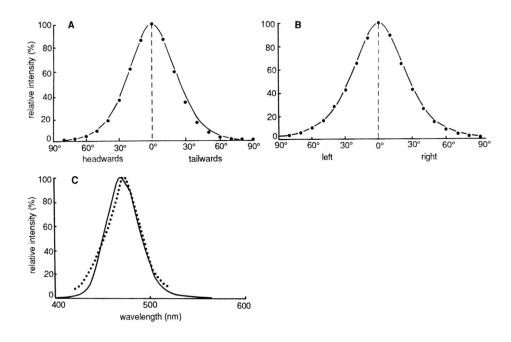

Figure 16.6. Comparison between the bioluminescence of the hatchetfish *Argyropelecus* and the ambient light in the sea. (A,B) Angular distributions of bioluminescence in the longitudinal (A) and transverse (B) planes for *A. affinis*. Solid line is a computed curve for daylight in the sea. (C) Spectral distributions of the bioluminescence of *A. aculeatus* (points) and light in the sea at about 500 m (solid line). From Denton *et al.* (1972, 1985).

There is as wide a range of light organ structures in fish with intrinsic systems as there is in cephalopods, and there may be several thousand organs on one individual. In very few cases has the luminescence even been observed, let alone in any behavioural framework. Even for the shallow water fish *Porichthys* there are only brief reports of a potential anti-predator flash response (Lane, 1967) and of its use by a male in a courtship display with a female whose luminescence had been artificially induced (Crane, 1965). Nevertheless the surprisingly close match between the spectral distribution of the luminescence of *Porichthys* and the spectral sensitivity of the fish (Figure 16.7) implies its use as an intraspecific communication system (Fernandez & Tsuji, 1976). The substantial database now available on the experimental physiology of *Porichthys* light organs is isolated from their behavioural employment. The latter cannot readily be extrapolated from the former.

One group of midwater fishes, the platytroctids (searsiids) have one type of light organ which can squirt particulate scintillating material into the water in what is a presumably defensive behaviour akin to that in some ctenophores. However, this is an extremely rare luminescent mode in fishes, many of which have numerous different types of light organ presumably for different purposes. Sexual dimorphism is common, for example in the

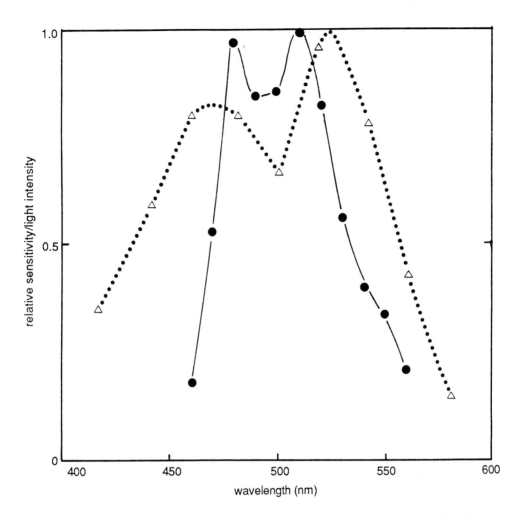

Figure 16.7. Comparison between the spectral sensitivity of the eye (150 μV criterion) of *Porichthys notatus* (dotted line; triangles) and the *in vivo* bioluminescence emission spectrum (solid line; filled circles). From Fernandez & Tsuji (1976).

caudal organs of myctophids and the postorbital organs of stomiatoids. Beebe & Vander Pyl's (1944) report that they obtained flash responses from the caudal organs of mycto- phids to the light of a luminous watch provides circumstantial support for potential intraspecific communication of this sort (Beebe, 1934). The specific distribution of patches of luminescent tissue, and of the light organs themselves, on myctophids and astronesthids (Figure 16.8) has prompted many suggestions concerning their use as intraspecific markers (*e.g.* Beebe & Vander Pyl, 1944; Goodyear & Gibbs, 1969) but there is still no behavioural support for the suggestions. Many of these organs have high frequency flashes (up to 10 Hz) (*e.g.* Barnes & Case, 1974) but there is no evidence for any specific information-coding in the flash frequencies.

Stomiatoid fishes have many different types of light organ. Several of these have been observed to luminesce in captured specimens, including the spherules present on the fins, tail and along the dorsal and ventral margins of the body. In this tissue it may appear as a glow or a train of flashes and is believed to be communicating a warning (perhaps of size) by outlining the fish's silhouette (O'Day, 1973). Among the organs responding in this way are those in the barbel. The elaborate luminous structures of the barbels in many stomiatoids, and some ceratioids, have long been interpreted as lures. There is no observational evidence for such a function but the complexity of specific barbels has

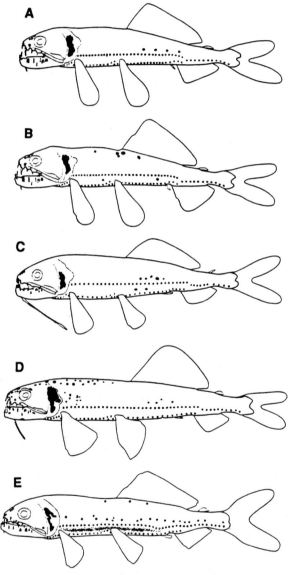

Figure 16.8. Distribution of luminous tissue patches (black) in four species of *Astronesthes*. (A) *A. cyaneus* - Pacific Ocean. (B) *A. cyaneus* - Indian Ocean. (C). *A. lamellosus*. (D) *A. macropogon*. (E) *A. micropogon*. From Goodyear and Gibbs (1969).

encouraged speculation on the potential for mimicry of particular prey species. A large suborbital light organ is characteristic of most stomatoids; when live animals are stimulated mechanically these organs may flash rhythmically with a slow (1-2 Hz) frequency (*e.g.* Harvey, 1952; Herring & Morin, 1978) and in animals in very good condition all the light organs may flash synchronously, providing a most dramatic response. Perhaps the most remarkable of the large light organs are the suborbital ones of *Malacosteus, Aristostomias* and *Pachystomias* whose light is limited to long wavelengths (λ_{max} 708 nm), and whose visual systems are unusually adapted to sense such wavelengths, thereby providing these fishes with a 'private' bandwidth with which to communicate with each other, or to locate red or orange prey species (Widder *et al.*, 1984; Bowmaker *et al.*, 1988).

GENERAL CONSIDERATIONS

Observations of luminous behaviour in marine organisms are exceedingly limited, unlike speculations. Many of the interpretations of the communicative(?) interactions are necessarily anecdotal, non-rigorous and anthropomorphic. This is inevitable in present circumstances, in which the paucity of *in situ* observations can lead to an exaggerated view of the equivalence of *in vitro* experimental data. Nevertheless there is a realistic basis for the belief that bioluminescence has an important communicative role in the sea.

Bioluminescent signals are only advantageous if they are received. The reception of the signals must be the main selective force for the development and maintenance of the visual systems of organisms in the sea at depths below the penetration of significant surface light. The adaptations of such visual systems may therefore provide clues to the signal value of bioluminescence in this environment. The general problems of vision underwater are dealt with elsewhere in this volume (Denton, Chapter 8; Land, Chapter 9) and it is sufficient to reiterate here that any light signal will be degraded by the combined effects of absorption and scattering, both of which are spectrally selective. Blue-green light is most effectively transmitted through clear oceanic water. These wavelengths are therefore best employed for long-range communication. Signal loss at short range is dominated by inverse square law attenuation, which is not spectrally selective. A wider range of wavelengths can therefore be effectively employed for communication at short range, either in the form of a broad bandwidth emission or as a narrow bandwidth of longer or shorter wavelengths. The ecological effectiveness of a light signal depends on its reception, absorption and transduction in the receiver's eye (and brain). The absorption of the visual pigments should be high over the waveband of the signal, though the correlation between the spectral characteristics of the signal and the absorption of the pigment need not be exact, particularly when the density of the visual pigment is high. Thus Hiller-Adams *et al.* (1988) calculate that the visual pigment of the deep-sea shrimp *Systellaspis debilis* (A. Milne-Edwards) (λ_{max} 493 nm) would absorb 99% of axially incident blue light (475 nm), whereas an equivalent visual pigment with λ_{max} 515 nm would absorb only 2% less. Precise correlation is clearly not essential. The sensitivity of an organism to a bioluminescent signal is also related to the retinal illuminance, which, in turn (for a point source) is a function of the aperture (= eye) size for both fish and cephalopod eyes and for superposition compound eyes. Aperture size should therefore be enhanced in animals to whom the reception of weak bioluminescent signals is important, and is indeed a common characteristic of many deep-sea species (see Denton, Chapter 8; Land, Chapter 9). These considerations apply equally to non-specific signals, but a finer tuning of visual and bioluminescent adaptations might be expected for intraspecific interactions. This has only rarely been identified. The spectral correlations in e.g. *Odontosyllis*, *Malacosteus* and *Porichthys* provide one indication. Others are the enlargement of the eyes of male signalling ostracods and the correlation between eye size and the presence of cuticular light organs in some oplophorid decapods. Similar correlations can be found among fireflies (Case, 1984) where the spectral subtleties have been more clearly defined (Seliger *et al.*, 1982; Eguchi *et al.*, 1984).

The complexity of luminous signals in fireflies has been correlated with high population densities or with the local presence of several different species, while the remarkable phenomenon of synchrony in fireflies may be a natural consequence of crowding (Case, 1984). Similar arguments can be used to explain the diversity of ostracod displays.

One area of particular interest is the visual propagation of bioluminescent signals. Waves of bioluminescence have been observed running through populations of fireflies and ostracods and there have been numerous accounts of how individuals of many groups can be stimulated to luminesce by a light flash. This response is almost routinely encountered by observers in submersibles and has been recorded remotely by Neshyba (1967) who investigated the effects of colour and pulse rate on the responses at depths of 30-700 m. Experiments are needed to establish how these responses relate to interactions involving normal bioluminescent signals, and the range over which such responses occur.

The subtleties of luminous communication in marine organisms have hardly begun to be unravelled. Most explanations are still based on a simple protean or 'startle' defensive reaction to a luminous flash, yet even in this situation the relative effectiveness of *e.g.* rapid repeated flashes compared to single flashes with a slower decay, have not been investigated. Hailman (1977) includes startle deception under the general heading of visual ambiguity in photic communications and most of his other categories of deception also have bioluminescent counterparts in the sea. Only mimicry, so dramatic a component of the repertoire of luminous deception in fireflies (Lloyd, 1984), has yet to be definitively identified in any marine species. A probable luminous dialogue between the sexes has so far been observed only in the anomalopids and *Odontosyllis*. The potential for complex luminous responses such as these is widely available in marine animals and these two examples must be merely the tip of a very large iceberg.

Chapter 17

THE CHEMISTRY OF BIOLUMINESCENCE: ORIGINS AND MECHANISM

FRANK MCCAPRA

School of Chemistry and Molecular Sciences, University of Sussex, Brighton, BN1 9QJ

The structures of the luciferins so far discovered make chemiluminescent reactions very likely. The luciferases seem suited to relatively simple transformations such as removal of a proton. It is argued that these facts, together with the ease with which the probable biosynthesis of the luciferins is modelled by chemical reactions, point to a menu of inherently chemiluminescent reactions from which bioluminescent reactions are selected. These reactions in certain cases are bright enough to be selectable in a Darwinian sense.

Light emission from bacteria is thought to be an accidental phenomenon accompanying a common redox reaction, and is merely an accompaniment to the main event. These observations predispose us to believe that, in many ways bioluminescence is, or originated as, an accompaniment to other metabolic activities. It is to be hoped that the genetic techniques now being used to give DNA sequences for the luciferases will show how enzymes with functions other than the catalysis of bioluminescence became associated with small molecules with an inherent tendency to chemiluminesce.

INTRODUCTION

Most bioluminescent species are marine and their light contributes substantially to the ecology of this environment. Chemical considerations of the mechanisms and origins of bioluminescence necessarily also encompass the additional diversity provided by terrestrial and freshwater species and the special insights they can provide.

The emission of light from a living organism is now clearly established to be the result of an oxidative chemical reaction. The phenomenon of chemiluminescence has received much attention in its own right, and the conclusion that bioluminescence is simply a biochemically controlled version of the chemical event is now inescapable. Since the proposal, and subsequent validation, of the theories governing the emission of light from organic compounds some 20 years ago, there has been extensive development of the basic ideas. This evolution in mechanistic interpretation has implications for bioluminescence, and part of the purpose of this contribution is to bring the process up to date, rather than to produce a comprehensive review. The other intention is to address one of the most intriguing problems in the field, that of the origins of bioluminescence. This must be a tentative exercise, for two reasons. It is done from the standpoint of the chemist, and

Herring, P.J., Campbell, A.K., Whitfield, M. & Maddock, L. (ed.). *Light and Life in the Sea.*
© 1990. Cambridge University Press.

is thus a partial account. Secondly, it quickly becomes apparent that we have access to only a fraction of the information required to make definitive statements about such a significant event. Origins, then, must be taken to mean only the molecular basis of the process. Nevertheless it should help to define the areas in which further investigation is most likely to be fruitful. One of these is already apparent in the rapid growth in the application of complement-DNA techniques to the structure, and uses, of the luciferases. A very valuable interaction between the chemistry of small molecules and the information obtained from genetic methods is in prospect.

The overall mechanism can be described as in Figure 17.1.

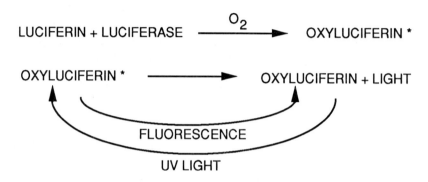

Figure 17.1. Oxidation of luciferin results in the excited singlet state of oxyluciferin. This product can be made to fluoresce with the same spectrum.

THE ORIGINS OF BIOLUMINESCENCE

There have been several speculative attempts to understand how light emission may have arisen in the course of evolution (Buck, 1978; Seliger, 1975; Hastings, 1983). The only proper context for the discussion of the primary event is Darwinian evolution, and the chemist with the relatively few variables in his material, is not a major player in this game. The most balanced and thought provoking of the essays in this area is that by Buck. There are many warnings implied in that work about the doubtful nature of extrapolations from our present inadequate knowledge, back to origins. Yet, as Buck himself made plain, there is a need to press ahead in the hope that a fusion of the scattered evidence, as it becomes available, will emerge.

An early, and often repeated, suggestion is that there is some detoxification aspect to the removal of oxygen as a result of bioluminescence, especially at the point where anaerobes had to contend with increasing concentrations of oxygen (McElroy & Seliger, 1962). The light was seen as an energy 'sink', involving as it must, quantised energy changes of around 300 kJ. This was never a good idea, in that most bio- and chemiluminescent reactions are not completely efficient, and the bulk of the energy appears as vibrational energy, *i.e.* heat, in any event. It is also suggested later, that the chemical evidence, such as it is, points towards light production arising as a *by-product* of a normal oxidation reaction. A related proposal (Seliger, 1975) is that the luciferases evolved to utilise oxygen directly as an electron acceptor, and that this provided a major selective advantage in the resulting metabolism of hydrocarbon materials by the organisms concerned. This proposal lacks chemical insight. The most widespread and studied of the luciferases, although designated as oxygenases (or sometimes dioxygenases) by virtue of the resulting products, are rather ineffectual oxidising agents. It is common knowledge (Walsh, 1979) that even the flavin-based bacterial system is closely related to that class of oxygenase which has to choose 'soft targets', such as electron-rich molecules, as opposed to the metalloenzymes such as P450, which, from the chemical point of view, are the only enzymes which are a match for compounds with low ionisation potentials such as hydrocarbons. Yet again, the view that light emission is a by-product carries with it, as will be seen, the implication that the oxygenation is virtually accidental, and had not been taken to any great level of efficiency (at least with respect to the type of substrate handled) before selective pressures ensured their survival because of the useful emission of light. It is argued, on the basis of the unfortunately few well-studied examples, that the oxidation is a function of *substrate* structure, and that some of the luciferases should perhaps not be classified as oxygenases at all, in that they lack a means of catalysing the use of the oxygen molecule in a way typical of the usual oxygenase.

Hastings (1983) has estimated, on the basis of common biochemical features, that bioluminescence has evolved as many as 30 different times. He is of the opinion that the luciferases evolved late, explaining the unexpected and difficult to rationalise phylogenetic distribution, and that efficient light emission appeared after the development of vision. Buck (1978) also believes that the properties of the luciferases would have to be fixed in the evolutionary progression by an unrelated, valuable function, until the evolution of the other components of the light reaction. This conclusion can be supported, at least in part, by the chemical mechanisms discussed here.

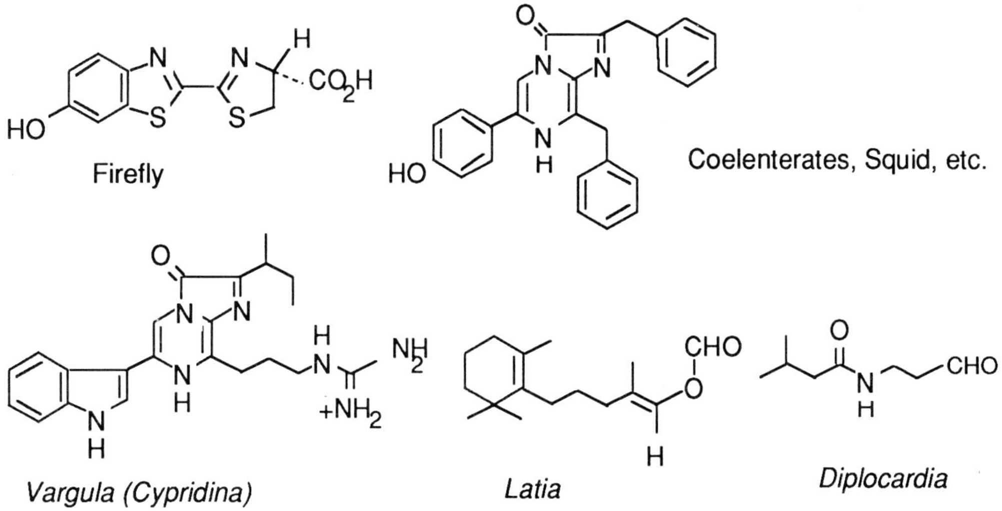

Figure 17.2. The structures of various luciferins.

BIOSYNTHESIS OF THE LUCIFERINS

The luciferins so far discovered are unique molecules, not found elsewhere in nature. Some structures (Figure 17.2) have origins which appear reasonable, by inspection.

For example, *Latia* luciferin (Shimomura & Johnson, 1968), as an enol formate of an aldehyde, is obviously an oxidative product derived from a terpene, and dinoflagellate luciferin (Dunlap *et al.*, 1981), together with Substance F, from luminescent euphausiid shrimps (Shimomura, 1980), can similarly be seen as metabolic products of chlorophyll, as indicated in Figure 17.3.

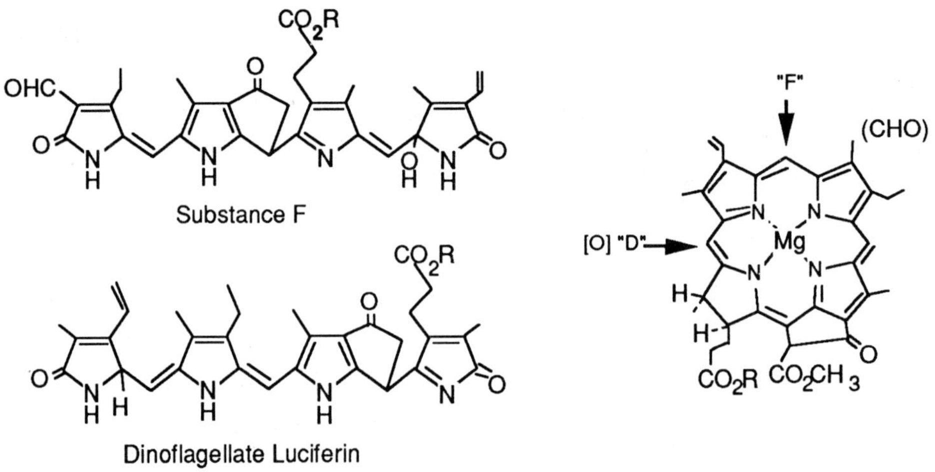

Figure 17.3. Oxidative cleavage (at F or D) generates the components of the light reactions.

The long chain aldehyde in the bacterial system is not strictly a luciferin, but its origin is straightforward, since fatty acids serve as the precursors (Ulitzur & Hastings, 1978). Nothing is known of the biosynthesis of *Diplocardia* luciferin (Wampler, 1981). This is also an aldehyde, and the corresponding luciferase could hardly be more different from the flavin-requiring bacterial luciferase, since it is a copper-containing protein. The origins of the remaining luciferins present much more of a challenge, and current knowledge is reviewed. The remarkable diversity in chemical provenance of all of them is strong support for the late, and indeed adventitious, appearance of bioluminescent phenomena.

FIREFLY LUCIFERIN

Although obviously a terrestrial organism, such is the unity of the bioluminescence field, that the title of this volume is no bar to its inclusion in the contents! There is no difficulty in identifying (McCapra & Razavi, 1975) cysteine as a component of the structure, present in the right hand or thiazoline ring of the luciferin (1) (Figure 17.4).

Indeed the last step of the synthesis involves the addition of D-cysteine to 2-cyano-6-hydroxybenzothiazole (2). The benzothiazole presented more of a challenge, but with the

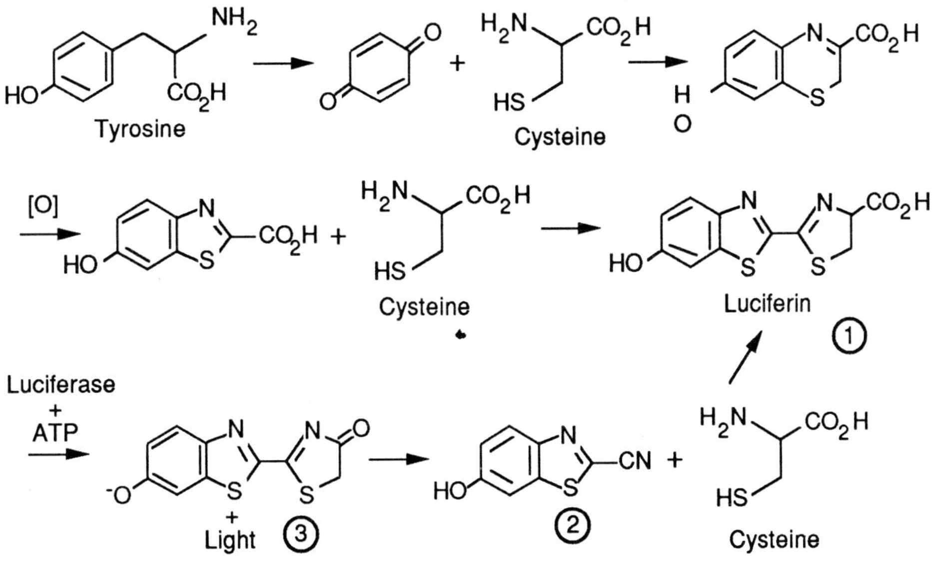

Figure 17.4. Biosynthesis of firefly luciferin.

presence of sulphur and nitrogen, we were tempted to look for cysteine here also. The atoms are not arranged as in cysteine, but the oxidative rearrangement shown provides the necessary modification. We attempted a biomimetic chemical synthesis from ben-zoquinone, which achieved just such a ring contraction, suggesting that the firefly might also adopt a *chemically* predestined path. We also suggested that the benzoquinone precursor would be tyrosine. This all-aminoacid source of the luciferin was confirmed by radioactive tracer studies on the firefly by Japanese workers (Okada *et al.*, 1976), and in the click beetle *Pyrophorus pellucens* in our laboratory (McCapra & Razavi, 1976).

A remarkable corollary to this investigation is that the oxidation reaction which gives light emission, and produces oxyluciferin (3), results, after hydrolysis in the firefly, in the same synthetic intermediate as used by the chemist, namely the cyanobenzothiazole (2). This compound is recycled, and combination with cysteine regenerates firefly luciferin (Okada *et al.*, 1974). Thus the only 'fuel' to be consumed in the generation of light is the readily available cysteine. A similar pattern is postulated later for coelenterazine, and by extension, *Cypridina* luciferin, although no experiments have yet been carried out to test this.

COELENTERAZINE

This is the name given by Shimomura to the luciferin originally discovered in the coelenterates *Aequorea* and *Renilla* (Shimomura *et al.*, 1980; Campbell, 1988, 1989). Its distribution has now been shown to be extraordinarily widespread. It was first found in luminescent animals ranging from shrimp through squid to fish. Subsequently it has

been found in a variety of *non-luminescent* marine organisms such as herring (Shimomura, 1987). Its dissemination by the food chain seems clear. The full story is yet to
be written, but it is probable that several organisms at the bottom of the food chain are
the originators of the compound, and that many luminescent higher organisms obtain
their supplies of this essential material in their diet. To my mind, this reinforces the view
that the luciferase, which must already be widely distributed, either serves another
metabolic purpose, or did so in the very recent past. Whether the luciferin (coelenterazine) also plays a non-luminescent metabolic role remains to be seen.

A reasonable suggestion for the biosynthesis of this heterocycle (Figure 17.5) is that it
is derived from three aminoacids, two tyrosines and a phenyl alanine. In the same way,
Cypridina luciferin can be constructed from tryptophan, isoleucine and arginine. The
aminoacids are not unmodified, and the actual precursor is more likely to be a dehydro-
aminoacid tripeptide. To test this hypothesis, such a precursor was synthesized, and the
cyclisation occurred, under mild conditions, in quantitative yield (McCapra & Roth,
1972). The conclusion we wish to draw from this, is that the post-translational modification of the original peptide (perhaps as part of a larger protein) predisposes a commonly
occurring type of compound to form a structure which is highly likely to *autoxidise* (see
later). The theme of adventitious formation of compounds capable of undergoing
chemiluminescent oxidation is thus enhanced.

The oxidation of coelenterazine during the luminescent reaction produces the emitting
molecule, the oxyluciferin (4, Figure 17.6) in the excited state. It is known to be easily
hydrolysed, and the product (5) is often found in the spent reaction mixture.

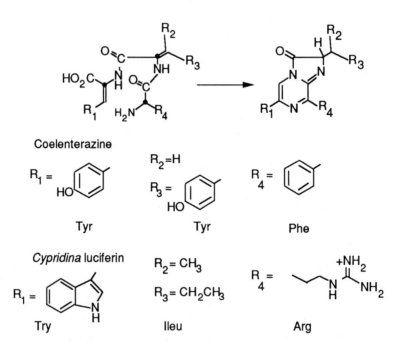

Figure 17.5. Biosynthesis of imidazolopyrazine luciferins from peptide precursors.

A synthesis of the luciferin starts with (5), using *p*-hydroxyphenylpyruvic acid (Inoue *et al.*, 1980). The compound formed by this combination is actually found in the livers of certain luminescent squid (*Watasenia*), and since *p*-hydroxyphenylpyruvic acid is in metabolic equilibrium with tyrosine, a re-cycling mechanism very similar to that in the firefly presents itself. In this case the 'fuel' for the light reaction is the aminoacid tyrosine rather than cysteine. No experiments have yet been carried out to confirm or disprove this hypothesis.

The story does not end here however. Certain coelenterates (*Aequorea, Renilla* and *Obelia* for example) increase the light yield and alter the colour of the initial chemiluminescent reaction by employing an electronic energy transfer process. This is accomplished by using a green fluorescent protein (GFP) which binds to the luciferase (or photoprotein - see below) to meet the requirements for close proximity between donor and acceptor molecules (Ward & Cormier, 1978). This same protein has had some of its aminoacids modified in a way similar, but not identical to that found in the biosynthesis of the luciferin. The aminoacid sequence involved in producing the chromophore also contains the dehydro-aminoacid, dehydro-tyrosine, and is now known to be as shown in Figure 17.7 (Ward, personal communication; serine is the reduced form).

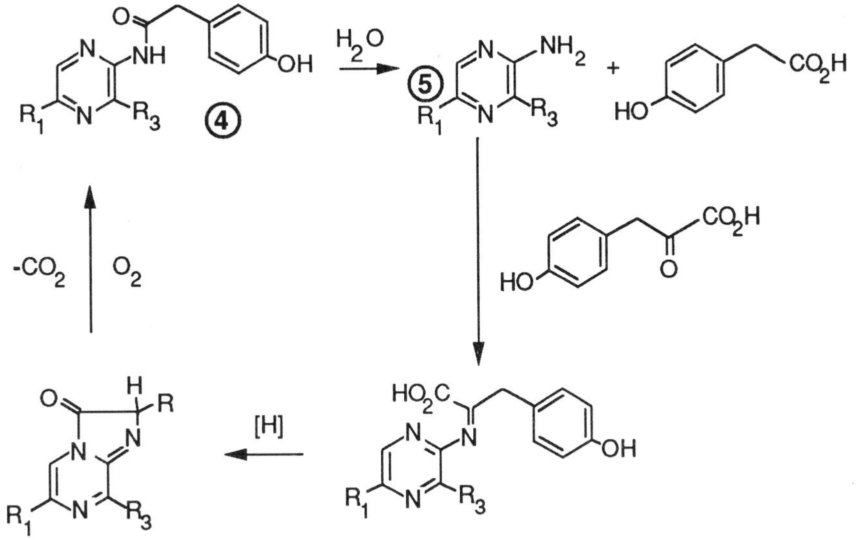

Figure 17.6. Proposed recycling mechanism for coelenterazine.

Although this has been confirmed by sequencing the gene for its biosynthesis, the exact properties of the chromophore are not yet established, since the chromophore is not fluorescent unless it is part of the fully renatured protein. Yet again one has the feeling that a common process, perhaps designed for other purposes, is being pressed into service in the interests of better bioluminescence. Since the genes for the luciferases, photoproteins and the fluorescent protein have been, or are being, cloned, the means to answer some of these questions is to hand.

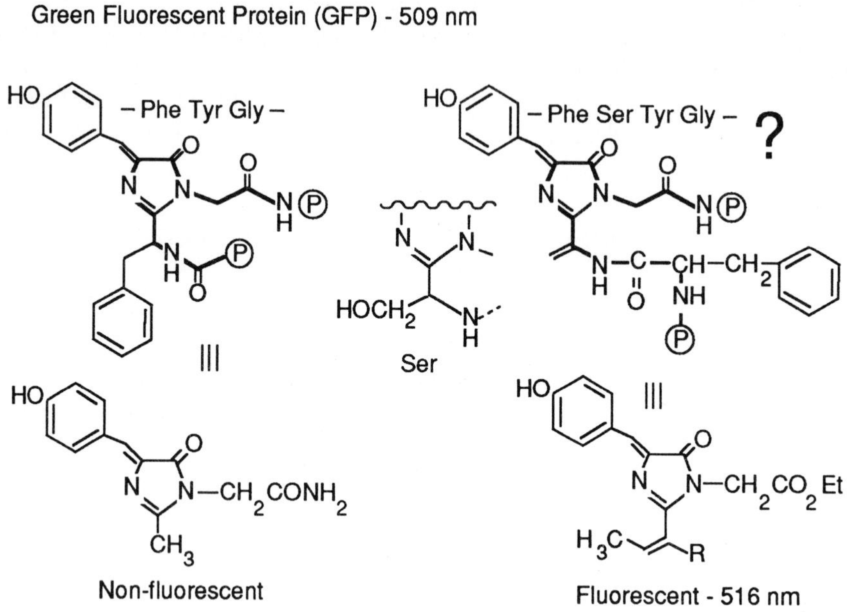

Green Fluorescent Protein (GFP) - 509 nm

Figure 17.7. Green fluorescent protein is derived from modified amino acids.

THE ROLE OF OXYGEN

Many comparisons between bioluminescence and respiration have been made. It is certainly true that all of the bioluminescent organisms examined to date use molecular oxygen in the final chemiluminescent reaction. This universality was obscured for some time, when it seemed that a class of reaction, involving what were called photoproteins, could be made to emit light in the absence of oxygen (Shimomura & Johnson, 1978). The trigger for the reaction was the addition of calcium ion.

Reference has already been made to the imprecision of the term oxygenase when applied to the action of a luciferase. It is true that some organisms, with an undefined chemistry, with unknown structures for their luciferins and luciferases, do show peroxidase activity (Cormier, 1978). Like many peroxidases, their substrate specificity does not seem to be high, and their existence, if finally substantiated, should not invalidate the thesis proposed. That is, the emission of light is a property more of the substrate than of the enzyme. By that I mean that the luciferases, in the case of all of the many tens of organisms which use coelenterazine, the five or so which use *Cypridina* luciferin and the luminescent beetles, do not actually catalyse the oxidation step itself. If catalysis is involved, then it is not by a means already discovered for the large number of classical oxygenases. In all of these examples, the luciferin is so constituted as to form a carbanion readily, and to undergo *spontaneous oxidation*. Such autoxidations usually occur at a diffusion controlled rate (Russell *et al.*, 1967). Although the exact kinetic properties of the oxidation step have not been fully examined for the luciferases mentioned, there is no evidence as yet to contradict this statement.

Model chemiluminescent compounds (McCapra & Perring, 1985) do indeed show this diffusion controlled oxidation, with light-emitting efficiencies not far removed from that of the natural reaction (Figure 17.8).

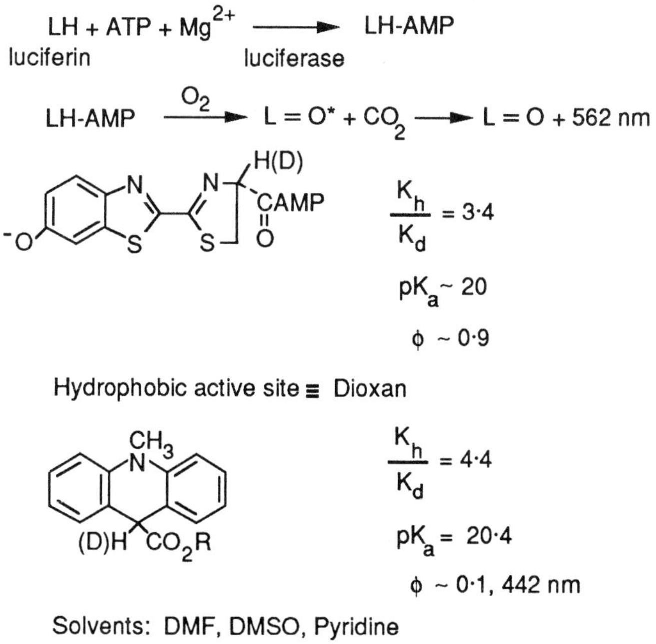

$$\text{LH} + \text{ATP} + \text{Mg}^{2+} \longrightarrow \text{LH-AMP}$$

luciferin   luciferase

$$\text{LH-AMP} \xrightarrow{\text{O}_2} \text{L} = \text{O*} + \text{CO}_2 \longrightarrow \text{L} = \text{O} + 562 \text{ nm}$$

$$\frac{K_h}{K_d} = 3\cdot4$$

$$pK_a \sim 20$$

$$\phi \sim 0\cdot9$$

Hydrophobic active site ≡ Dioxan

$$\frac{K_h}{K_d} = 4\cdot4$$

$$pK_a = 20\cdot4$$

$$\phi \sim 0\cdot1, 442 \text{ nm}$$

Solvents: DMF, DMSO, Pyridine

Figure 17.8. Luciferin bioluminescence and a model.

Part of the reason for this circumstance is the very high radical stability conferred by the nature of the heterocycles involved. The radical is exemplified in Figure 17.9, and superoxide ion is the expected other product of the interaction of the carbanion shown with molecular oxygen. Thus the function of these particular enzymes is to remove a proton, and provide a largely non-aqueous environment for maximum efficiency in succeeding steps, particularly the electron transfer to oxygen. There is ample precedent for this type of catalyst-free oxidation in carbanions of many types (Russell *et al.*, 1967). The probability that such a set of enzymes have, or had, a role other than that in bioluminescence is thus very high. It is noteworthy that these luciferases have no oxidative co-factors, yet another reason for reluctance in using the description oxygenase. *I further believe that these properties provide a mechanism for the control of the flashing behaviour of the firefly.* All that is required of the luciferase is that, in response to a nervous impulse, the enzyme changes conformation by any appropriate mechanism, such that the acid-base reaction is repressed. This will occur well within the time frame observed in the flashing regime. The protonated (*i.e.* unchanged) luciferin will certainly not oxidise, whereas the rate of oxidation of the anion is among the fastest known reactions.

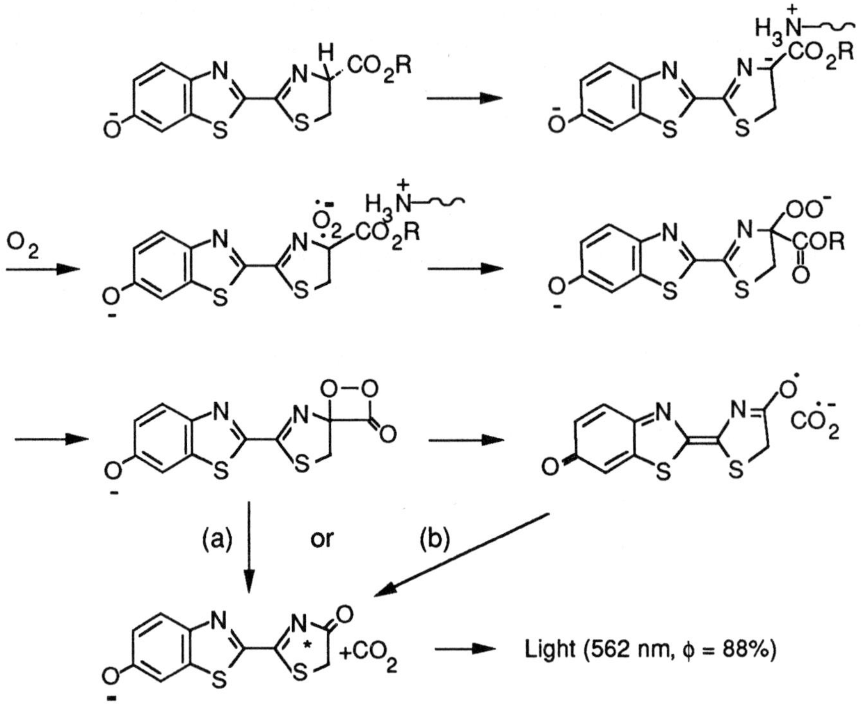

Figure 17.9. Mechanism of firefly luciferin luminescence. The direct route (a) is now favoured.

It is certain that many other chemiluminescent reaction types will emerge in bioluminescence, but it is nevertheless likely that, as with the peroxidases, we will be able to say that the degree of specialisation is low enough to support the contention that the luminescence arose as part of an organisation in pursuit of other evolutionary goals. Thus the evidence adduced from the chemical mechanisms is very much in support of the ecological record of diversity. In effect, the rules governing the occurrence of *chemi*luminescence are sufficiently flexible as to allow many biochemical systems the opportunity to develop a large variety of *bio*luminescent types.

There have been suggestions that oxygen can be 'stored' in the proteins of the coelenterate luciferases (Cormier *et al.*, 1977). Of course, some of these do form a complex with oxygen, luciferin and luciferase, known as a photoprotein. However, the oxygen is present as a peroxide of the luciferin, and there is no evidence for any other binding group. There is an often repeated error in the representation of the so-called 'bound' oxygen and the luciferin, to be found, for example, in the otherwise excellent contribution by Musicki *et al.* (1986). Here, the bound oxygen is at the peroxide *i.e. reduced* level, as is the luciferin. No reaction is possible unless one of these partners is at a higher oxidation level. That is, the oxygen is usually at the oxidation level of *molecular* oxygen, not peroxide. The binding of oxygen gas thus becomes a *major* problem, without a co-factor. There are no such co-factors.

In chemiluminescence there are many examples of compounds which form peroxides such as the luciferin peroxide, needing only a change of pH to trigger the reaction (McCapra & Perring, 1985). The conformational change in the binding of calcium is the equivalent step for the photoprotein.

THE LIGHT-GENERATING MECHANISM

Although the mechanism of the light-emitting step was set out in an essentially correct form many years ago (McCapra & Richardson, 1964; McCapra & Perring, 1985) there has been a slow evolution in interpretation. After the confirmation of the dioxetanone route shown in Figures 17.10 and 17.11, several groups adopted the idea of intramolecular electron transfer, given the label CIEEL (Chemically Induced Electron Exchange Luminescence) by Schuster (Schuster, 1979).

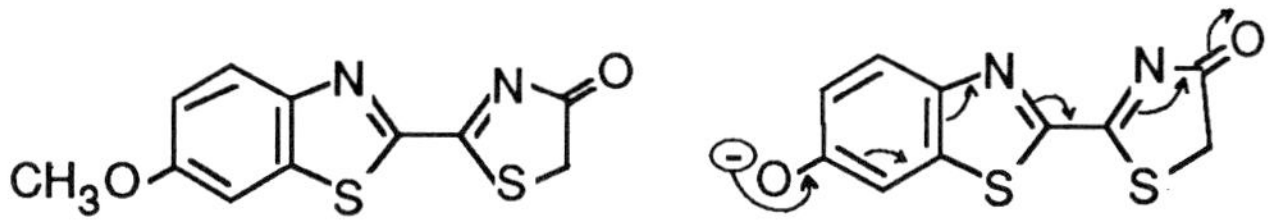

Figure 17.10. Charge transfer is required for singlet state formation.

There are many examples of this route to chemiluminescence, especially in electrically generated light emission. However, there are some problems in accepting that discrete electron transfer is involved. In theoretical terms, it is simpler to see the population of the excited state as a smooth transition into the new state without the generation of high energy intermediates such as radical ions. Indeed, these radical ions are just one extreme of a type of transition state known as a charge transfer state. This idea was part of the earliest proposals on the nature of dioxetan decomposition, and has been reconsidered in some recent publications (Gundermann & McCapra, 1987; Catalani & Wilson, 1989).

In addition, we have attempted to detect such radical ions in several reactions which should produce them, by a method of exquisite sensitivity (McCapra et al., 1981). Although negative experiments are never the last word, in combination with other evidence, we feel that the pendulum has swung back to the notion that dioxetan decomposition, with the emission of light, is a concerted process. This is especially true of the intramolecular charge transfer type, exemplified by the luciferins discussed above. The necessary overlap between the orbitals of the electron-rich donor part of the molecule and the lowest unoccupied orbitals of the peroxide portion is hard to envisage.

Recently there has been a re-examination of the compounds which gave rise both to comparisons with the luciferins, and to the original coinage of the CIEEL acronym (Catalani & Wilson, 1989). These contributions point in the new direction discussed above. The characteristics of the structure of firefly luciferin are mirrored by several synthetic chemiluminescent compounds, and these also show evidence of a charge transfer interaction, summarised in Figure 17.10 by the flow of arrows in the formulae.

It is known that blocking this flow results in almost complete repression of the luminescence, even where the necessary fluorescence of the oxidised product is maintained.

An interesting and important practical consequence of this interpretation is seen in the bioluminescence of the coelenterates. The emission of light from the oxyluciferin derived from coelenterazine is that of the amide anion (6) shown in Figure 17.11 (McCapra & Manning, 1973). This anion can only be formed direct from the transition state, for this is the only way in which the amide anion (pKa estimated as 25) can be formed in solutions of pH as low as 6·0. There is an internal marker for this in that the phenolic hydroxyl, whose ionisation is readily detectable by changes in the fluorescence spectrum, is unionised at these low pH values., yet its pKa is the expected 8-9. These events are essential for the colours of the light observed from the organism, and for the transfer of energy to the green fluorescent protein which is clearly an enhancement of value to the organisms.

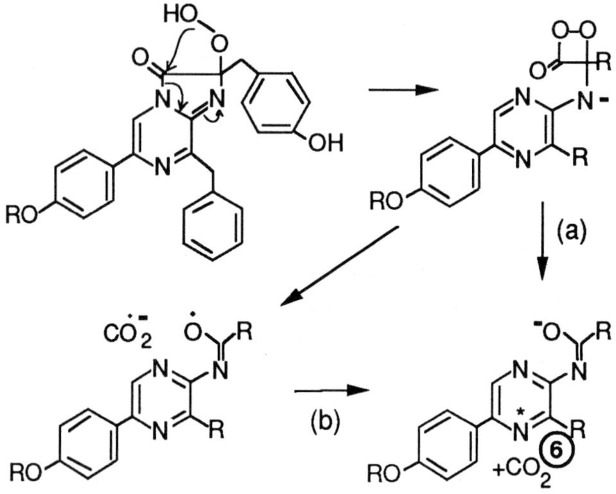

Figure 17.11. Mechanism of coelenterazine luminescence. The direct route (a) is now favoured.

BACTERIAL BIOLUMINESCENCE

Bacteria, the most investigated, but still puzzling, luminescent organisms have been left to the end of this discussion because what is known about them summarises much of the discussion on both origins and mechanism (Hastings, 1983).

The light has been described in terms ranging from totally adventitious to essential for natural selection. I will leave the evolutionary arguments to those very much better qualified to make them. However, my own feeling is that we *do* have an example here of light emission which is virtually irrelevant to the main functions of the organism. Paradoxically this makes the phenomenon particularly important, since it may serve as a model for the many examples of seemingly meaningless (as bioluminescence) low-level light emission such as that which occurs during phagocytosis (Allen, 1982).

There are two reasons for believing the above. Hastings (1983) has come to the conclusion that bacteria (and similar 'primitive' systems such as those of the luminescent fungi) are different from the more obviously developed examples. He has described their ecology in detail, and the way in which this supports such a view. The second reason, also given by Hastings, is that it is not at all unreasonable to suppose that the primary function of this enzyme is the oxidation of long chain aldehydes to the corresponding fatty acid. It may be that there are other substrates *in vivo* which are also oxidised, although none has been found to give light with the enzyme *in vitro*. The chemical and biochemical properties of the enzyme cyclohexanone oxygenase (Latham & Walsh, 1987) (from *Pseudomonas* spp.) are very similar, down to the classification of the reaction catalysed as a Baeyer-Villiger type (Figure 17.12). The emission of light is, after all, the property of the substrate molecule, and it is not difficult to imagine that a certain mechanism, refined by nature for other purposes, carries with it the possibility of raising one of the products to an excited state, provided sufficient energy is released for the emitted quantum.

If this *light-emitting feature* of the reaction is not selected for, then the light reaction will merely trail in the wake of the metabolically essential reaction. It must be pointed out that most reactions of peroxides are chemiluminescent, and that many of the model reactions examined in an attempt to explain bacterial luminescence are also hard to understand.

The bacteria also show that the energy of the original light can be transferred to acceptor molecules. Yellow mutants, which use a protein-bound flavin as the energy acceptor, and particularly the presence of a blue fluorescent protein (BFP) with lumazine as its chromophore (Lee, 1985), all indicate that the light generating reaction is not so much an FMN-based light-emitting reaction, but a more general peroxide chemiluminescence. The BFP can be added to a normal reaction, and its influence demonstrated, especially by a change in the colour of the emitted light from blue-green to blue. The proven emitter in the unperturbed reaction is the hydroxyflavin (7) in Figure 17.12.

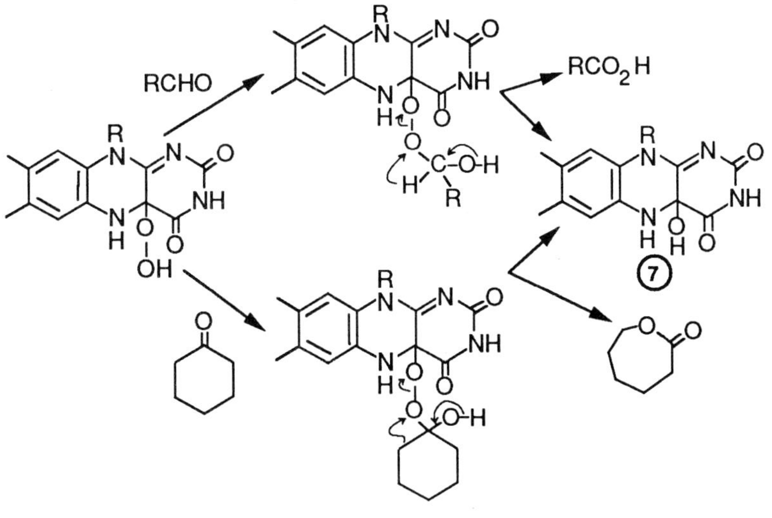

Figure 17.12. Comparison of bioluminescent pathway with dark oxidation of cyclohexanone.

This molecule is probably acting as an acceptor in the same way as the other fluorescent acceptors described above. The details are still not satisfactorily described, but the literature is full of examples of peroxides of almost every structural type, reacting with fluorescers to give chemiluminescence by a variety of mechanisms. The various attempts to place this reaction in the CIEEL category are all flawed, and a very recent publication has shown how very inefficient these reactions are by re-measuring what was thought to be the best of the examples, especially in relation to bacterial bioluminescence (Catalani & Wilson, 1989). No chemiluminescent Baeyer-Villiger reactions are known. We have still to discover the mechanism of light emission in this luminescent organism.

Chapter 18

LIVING LIGHT: FROM THE OCEAN TO THE HOSPITAL BED

ANTHONY K. CAMPBELL

Department of Medical Biochemistry, University of Wales College of Medicine, Heath Park, Cardiff, CF4 4XN

Bioluminescent proteins and their luciferins can be detected down to 10^{-18}-10^{-21} mol. This is sufficient to measure and locate intracellular signals such as Ca^{2+}, energy supply in the form of ATP, and the end response of cells responding at their plasma membrane to activators, drugs or pathogens. Methods have been developed using liposome cell fusion, or transfection of cDNA or mRNA, to incorporate these bioluminescent indicators into living cells. This has led to the identification of two distinct mechanisms responsible for activating cells such as human neutrophils, one requires a rise in cytosolic Ca^{2+} the other does not. A mechanism of reversible cell injury induced by membrane pore forming proteins has also been discovered which appears to play an important role in inflammatory diseases such as rheumatoid arthritis and multiple sclerosis. The ability to clone and genetically manipulate bioluminescent proteins has led to their application as reporter genes and, at the patient's bedside, to their use as replacements for ^{125}I and ^{32}P in immunoassay and recombinant DNA technology.

INTRODUCTION

Advances in patient care depend critically on three factors; firstly, increased understanding of the disease process, in modern terms at the molecular level, secondly, the production of new reagents for use in therapy aimed at susceptible targets identified as a result of the known disease mechanism, thirdly, the development of new technology enabling new diagnostic tests to be made available to improve the speed, sensitivity and accuracy of identifying a particular disease in an individual patient. Three marine luminous organisms (Figure 18.1) illustrate this philosophy well. The luminous hydroid *Obelia* had led to the discovery of a mechanism for demyelination in multiple sclerosis arising from a rise in intracellular calcium. The piddock, *Pholas dactylus*, enables oxygen metabolite release to be measured from white cells isolated from the joints of patients with rheumatoid arthritis, whereas the genes responsible for light emission in oceanic luminous bacteria have enabled a uniquely sensitive method for detecting *Salmonella* contamination in food to be developed.

Living light has been exploited by man in some remarkable ways (Table 18.1). All require a regular source of luminous organisms, but the scientific applications depend on two other factors. Firstly the chemistry and biochemistry must be defined so that any

Herring, P.J., Campbell, A.K., Whitfield, M. & Maddock, L. (ed.). *Light and Life in the Sea.*

Figure 18.1. Three British luminous organisms with application in cell biology: (A) the glow-worm, *Lampyris noctiluca* for ATP, (B) the hydroid, *Obelia geniculata* for cytosolic free Ca^{2+}, (C) the common piddock, *Pholas dactylus* for O_2^-. From Campbell (1988) with permission of Ellis Horwood.

Table 18.1. *Exploitation of chemiluminesence by man*

Application	Example
1. Night - light	Wartime use of luminous ostracods for map-reading by Japanese soldiers
	Illumination of combustible material by luminous fungi
	Night lights for modern mariners
2. Decoration and ritual	Fireflies and cucujos in head-dresses of Mexican women
3. Recreation	Luminous golf balls
4. Fishing	Pieces of luminous squid on Portuguese fishing lines
5. Military	Detection of nuclear submarines
6. Scientific	Model for the chemistry of the excited state
	Model for anaesthetic and drug action
	Model for biochemical evolution
	Probe for the genetic engineer
	Analytical tool in biochemistry, microbiology and biotechnology

unique features capable of exploitation can be identified. Secondly, instrumentation is required so that the parameters critical for exploiting it can be quantified, in particular the kinetics and spectral characteristics of the reaction responsible for the light emission.

Although bioluminescence is found in some 700 genera representing 16 phyla (Herring, 1987; Campbell, 1988) in only a few have the chemistry and biochemistry been characterised sufficiently for them to be exploited. One of the most widely used analytical techniques based on bioluminescence is the measurement of ATP using firefly luciferin-luciferase. This generic term was first coined by Dubois in 1887, but has resulted in some confusion regarding the variety of biochemical compounds responsible for living light. The luciferins which have been identified so far can be grouped conveniently into five 'families' (Figure 18.2). However it is important to remember that the luciferase, the enzyme responsible for catalysing luciferin oxidation to generate light, has a unique amino acid sequence in each luminous species. This will be true even in a taxonomic luminous family using the same chemical reaction, though strong sequence homology is to be expected, particularly in the protein domains responsible for catalysis. There is no obvious biological pattern to the occurrence of the various luciferins. For example coelenterazine is responsible for bioluminescence in organisms in six distinct phyla (Shimomura et al., 1980; Campbell & Herring, 1989).

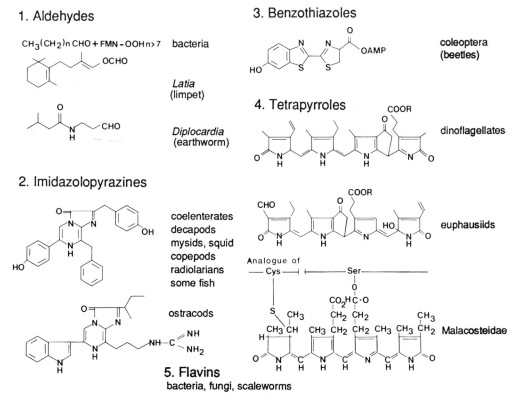

Figure 18.2. Five known chemical 'families' in living light. The structure for Malacosteidae is only proposed (Campbell & Herring, 1987), the precise structure has yet to be determined. From Campbell (1988) with permission of Ellis Horwood.

A further confusion has arisen regarding the role of ATP in bioluminescence. All living cells, including luminous ones, require ATP to remain alive and to carry out their specialised functions. However, only in luminous beetles and some other insects is hydrolysis of ATP involved directly in the chemiluminescent reaction. Even in these cases ATP hydrolysis is not the energy source for light emission, this comes from the breakdown of the dioxetanone intermediate formed by oxidation of the luciferin (McCapra, Chapter 17).

BIOLUMINESCENCE AS AN ANALYTICAL TOOL

The unique potential of bioluminescence, both as a research tool in biology and in analysis at the patient's bedside, depends on three key features:

1. Sensitivity: just a few hundred molecules reacting per second can be detected.
2. Intracellular probe: chemical reactions can be quantified and located from within living cells.
3. Non-radioactive: chemiluminescence compounds can be detected with the same, or better, sensitivity than ^{125}I or ^{32}P, yet without the appropriate triggering conditions, and the chemi- or bio-luminescence substance is stable.

The exploitation of these three features in turn depends critically on the development of instrumentation to detect and quantify very low light intensities, to visualise these and to provide a spectral analysis

Over the past century many researchers into chemiluminescence have realised that by coupling one of the components of a chemiluminescent reaction to a substance of biological interest, the latter can be measured easily down to a picomol-femtomol (10^{-12}-10^{-15}), often down to an attomol (10^{-18}) and sometimes down to 10^{-21} mol, known affectionately in our laboratory as a tipomol (from Welsh tipyn = little). 10^{-24} mol is an impossomol! It is this sensitivity, achievable because of the low chemical and instrument 'noise', which can make chemiluminescence analysis superior to its rivals absorbance, fluorescence, nmr or radioactivity. Beijerinck (1902) was able to measure oxygen released by photosynthesis using luminous bacteria. Harvey, in 1916, whilst studying the catalysis of pyrogallol + H_2O_2 chemiluminescence by plant peroxidase, or oxidases as he called them, commented on the exquisite sensitivity of detecting these enzymes. McElroy (1947), following his discovery of the ATP requirement in the firefly chemiluminescent reaction, soon established a sensitive assay for this key metabolite in energy metabolism.

Bioluminescent reactions require a minimum of three and a maximum of six components to generate the light. Any one of these components may be coupled to a substance or reaction of biological interest (Table 18.2):

1. The luciferin: *e.g.* as a way of detecting the label in immunoassay or a DNA probe.
2. The luciferase: *e.g.* as a reporter gene in genetic engineering.
3. The oxidant: *e.g.* O_2, O_2^- or H_2O_2 measurement.
4. A cofactor: *e.g.* ATP, NAD(P)H, FMN.
5. A cation: *e.g.* Ca^{2+}.

6. A fluorescent energy transfer acceptor: *e.g.* homogeneous immunoassay or DNA assay.

But do we really need the sensitivity achievable using these bioluminescent assays? Has anything new about the physiology or pathology of cells been discovered by applying them? Have they real application at the patient's bedside in aiding clinical diagnosis?

Clinically important substances in plasma range from about 140 mM for Na^+ to a few pM for some free hormones (*e.g.* thyroid). Measurements in a few millilitres of blood demand an assay sensitivity of pmol-fmol (10^{-12}-10^{-15} mol). The ATP content of one eukaryotic cell is usually between 0·1 and 10 fmol (10^{-16}-10^{-14} mol), but in a bacterium may be as little as an amol (10^{-18} mol). The successful detection of *Salmonella* contamination in food demands an assay capable of detecting one live organism in 25 g of food.

A major problem in contemporary cell biology has been to develop techniques for measuring, and locating, cations such as Ca^{2+} and H^+, and nucleotides, enzymes and other proteins in living cells. Conventionally, biochemical analysis of tissue or cell

Table 18.2. *Some analytical applications of chemiluminescence*

Chemiluminescent system	Application
A. Synthetic compounds	
1. Phthalazine diones (luminol, isoluminol & derivatives)	reactive oxygen metabolites , phagocyte activation haem compounds, transition metals immunoassay label, peroxidase as a label, NO, NO2.
2. Acridinium salts (lucigenin & acridinium esters)	O_2^-, immunoassay label
3. Adamantylidene adamantane (-P)	immunoassay label
4. O_3	NO, NO_2, NO_3^-
B. Bioluminescent compounds	
1. Firefly (*Photinus*) luciferin-luciferase	ATP and related metabolites Enzymes producing or degrading ATP, alkaline phosphatase as a label in immunoassay and DNA probe; reporter gene in genetic engineering
2. Bacterial (*Photobacterium* or *Vibrio*) oxidoreductase-luciferase	NAD(P)H, FMN and related metabolites, enzymes producing or degrading NAD(P)H and $FMNH_2$, long chain aldehydes, immuno-assay label detection, O_2, reporter gene in genetic engineering
3. Coelenterate (*Aequorea* or *Obelia*) photoprotein	intracellular free Ca^{2+}, label in immunoassays or DNA probe
4. Sea pansy (*Renilla*) luciferin-luciferase	phosphoadenosine phosphate (SO_4)
5. Piddock (*Pholas*) luciferin	O_2^-, peroxidase, phagocyte activation
6. Earthworm (*Diplocardia*) or acornworm (*Balanoglossus*) luciferin-luciferase	H_2O_2
7. Polychaete worm (*Chaetopterus*)	Fe^{2+}
C. Ultraweak chemiluminescence	endogenous peroxides

For references see: DeLuca (1978, 1987), Campbell (1988)

preparations has followed the 'grind and find' approach involving the extraction of these substances from thousands or millions of cells. However this not only destroys the hetereogeneity between different cells and the polarity within each cell, but also prevents some parameters being measured at all. The concentration of cytosolic free Ca^{2+} is about $0.1\ \mu M$ in a resting cell and may rise to $1-5\ \mu M$ following activation by a stimulus, drug or pathogen. Yet the total cell Ca^{2+} remains at some 1-5 mmoles per kg cell water. Changes in cytosolic Ca^{2+} can thus be studied only whilst the plasma membrane remains intact. Furthermore many examples of cell activation or injury involve individual cells crossing a series of thresholds. These include end responses such as cell movement, transformation, division or death. It is therefore essential that the intracellular signals, energy supply and covalent modification of the proteins responsible for eliciting an end response can be detected, and located in a particular site, within individual cells. This demands a sensitivity of detection in the $10^{-15}-10^{-20}$ mol region.

Radioactive probes have provided the research and clinical scientist with sensitive and specific probes for unravelling metabolite pathways, and as labels in new technologies such as immunoassay and recombinant DNA technology. However radioisotopes such as ^{125}I and ^{32}P are hazardous and short-lived. Their decay can damage the biological molecules surrounding them, their detection often requires cumbersome or time-consuming devices not suitable for automation e.g. autoradiography, and they cannot be adapted to provide 'homogeneous' assays where no separation steps are required, thereby introducing imprecision and problems with automation. Chemiluminescent substances can now be used either directly or indirectly as alternatives to ^{125}I and ^{32}P, detectable with the same or better sensitivity ($10^{-18}-10^{-17}$ mol) within a second or so, are stable for years and can be used to develop homogeneous ligand-ligand assays (Campbell & Patel, 1983).

But before examining how living light has been used to measure substances or processes of real biological or clinical interest (Table 18.3) we must first decide how the light emanating from the reaction is to be detected and quantified.

Table 18.3. *The major biomedical applications of bioluminescence*

Analyte	Bioluminescent system
1. Enzyme and metabolites	firefly luciferin-luciferase
	bacterial oxidoreductase-luciferase
2. Reactive oxygen metabolites	pholasin
	earthworm luciferin-luciferase
	acorn-worm luciferin-luciferase
	ultraweak chemiluminescence
3. Intracellular free calcium	aequorin
	obelin
4. Non-isotopic labels in immunoassay and DNA technology	firefly luciferin-luciferase
	bacterial oxidoreductase-luciferase
	aequorin
5. Reporter genes in genetic engineering	bacterial lux genes
	firefly luciferase
	aequorin

DETECTION AND QUANTIFICATION

Most bioluminescent reactions can be described by the following equation:

$$LH_2 + oxidant + cofactor(s) + cation(s) \rightarrow products + L^* \rightarrow light \qquad (18.1)$$

At constant oxidant and cofactor concentration the reaction becomes first order and exhibits a typical exponential decay:

$$light\ intensity\ (I) = dh\nu/dt = \varnothing_{CL} . (LH_2)_o . k . e^{(-kt)} \qquad (18.2)$$

where $\varnothing_{CL}$ = overall chemiluminescence quantum yield (usually 0·1-1 for bioluminescence), $(LH_2)_o$ = amount (*not* concentration) of active luciferin at time zero, k = exponential decay rate constant. k will depend on, and change with, the concentration of the reactants other than the luciferin. Strictly $(LH_2)_o$ is not the total amount of luciferin present but rather that bound to the enzyme luciferase. Only when luciferase » luciferin will $(LH_2)_o$ = total luciferin present in equation (18.2). The analyte, the material or gene expression to be measured, may then alter $\varnothing_{CL}$, $(LH_2)_o$, or k, or it may alter two other parameters not represented in equation 1, colour and polarisation. An enzyme or metabolite which changes the rate of a bioluminescent reaction can then be quantified by monitoring intensity (I), the rate constant (k) or the peak height ($\varnothing_{CL} . (LH_2)_o . k$). Sometimes when the decay rate is slow, a mini-integral, *e.g.* 10 s, is sufficiently precise. In contrast when quantifying labels in immunoassay or following DNA hybridization, a measure of the total amount of label is required. If the label is the chemiluminescent substrate itself then an end point assay is used measuring total light ($\varnothing_{CL} . (LH_2)_o$). However if the label is a catalyst, *e.g.* luciferase, peroxidase or alkaline phosphatase, then the assay conditions are usually adjusted so that light intensity, rather than total light yield, is proportional to the amount of label present.

Equation (18.2) also helps us to identify the parameters which must be optimised to gain maximum potential from a bioluminescent assay:

1. The substance to be measured must itself, or after coupling it to one of the components of the bioluminescent reaction, alter one or more of the physical characteristics of the reaction: rate of light emission, total light yield, colour or state of polarisation of emitted light.
2. The physical conditions (*e.g.* pH, temperature, ionic strength) must be compatible with the stability of the substance or biological system being studied. This is particularly relevant when measuring things inside live cells.
3. The quantum yield $\varnothing_{CL}$ should be high, ideally 0·1-1, for maximum sensitivity, and not be susceptible to alterations by assay conditions, solvents or serum.
4. The kinetics of the reaction must be appropriate. A slow reaction taking many hours to generate a measurable signal would reduce its value as a replacement for radiolabels, but might be particularly suitable for studies inside cells where complete consumption of the chemiluminescent substrate in a fast reaction would mean that the biological phenomenon could be observed for perhaps only a few seconds.

Four types of detectors have been developed to quantify light intensity:
1. Thermal (bolometers and thermopiles).
2. Chemical (photographic paper).

3. Biological (the eye).

4. Electrical (photoelectric devices and photomultiplier tubes).

The three detectors in current use are highly sensitive photographic film (*e.g.* 20000 ASA), photomultiplier tubes, and solid state detectors such as photodiodes and charge-coupled devices. To detect very low intensities of just a few hundred photons per second a photomultiplier with a low dark current must be used. This will saturate at about 10^{-7} photons s^{-1} and is blue sensitive. Thus the firefly luciferin-luciferase reaction needs to generate some five times as many photons as the Ca^{2+}-activated photoprotein aequorin to generate the same number of pulses from a bialkali photomultiplier. A typical, high sensitivity chemiluminometer consists of four parts:

1. A temperature-controlled, completely light-tight housing, enabling solutions to be injected to trigger or alter the chemiluminescent reaction being observed.

2. A highly sensitive, cooled photomultiplier tube maintained at some 1000 v (or a photodiode, red sensitive, in hand-held equipment).

3. A signal processor, with discriminator for pulse counting (more sensitive in terms of signal:noise than wholly analogue devices), and an analogue converter for displaying traces.

4. Data processor and display: scaler, chart recorder, oscilloscope and interfaced to computer + VDU + printer.

Several excellent commercial chemiluminometers are now available, including microtitre plate readers. Homogeneous assays based on colour shifts and/or energy transfer require a dual photomultiplier system (Campbell *et al.*, 1985). Assays in microtitre plates or southern, northern and western blots can be visualised on Polaroid film (Kricka & Thorpe, 1987), quantification being achieved by eye, with or without a neutral-density filter system for standards. However for low light intensity assays or for visualising bioluminescent genes expressed in cells or whole organisms, an image intensifier is required. The four-stage electromagnetically focussed system pioneered by Reynolds (Reynolds, 1978; Eisen & Reynolds, 1985) is probably still the most sensitive, as good as a photomultiplier tube, but is no longer commercially available. Liquid N_2 cooled charge-coupled devices providing a video signal easily read off pixel by pixel by computer, offer exciting prospects for the future. The most sensitive and flexible systems for biologists usually combine a sensitive light detector with an electron multiplier, followed by computer analysis to produce a pseudo-colour image or contour map, or TV recording. Popular combinations include: intensified silicon intensified TV tubes or image photon detecting systems where the primary detector is a photocathode, the electrostatically focussed electrons are then multiplied by a microchannel plate and the resulting image on the phosphor processed by a charge-coupled device. These devices enable luminescent events to be located and quantified within single cells.

CELL PHYSIOLOGY AND PATHOLOGY

A key feature of all animal and many plant cells is their ability to change state or behaviour in response to an initial event at the plasma membrane. The primary stimulus may be physical, such as an action potential or touch; chemical, such as neurotransmitter or hormone binding to a receptor; or biological, such as a bacterium or virus. In addition, the plasma membrane is the primary target of many pathogens, not only microorganisms but also toxins released by them, as well as cellular and humoral components of the body's defence system such as phagocytes, T lymphocytes, antibodies and complement. The end result is the initiation of an event within the cell leading to phenomena including cell movement, division, transformation, secretion, cell leakage or even death. Furthermore pathogens can also initiate defence pathways as a result of chemical or morphological events which begin inside the cell, well away from the plasma membrane. All these processes require the generation of intracellular signals, initially at the plasma membrane, which can reach an appropriate target within the cell. The signal can then either directly trigger the event following binding to a particular macromolecule, or induce covalent modification such as phosphorylation of proteins, which are then responsible for provoking the cell's end response. Energy in the form of ATP is also required.

Three groups of intracellular signals have been identified so far:

1. Cations *e.g.* Ca^{2+}, H^+.
2. Nucleotides *e.g.* cyclic AMP, cyclic GMP.
3. Phospholipid derivatives *e.g.* inositol phosphates and diacyl glycerol.

The ultimate objective of the molecular cell biologist is to identify the primary and secondary signals involved in controlling cell activity, and to define the molecular basis of the chemisymbiosis between these signals, energy supply and protein modification determining if and when a particular cell passes an end-response threshold (see Campbell, 1983, 1988). In order to achieve this it is essential that intracellular signals, ATP, protein modification and the end response can be quantified and located temporally in individual cells.

Ridgway & Ashley (1967) injected the Ca^{2+}-activated photoprotein aequorin into the giant muscle fibre of the barnacle *Balanus nubilis*, and for the first time were able to correlate cytosolic free Ca^{2+} with the onset and development of tension. Since then, aequorin, and its British relative obelin (Campbell, 1974) obtained from the hydroid *Obelia geniculata* in Plymouth, have been incorporated into a wide range of invertebrate and vertebrate cells (Ashley & Campbell, 1979; Campbell, 1983, 1988). We, and others, have had to develop special ways of incorporating these proteins and their mRNA into small mammalian cells (Campbell & Hallett, 1983; Campbell *et al.*, 1988; Scolding *et al.*, 1989). pH or temperature sensitive liposomes targetted onto a cell with a monoclonal antibody enables active photoprotein, apoprotein or mRNA to be incorporated into phagocytic cells such as neutrophils and non-phagocytes such as oligodendrocytes. Native photoprotein is formed from apoprotein by addition of the prosthetic group coelenterazine (Figure 18.3). Aequorin and obelin bind $3Ca^{2+}$, and their chemiluminescence is related to absolute free Ca^{2+} by a rate constant. Shimomura *et al.* (1988) have synthesized a coelenterazine analogue which results in a bimodal spectrum for aequorin. The ratio of chemiluminescence at 410-420 to 460-480 nm is linearly related to free Ca^{2+}. Obelin behaves in a similar manner (Campbell, Kishi & Shimomura, unpublished). However the new photoprotein is much less stable than the natural protein. The particular advantages photoproteins offer for measuring cytosolic free Ca^{2+} compared with metallochromic dyes, microelectrodes, nmr or fluorescent probes are: selectivity for Ca^{2+}, they cover the entire range of cytosolic free Ca^{2+} (0·1-100 µM), *i.e.* resting - stimulated - pathological, they can amplify the Ca^{2+} signal without disturbing the cell's Ca^{2+} balance, and are less susceptible to internal redistribution or leakage during tissue development or cell injury.

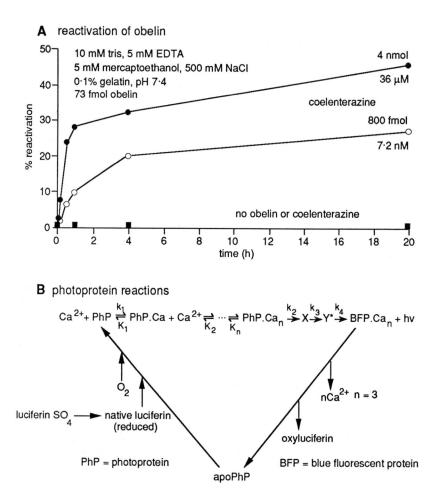

Figure 18.3. Reactivation of obelin from apoobelin from Campbell (1988) with permission of Ellis Horwood

Important discoveries arising from the use of photoproteins have been the identification of primary cell stimuli, either dependent or independent of a rise in cytosolic Ca^{2+}; identification of secondary regulators acting on the Ca^{2+} transient; the requirement for internal Ca^{2+} stores; definition of the range of cytosolic free Ca^{2+}; identification of gradients (Eisen & Reynolds, 1985) and oscillations (Woods *et al.*, 1986) in cytosolic free

Ca^{2+} as a possible mechanism for maintaining cell thresholds; and the distinction between rises in Ca^{2+} as friend or foe, cause or consequence, following injury to a cell. These are well illustrated by the activation of neutrophils (Figure 18.4), the cell on which we have focussed.

We need oxygen to live, yet oxygen can be toxic. Neutrophils are the dominant white cell in the blood and our first defence against microorganisms. Part of the process for killing invading pathogens involves production of O_2^-, then H_2O_2 and OCl^- in the phagosome. Release of these highly reactive metabolites causes damage to surrounding tissue in rheumatoid arthritis, myocardial infarction, kidney and lung disease. Chemi-

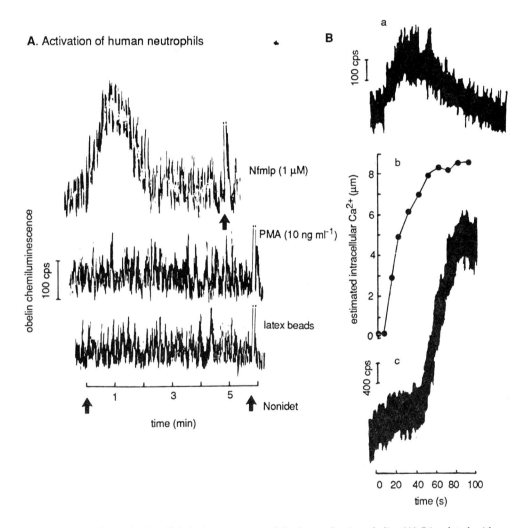

Figure 18.4. Cytosolic free Ca^{2+} in human neutrophils detected using obelin: (A) Stimulated with chemotactic peptide (Nfmlp), phorbol ester (PMA) or particles (latex beads) from Campbell (1988) with permission of Ellis Horwood. (B) Effect of the membrane attack complex of complement. a. Obelin chemiluminescence, b Estimated intracellular free Ca^{2+}, c. Oxygen metabolite production estimated using luminol dependent chemiluminescence. Complement was activated at the cell surface at time zero. From Hallett & Campbell (1982) with permission of MacmillanPublications Ltd.

luminescence provides a sensitive method for detecting release of reactive oxygen metabolites (luminol for H_2O_2 in the presence of peroxidase; lucigenin or pholasin down to a single cell (Robert *et al.*, 1987) for O_2^-; vinyl pyrenes for $^1O_2$). Using obelin to monitor cytosolic free Ca^{2+} in neutrophils we have shown that chemotactic stimuli activate the O_2^- producing pathway via a rise in cytosolic Ca^{2+}, whereas phagocytic stimuli and tumour promoters activate via mechanisms independent of a rise in cytosolic free Ca^{2+} (Campbell & Hallett, 1983). Adenosine inhibits O_2^- release only by those stimuli dependent on a rise in cytosolic free Ca^{2+}, and can abolish the Ca^{2+} transient.

We have also used neutrophils, and oligodendrocytes, to identify a new mechanism of reversible cell injury which we propose plays a key role in the release of oxygen metabolites in the joints of patients with rheumatoid arthritis, and in inducing demyelination in multiple sclerosis (Campbell & Luzio, 1981; Hallett *et al.*, 1981; Scolding *et al.*, 1989). Soluble, globular proteins from the complement pathway, in particular the terminal component C9, insert themselves into the plasma membrane. The 'British doughnut' shaped molecule induces a rapid influx of Ca^{2+} and a release of Ca^{2+} from internal stores. This Ca^{2+}, together possibly with a protein kinase C pathway, activates a defence mechanism against this potentially lethal protein in the membrane. Unless it is removed the cells will eventually undergo osmotic lysis. The membrane attack complexes containing C9 are channelled into a vesicle as inactive 'American donut' shaped 'pores', and are removed by budding and endocytosis. These cells recover rapidly and can respond immediately to chemotactic and phagocytic stimuli. However, many cells are not able to protect themselves in time and undergo a second permeability threshold to molecules of about 400-1000 molecular weight, probably as the result of aggregation of the C9 'doughnuts' in the plasma membrane. These cells can still protect themselves, but once the C9 complexes have been removed it takes an hour or so for the cells to recover completely. Some cells cannot protect themselves in time and these lyse. Support for the pathological significance of this mechanism comes from the detection of vesicles containing complement complexes in several immune based and inflammatory diseases. Similar, structurally related proteins are released as perforins by T cells, toxins by bacteria and some invertebrates, and are found in viral coats. Thus the measurement of free Ca^{2+} using a bioluminescent protein appears to have led to the discovery of a cellular defence system of wide biological significance.

NEW LAMPS FOR OLD AT THE HOSPITAL BED

Immunoassay is a widely used technique in clinical diagnosis for measuring steroid, thyroid and polypeptide hormones, vitamins, cyclic nucleotides, cancer related proteins, autoantigens, microbial antigens and antibodies, and drugs. The technique depends on labelling either a standard antigen or antibody, and then mixing antibody, antigen and sample. A variety of 'competitive' and titrametric strategies using single-site, two-site (sandwich) methods and either polyclonal or monoclonal antibodies have been developed. The commercial markets world-wide are thousands of millions of dollars per year, such is the scale of their use in hospitals and elsewhere. Three chemiluminescent labels have now been used widely, and in laboratories where real clinical decisions are taken:

1. Isoluminol derivatives, *e.g.* aminobutyl ethyl isoluminol (Campbell & Patel, 1983).
2. Acridinium esters (Weeks *et al.*, 1983; Weeks & Woodhead, 1987).
3. Peroxidase detected using luminol chemiluminescence enhanced by compounds such as iodophenol (Whitehead *et al.*, 1983).

The precise detection limited for these labels has been the source of some controversy. All of these labels can be detected down to 10^{-15} mol, but the signal:chemical-noise ratio appears best with acridinium esters, which are detectable down to 10^{-17} mol, equivalent to ^{125}I. Recently two ingenious methods for exploiting the bacterial and firefly luciferases have been developed. In the latter luciferin phosphate (Miska & Geiger, 1983) is hydrolysed by alkaline phosphatase, the immunoassay or DNA probe label. The luciferin produced is then able to generate light in the presence of the luciferase. Light emission is prolonged and like luminol-enhanced peroxidase chemiluminescence can be detected in microtitre plates or nitrocellulose by a camera luminometer (Geiger *et al.*, 1989). Similar phosphorylated derivatives of adamantane dioxetans have also been developed (Hummelen *et al.*, 1987).

Bacterial luciferase has been coupled to a homogeneous immunoassay protocol (Terouanne *et al.*, 1986) whereby an antibody to the antigen being measured, bacterial oxido-reductase and luciferase are covalently attached to solid phase. NAD glucose 6 phosphate dehydrogenase is used as the label on the antigen. As more free antigen is added, the less the antigen-dehydrogenase is bound to solid phase. Only NADH generated locally at the solid phase surface is detected by the luciferase. We developed an alternative homogeneous immunoassay strategy using chemiluminescence energy transfer (Campbell & Patel, 1983; Campbell *et al.*, 1985). Antigens of molecular weight 180-150000 can be detected by a change in colour of the light emission without the need to separate bound from free antigen or antibody. This principle can equally be applied to the detection of DNA-DNA or DNA-RNA recombinants without the need for blotting. Peroxidase, acridinium esters and alkaline phosphatase can be linked to DNA probes by coupling to the 3' or 5' ends, or by biotinylation/avidin and nick translation. Microbial DNA and RNA markers, and restriction fragment length polymorphisms have been detected using both solid phase reagents and southern or northern blotting with peroxidase and acridinium ester labels, but these are not as sensitively detectable as ^{32}P (Matthews *et al.*, 1985). However Ca^{2+} activated photoproteins, and alkaline phosphatase detected using firefly luciferin-phosphate, can be measured down to 10^{-20} mol, some one-two orders of magnitude more sensitive than ^{32}P. The rapidly expanding use of the polymerase chain reaction is making many applications of ^{32}P with blotting redundant. The need now is for a bioluminescent DNA/RNA probe which provides a homogeneous method of detecting specific genes or alleles, without the need for electrophoretic separation or blotting.

GENETIC MANIPULATION OF BIOLUMINESCENCE

Messenger RNA coding for bioluminescent proteins can be detected by translation with rabbit reticulocyte lysate to form active light-emitting protein (Figure 18.5). Three bioluminescent proteins have been cloned and sequenced: the bacterial luciferase α and β subunits (Cohn *et al.*, 1983; Miyamoto *et al.*, 1987), the firefly luciferase (DeWet *et al.*, 1985) and aequorin (Inouye *et al.*, 1985; Prasher *et al.*, 1985). We have used the pCDV1 primer plus Honjo linker (Nomo *et al.*, 1986) to clone firefly luciferase, active luciferase being detected by light emission in the transformed *Escherichia coli* (Sala-Newby, Kalsheker & Campbell, unpublished). All three genes have been expressed in *E. coli* to generate light (DeWet *et al.*, 1985; Engebrecht & Silverman, 1987; Inouye *et al.*, 1985). Firefly and bacterial luciferase cDNA's have also been expressed in plant and mammalian cells, and these genes have been linked to promoters such as SV40 and auxin. Visualisation of light-emitting cells in whole tissues or plants using image intensifiers enables promoter activation to be located (Dilella *et al.*, 1988; Ow *et al.*, 1986; Riggs & Chrispeels, 1987). Firefly luciferase contains a sequence targetting it into peroxisomes (Keller *et al.*, 1987). The intracellular location of this luciferase formed from cDNA has not yet been established. Photoproteins can also be formed in mammalian cells from mRNA (Campbell *et al.*, 1988).

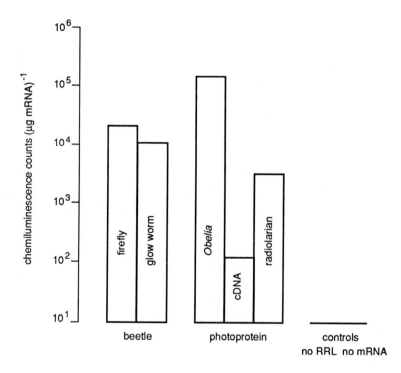

Figure 18.5. Formation of bioluminescent proteins from mRNA. RRL: rbbit reticulocyte lysate. 5μl RRL + 2μl mRNA for 1 h at 30°C.

An ingenious method for detecting very small numbers of pathogenic bacteria such as *Salmonella* has been developed by Ulitzur & Kuan (1989). A phage selective for the particular bacterium is genetically engineered containing the operon for bacterial luciferase and the other proteins (known as lux genes A, B, C, D, E, R, and I) for light emission. Only when this phage infects a living bacterium is light generated.

FUTURE

Much has been learned about the physiology and pathology of cells using the bioluminescent proteins aequorin, obelin and firefly luciferase. However the clinical exploitation of bioluminescence, as opposed to chemiluminescence from synthetic compounds, has been restricted by the lack of availability of bulk reagents with the necessary quality control. A few litres of transformed *E. coli* can yield several hundred milligrams of luciferase. Thus cloning and genetic engineering offer exciting prospects not only for generating sufficient quantities of bioluminescent proteins, but also for producing proteins with new properties so heralding a new era in the study of the chemistry of living light. The coupling of bioluminescent genes to specific promoters together with their incorporation into transgenic organisms provides a unique technique for studying the physiology and pathology of gene expression during development. The discovery from thermal vents or the genetic engineering of heat stable luciferase and photoproteins will have valuable application in procedures such as the thermal cycling in the polymerase chain reaction.

The commercial markets for these new reagents is huge, but these new technologies also now offer the scientist an opportunity to probe what is perhaps the central problem in contemporary biology, how step by step changes in DNA throughout evolution led to the appearance of chemical events which only then became susceptible to the forces of Darwinian-Mendelian selection.

I thank members of my research group for many years hard work. I thank the Director and staff of the Marine Biological Association and the late Dr L.H.N. Cooper FRS for much support, and Dr P.J. Herring at IOS, Deacon Laboratory for the opportunity to study living light on RRS 'Discovery'. I also thank the MRC, AFRC, Arthritis and Rheumatism Council, the Multiple Sclerosis Society and Brown and Maurice Hill Bequest Fund of the Royal Society for financial support

Summary and Perspectives

It seems appropriate at the end of a volume such as this to provide a brief summary of the wide variety of topics discussed in the chapters, and to speculate about some possible future developments.

The primary impact of light on life in the sea is in the surface layers that are penetrated by radiation from the sun. In this photic zone, the light regime is complex and it is not until we reach a depth of 200 metres that the spectral characteristics of the incident light become uniform and predictable.

The spectral distribution of the energy available to organisms in the photic zone depends on the characteristics of the sun's radiation and on the scattering and absorption properties of sea water, of phytoplankton cells and of suspended particles (Sathyendranath & Platt, Chapter 1). Photosynthetically active radiation (PAR) is largely confined to the wavelength band 400 to 700 nanometres and constitutes approximately one-third of the photon flux incident on the sea surface. Less than 10% of the PAR is effectively trapped by photosynthesising organisms. Fluorescence and bioluminescence contribute only a small fraction of the total light energy available in the photic zone but can be used to good effect in investigating the distribution of plants in the sea, since the spectral characteristics act as fingerprints for the organisms involved.

The light emitted by the sea can be detected by remote sensing devices and its spectral quality is usually dominated by the scattering and absorption characteristics of phytoplankton cells. Ocean colour can therefore be exploited to determine the distribution of phytoplankton biomass on a global scale and to relate its patchiness to the physical processes controlling the stability of the surface layers and the supply of nutrients (Robinson, Chapter 2). Investigations now underway are seeking to extend the information that can be obtained from remote sensing by linking the spectral data to other factors such as temperature, location and season to provide measures of gross productivity on a global scale. The synoptic view provided is of vital importance to studies of the health of phytoplankton communities in the ocean and their influence on climatic change.

To interpret the information received from passive (satellite) or active (air-borne lasers) measurements of ocean colour it is necessary to make extensive corrections for the atmospheric absorption and scattering of light and for the presence of suspended sediments in the water that can influence the light regime by scattering and absorption and thereby obscure the signal attributable to the phytoplankton (Robinson, Chapter 2). It is also necessary to allow for variation in the vertical distribution of phytoplankton cells below the integration depth of the ocean colour measurements (usually less than 10 m). Consequently, the remotely-sensed data have to be calibrated against measurements made in the water column. This is not a trivial matter and elaborate undulating instruments have been developed to obtain information on the spectral characteristics of

Herring, P.J., Campbell, A.K., Whitfield, M. & Maddock, L. (ed.). *Light and Life in the Sea.*
© 1990. Cambridge University Press.

upwelling and downwelling radiation within the photic zone on time and space scales compatible with those of remotely sensed data (Aiken & Bellan, Chapter 3).

The photic zone itself may be naively considered as the depth to which phytoplankton can effectively harvest the incoming radiation. In reality phytoplankton growth can be sustained only by maintaining a favourable balance of photosynthesis over respiration (Tett, Chapter 4). The 'compensation point' is defined as the depth below which algae consume more energy in respiration than they receive from photosynthesis and corresponds approximately to an incident radiation of 1% PAR. In the open ocean, turbulence will constantly move the phytoplankton into and out of the photic zone and a critical depth must be defined over which the integrated values of photosynthesis and respiration exactly match. The assessment of the critical depth is complicated because it involves the determination of factors controlling both the efficiency of utilisation of light energy in the production of phytoplankton biomass and the rate of breakdown of the accumulating biomass by respiration and decay. However, it is essential that these factors should be resolved since the response of the critical depth to changes in global climate is now of vital concern.

One important factor is the efficiency of the initial photon-capturing step. Marine plants contain an array of accessory pigments that trap the incoming photons and facilitate the transfer of the absorbed light energy to assemblies of chlorophyll *a* molecules where the photosynthetic process begins (Dring, Chapter 5). These accessory pigments confer different wavelength sensitivities on the various species of phytoplankton and also provide a means for identifying their presence from pigment analyses and possibly from ocean colour measurements using appropriate spectral resolution. Despite the variety of pigmentation observed, the balance of evidence suggests that it is the total irradiance, rather than its spectral quality, that controls the depth distribution of phytoplankton and benthic algae in the sea. The relationships are, however, complex and cyanobacteria, which possess a pigment (phycoerythrin) which is adapted to collecting light at the blue/green wavelengths and low intensities predominant at depths of a few hundred metres, are often more abundant and more productive in surface waters (Joint, Chapter 6).

In the multicellular marine plants found in estuaries and coastal waters, light also influences development by controlling the direction of growth (phototropism) and by influencing patterns of cell division (photomorphism). The way in which the directional light signal is interpreted and converted into spatial information depends on the nature and distribution of photopigments within the cell (Brownlee, Chapter 7). The research into this aspect of the impact of light on plant life has many parallels in the investigation of the perception and interpretation of light fields by animals. The mechanism of photoreception in the vertebrate eye starts with a messenger of transduction (cyclic GMP) in the rods and cones of the retina (Matthews, Chapter 11), whereas a different biochemical mechanism is involved in the adaptation of the eye to light, based on the influx and efflux of calcium ions. An invertebrate with very highly developed vision, the squid, has highly specialised and structurally ordered photoreceptors which provide a favourable system for the study of the enzyme pathways in photoreception (Saibil, Chapter 12).

Only in coastal and reef conditions does the light environment of the sea approach the complexitiy of that on land. Local differences in the quantity and quality of the light (on which are superimposed the cyclical day/night fluctuations) can be exploited by animals in these environments by retreating into shadows or by using behavioural or chromatic patterns to increase or decrease their visibility (Potts, Chapter 14). The optical complexity of this environment renders extreme specialization of one element of the visual system, at the expense of others, a highly evolutionary risk - unless alternative sensory systems can compensate. In the open sea the illumination is much more uniform and predictable, particularly at depths below the photic zone (Denton, Chapter 8). Acute changes in the temporal and spatial nature of the radiance distribution are primarily caused by bioluminescence. The visual tasks within such an environment allow a greater commitment to particular specializations of the visual systems of the animals concerned, with less risk of temporary visual handicap by exposure to local fluctuations or patchiness in the light conditions.

The specializations can often be identified in the optical anatomy of the eye, which may, for example, be tubular or bilobed (Land, Chapter 9). Physiological specializations (Denton, Chapter 8; Land, Chapter 9; Partridge, Chapter 10) are less obvious, but no less significant, and in deep water species are frequently designed to maximise the photon capture efficiency. However, little is yet known about the neural and processing capabilities necessary to translate the identified physiological subtleties into successful behavioural patterns for survival. As Denton so clearly emphasises, the ultimate limitations on some aspects of vision in the sea may well be imposed by the processing systems rather than by signal reception and transduction.

Unlike the land, the sea is a vast three-dimensional environment of limited transparency to visible wavelengths. Horizontal excursions by an animal at any depth in the open sea will not take it into a significantly different light regime (Denton, Chapter 8). Vertical excursions, however, will have major consequences in the top 800 m, most notably in exponential changes in light intensity. Nevertheless by linking its vertical movements to changes in overhead daylight an animal can minimise the rate of change in its light environment. The diel vertical migrations of many species can certainly have this effect, optimizing the effectiveness of an eye adapted to a particular level of illumination (Blaxter and Batty, Chapter 13; Denton, Chapter 8). Rates of light adaptation in animals whose eyes lack a mobile iris are relatively slow and very little is yet known about the range of intensities to which different marine species can adapt or the time course of such adaptations. Surface light intensities can cause irreparable damage to the eyes of deep-living species, yet many species undergo an ontogenetic migration from the near-surface to deeper waters, during which their visual systems adapt to the changes in both intensity and spectral distribution (Partridge, Chapter 10).

As the light from the surface declines with depth (or time of day) so biologically generated light becomes of increasing significance. Bioluminescence provides the unpredictable temporal and spatial elements in an otherwise relatively uniform light environment. Its functional value depends on its efficient perception, either directly or by reflection, from a potential prey, predator, mate or rival. Many nocturnally active shallow marine species employ bioluminescence in a variety of ways (*e.g.* flashlight fishes and luminous brittle stars) and its role in deeper water species is even more extensive (Herring, Chapter 16). The visual adaptations of such species will be increasingly determined by the characteristics of the ambient bioluminescence. The emission by some fish of longwave red bioluminescence to which they are also uniquely sensitive is a case in point.

Many marine animals are attracted by a light, a response presumed to be exploited by the luminous lure of the anglerfish. This attraction has also been widely exploited for commercial purposes (Pascoe, Chapter 15). The varied responses of animals to the lights on a fishing net provide a possible analogue of their responses to similar, but lower intensity, light produced in a net by bioluminescence, with the consequent bias to the net's capture efficiency. The frequency with which bioluminescence has evolved in marine organisms (McCapra, Chapter 17) is an indication of its importance in the success of both nocturnal and deep-water species.

A knowledge of the molecular biology of phenomena such as photosynthesis, lens structure and vision, and bioluminescence has an obvious part to play in our understanding of the evolutionary pathways responsible for optimizing these processes. This knowledge also highlights one of the central problems in modern biology, the conflict between evolution at the molecular level and at the level of gene selection; the former seldom conforms to the principles of Darwinian-Mendelian selection. Bioluminescent indicators have been developed for probing the chemistry of living cells. This has led to new insights not only into the physiology of cells but also into what goes wrong with them in human disease. We have now seen how the identification of a common chemistry, the imidazolopyrazines, in luminous organisms from six phyla (Campbell, Chapter 18) has implications for studying food chains in the deep ocean.

Advances in technology and scientific methods promise to shed new light on many of the issues raised in the chapters of this volume. Developments in molecular biology are leading to further progress in understanding the relationship between light and the behaviour of marine organisms, and communication between them, by the identification and manipulation of the genes responsible for their development. The molecular approach is not a substitute for studies on individual organisms or whole ecosystems. Rather it gives the biologist a new pair of 'eyes' with which to view the phenomenon under investigation leading to new questions and a new experimental strategy for a wiser perspective, enhancing our perception of its natural beauty. At the opposite end of the spectrum new electronic instruments enable us to obtain, process and model vast amounts of information about the sea. Such instruments may be deployed on ships or submersibles, on moorings or satellites, and all provide data leading to new insights into different aspects of the study of light in the sea and how it is changed, used or produced by living organisms. The advent of the new research submersible offers exciting prospects for experimental observations *in situ*. Light is so fundamental a feature for, and of, animal life in the sea that an effective description and interpretation of the ecology of the oceans cannot be achieved without a full analysis of the integrated roles of light, colour and vision in this immense environment.

Bibliography

Numbers in square brackets, *e.g.* [Ch. 2], indicate the Chapter in this volume in which the reference is quoted.

Abbott, M.A. & Zion, P.M., 1987. Spatial and temporal variability of phytoplankton pigment off Northern California during Coastal Ocean Dynamics Experiment 1. *Journal of Geophysical Research*, **92**C, 1745-1755. [Ch. 2]

Aguilar, M. & Stiles, W.S., 1954. Saturation of the rod mechanism at high levels of stimulation. *Optica Acta*, **1**, 59-65. [Ch. 8]

Aho, A-C., Donner K, Hydén, E. Larsen, L.O. & Reuter, T., 1988. Low retinal noise in animals with low body temperature allows high visual sensitivity. *Nature, London*, **334**, 348-350. [Ch. 10]

Aiken, J., 1981*a*. The Undulating Oceanographic Recorder mark 2. *Journal of Plankton Research*, **3**, 551-560. [Ch. 3]

Aiken, J., 1981*b*. A chlorophyll sensor for automatic, remote, operation in the marine environment. *Marine Ecology - Progress Series*, **4**, 235-239. [Ch. 3]

Aiken, J., 1985. The Undulating Oceanographic Recorder mark 2. A multirole oceanographic sampler for mapping and modeling the biophysical marine environment. *Advances in Chemistry Series*, **209**, 315-332. [Ch. 3]

Aiken, J. & Bellan, I., 1986*a*. Synoptic optical oceanography with the Undulating Oceanographic Recorder. *Proceedings of the Society of Photo-optic Instrumentation Engineers*, **637**, 221-229. [Ch. 2, 3]

Aiken, J. & Bellan, I., 1986*b*. A simple hemispherical, logarithmic light sensor. *Proceedings of the Society of Photo-optic Instrumentation Engineers*, **637**, 211-216. [Ch. 3]

Alberte, R.S. & Andersen, R.A., 1986. Antheraxanthin, a light harvesting carotenoid found in a chromophyte alga. *Plant Physiology*, **80**, 583-587. [Ch. 5]

Alberte, R.S., Friedman, A.L., Gustafson, D.L., Rudnick, M.S. & Lyman, H., 1981. Light-harvesting systems of brown algae and diatoms. Isolation and characterisation of chlorophyll *a/c* and chlorophyll *a*/fucoxanthin pigment protein complexes. *Biochimica et Biophysica Acta*, **635**, 304-316. [Ch. 5]

Alberte, R.S., Wood, A.M., Kursar, T.A. & Guillard, R.R.L., 1984. Novel phycoerythrins in marine *Synechococcus* spp.: characterization, and evolutionary and ecological implications. *Plant Physiology*, **75**, 732-739. [Ch. 5]

Alfoldi, T.T. & Munday, J.C., 1978. Water quality analysis by digital chromaticity mapping of Landsat data. *Canadian Journal of Remote Sensing*, **4**, 108-126. [Ch. 2]

Ali, M. A., 1959. The ocular structure, retinomotor and photobehavioral responses of juvenile Pacific salmon. *Canadian Journal of Zoology*, **37**, 965-996. [Ch. 14]

Ali, M. A., 1976. Retinal pigments, structure and function and behaviour. *Vision Research*, **16**, 1197-1198. [Ch. 14]

Ali, M.A., 1984. *Photoreception and Vision in Invertebrates*. New York: Plenum. [Ch. 9]

Ali, M.A. & Anctil, M., 1976. *Retinas of Fishes: An Atlas*. Berlin: Springer-Verlag. [Ch. 10]

Ali, M.A. & Kobayashi, H., 1968. Electroretinogram-FFF in albino trout. *Experientia*, **24**, 454-455. [Ch. 10]

Allen, R. C., 1982. Biochemiexcitation: chemiluminescence and the study of chemical oxygenation reactions. In *Chemical and Biological Generation of Excited States* (ed. W. Adam and G. Cilento), pp. 309-344. New York: Academic Press. [Ch. 17]

Allen, R.D., Jacobsen, L., Joaquin, J. & Jaffe, L.F., 1972. Ionic concentrations in developing *Pelvetia* eggs. *Developmental Biology*, **27**, 538-545. [Ch. 7]

Andersen, R.A. & Mulkey, T.J., 1983. The occurrence of chlorophylls c_1 and c_2 in the Chrysophyceae. *Journal of Phycology*, **19**, 289-294. [Ch. 5]

Anderson, D.M., Nosenchuck, D.M., Reynolds, G.T. & Walton, A.J., 1988. Mechanical stimulation of bioluminescence in the dinoflagellate *Gonyaulax polyedra* (Stein). *Journal of Experimental Marine Biology and Ecology*, **122**, 277-288. [Ch. 16]

Anderson, J.M., 1983. Chlorophyll-protein complexes of a *Codium* species, including a light-harvesting siphonaxanthin-chlorophyll *a/b*-protein complex, an evolutionary relic of some chlorophyta. *Biochimica et Biophysica Acta*, **724**, 370-380. [Ch. 5]

Anderson, J.M., 1985. Chlorophyll-protein complexes of a marine green alga, *Codium* species (Siphonales). *Biochimica et Biophysica Acta*, **806**, 145-153. [Ch. 5]

Anderson, J.M., Evans, P.K. & Goodchild, D.J., 1987. Immunological cross-reactivity between the light-harvesting chlorophyll *a/b*-proteins of a marine green alga and spinach. *Physiologia Plantarum*, **70**, 597-602. [Ch. 5]

Andrews, C.C., Karl, D.M., Small, L.F. & Fowler, S.W., 1984. Metabolic activity and bioluminescence of oceanic faecal pellets and sediment trap particles. *Nature, London*, **307**, 539-541. [Ch. 16]

Angel, M.V., 1968. Bioluminescence in planktonic halocyprid ostracods. *Journal of the Marine Biological Association of the United Kingdom*, **48**, 255-257. [Ch. 16]

Anon., 1980. *Solar Radiation Data for the United Kingdom 1951-75*. Met.O.912, Bracknell: Meteorological Office. [Ch. 4]

Apel, J.R., 1987. Principles of ocean physics. In *International Geophysics Series*, vol. 38 (ed. W.L. Donn). London: Academic Press. [Ch. 1]

Apel, K., 1977. The light-harvesting chlorophyll *a/b*-protein complex of the green alga *Acetabularia mediterranea*. Isolation and characterization of two subunits. *Biochimica et Biophysica Acta*, **462**, 390-402. [Ch. 5]

Applebury, M. & Hargrave, P., 1986. Molecular biology of the visual pigments. *Vision Research*, **26**, 1881-1895. [Ch. 12]

Archer, S.N. & Lythgoe, J.N., in press. The visual pigment basis for cone polymorphism in the guppy *Poecilia reticulata*. *Vision Research*. Ch. 10]

Archer, S.N., Lythgoe, J.N. & Partridge, J.C., 1987. Visual pigment polymorphism in the guppy *Poecilia reticulata*. *Vision Research*, **27**, 1243-1252. [Ch. 10]

Arnone, R.A. & LaViolette, P.E., 1986. Satellite definition of the bio-optical and thermal variation of coastal eddies associated with the African current. *Journal of Geophysical Research*, **91**C, 2351-2364. [Ch. 2]

Ashley, C.C. & Campbell, A.K. (ed.), 1979. *Detection and Measurement of Free Ca^{2+} in Cells*. Amsterdam: Elsevier/North-Holland. [Ch. 18]

Atlas, D. & Bannister, T.T., 1980. Dependence of mean spectral extinction coefficient of phytoplankton on depth, water color, and species. *Limnology and Oceanography*, **25**, 157-159. [Ch. 4]

Attwell, D., 1986. Ionic channels and signal processing in the outer retina. *Quarterly Journal of Experimental Physiology*, **71**, 497-536. [Ch. 11]

Austin, R.W. & Petzold, T.J., 1981. The determination of the diffuse attenuation coefficient of sea water using the CZCS. In *Oceanography from Space* (ed. J.F.R. Gower), pp. 239-256. New York: Plenum Press. [Ch. 2]

Avery, J.A., Bowmaker, J.K., Djamgoz, M.B.A., Downing, J.E.G., 1983. Ultra-violet sensitive receptors in a freshwater fish. *Journal of Physiology*, **334**, 22-24P. [Ch. 10]

Baer, K.M. & Saibil, H.R., 1988. Light- and GTP-activated hydrolysis of phosphatidylinositol bisphosphate in squid photoreceptor membranes. *Journal of Biological Chemistry*, **263**, 17-20. [Ch. 12]

Bailey, K.M. & Batty, R.S., 1983. A laboratory study of predation by *Aurelia aurita* on larval herring (*Clupea harengus*): experimental observations compared with model predictions. *Marine Biology*, **72**, 323-327. [Ch. 13]

Baker, K.S. & Frouin, R., 1987. Relation between photosynthetically available radiation and total insolation at the sea surface under clear skies. *Limnology and Oceanography*, **32**, 1370-1377. [Ch. 4]

Baker, K.S. & Smith, R.C., 1982. Bio-optical classification and model of natural waters 2. *Limnology and Oceanography*, **27**, 500-509. [Ch. 2, 10]

Bannister, T.T., 1974. Production equations in terms of chlorophyll concentration, quantum yield, and an upper limit to production. *Limnology and Oceanography*, **19**, 1-30. [Ch. 4]

Bannister, T.T. & Weidemann, A.D., 1984. The maximum quantum yield of phytoplankton photosynthesis *in situ*. *Journal of Plankton Research*, **6**, 275-294. [Ch. 4]

Banse, K., 1976. Rates of growth, respiration and photosynthesis of unicellular algae as related to cell size - a review. *Journal of Phycology*, **12**, 135-140

Barale, V., McClean, C.R. & Malanotte-Rixxoli, P., 1986. Space and time variability of the surface color field in the northern Adriatic Sea. *Journal of Geophysical Research*, **91C**, 12957-12974. [Ch. 2]

Barkdoll, A.E., Pugh, E.N., Jr & Sitaramayya, A., 1989. The calcium dependence of the activation and inactivation kinetics of the light-activated phosphodiesterase of retinal rods. *Journal of General Physiology*, **93**, 1091-1108. [Ch. 11]

Barkmann, W.R. & Richards, K.J., 1989. Interactions between the ocean mixed layer and mesoscale eddies. *Annales Geophysicae*, SII2-6, 130. [Ch. 2]

Barlow, H.B., 1957a. Increment thresholds at low intensities as signal/noise discrimination. *Journal of Physiology*, **136**, 469-488. [Ch. 8]

Barlow, H.B., 1957b. Purkinje shift and retinal noise. *Nature, London*, **179**, 255-256. [Ch. 10]

Barlow, H.B., 1958. Temporal and spatial summation in human vision at different background intensities. *Journal of Physiology*, **141**, 337-350. [Ch. 8]

Barlow, H.B., 1988. The thermal limits of seeing. *Nature, London*, **334**, 296-297. [Ch. 10]

Barlow, R.G. & Alberte, R.S., 1985. Photosynthetic characteristics of phycoerythrin-containing marine *Synechococcus* spp. I. Responses to growth photon flux density. *Marine Biology*, **86**, 63-74. [Ch. 5, 6]

Barnes, A.T. & Case, J.F., 1972. Bioluminescence in the mesopelagic copepod *Gaussia princeps* (T. Scott). *Journal of Experimental Marine Biology and Ecology*, **8**, 53-71. [Ch. 16]

Barnes, A.T. & Case, J.F., 1974. The luminescence of lanternfish (Myctophidae): spontaneous activity and responses to mechanical, electrical and chemical stimulation. *Journal of Experimental Marine Biology and Ecology*, **15**, 203-221. [Ch. 16]

Barrett, J. & Anderson, J.M., 1977. Thylakoid membrane fragments with different chlorophyll *a*, chlorophyll *c* and fucoxanthin compositions isolated from the brown seaweed *Ecklonia radiata*. *Plant Science Letters*, **9**, 275-283. [Ch. 5]

Barrett, J. & Anderson, J.M., 1980. The P700 chlorophyll *a*-protein complex and two major light-harvesting complexes of *Acrocarpia paniculata* and other brown seaweeds. *Biochimica et Biophysica Acta*, **590**, 309-323. [Ch. 5]

Basch, L.V., 1988. Bioluminescent anti-predator defense in a subtidal ophiuroid. In *Echinoderm Biology* (ed. R.D. Burke *et al.*), pp. 503-515. Rotterdam: A.A. Balkema. [Ch. 16]

Bassot, J.-M., Bilbaut, A., Mackie, G.O., Passano, L.M. & Pavans de Cecatty, M., 1978. Bioluminescence and other responses spread by epithelial conduction in the siphonophore *Hippopodius*. *Biological Bulletin. Marine Biological Laboratory, Woods Hole, Mass.*, **155**, 473-498. [Ch. 16]

Bastian, B.L. & Fain, G.L., 1979. Light adaptation in toad rods: requirement for an internal messenger which is not calcium. *Journal of Physiology*, **297**, 493-520. [Ch. 11]

Batty, R.S., 1987. Effect of light intensity on activity and food-searching of larval herring, *Clupea harengus*: a laboratory study. *Marine Biology*, **94**, 323-327. [Ch. 13]

Batty, R.S., 1989. Escape responses of herring larvae to visual stimuli. *Journal of the Marine Biological Association of the United Kingdom*, **69**, 647-654. [Ch. 13]

Batty, R.S., Blaxter, J.H.S. & Libby, D.A., 1986. Herring (*Clupea harengus*) filter-feeding in the dark. *Marine Biology*, **91**, 371-375. [Ch. 13]

Baverstock, J., Fyles, J. & Saibil, H., 1989. Calcium inhibits GTP-binding proteins in squid photoreceptors. In *Receptors, Membrane Transport and Signal Transduction* (ed. A.E. Evangelopoulos *et al.*), pp. 76-84. Berlin: Springer-Verlag. [NATO ASI Series, vol. H29] [Ch. 12]

Baylis, L.E., Lythgoe, R.J. & Tansley K., 1936. Some new forms of visual purple found in sea fishes, with a note on the visual cells of origin. *Proceedings of the Royal Society* (B), **816**, 95-113. [Ch. 10]

Baylis, P.E., 1981. Dundee University meteorological satellite ground receiving and data archiving facility. In *Oceanography from Space* (ed. J.F.R. Gower), pp. 49-58. New York: Plenum Press. [Ch. 2]

Baylor, D.A. & Hodgkin, A.L., 1973. Detection and resolution of visual stimuli by turtle photoreceptors. *Journal of Physiology*, **234**, 163-198. [Ch. 11]

Baylor, D.A. & Hodgkin, A.L., 1974. Changes in time scale and sensitivity in turtle photoreceptors. *Journal of Physiology*, **242**, 729-758. [Ch. 11]

Baylor, D.A., Lamb, T.D. & Yau, K.-W., 1979. The membrane current of single rod outer segments. *Journal of Physiology*, **288**, 589-611. [Ch. 11]

Baylor, D.A., Matthews, G. & Yau, K.-W., 1980. Two components of electrical dark noise in toad retinal rod outer segments. *Journal of Physiology*, **309**, 591-621. [Ch. 11]

Beatty, D.D., 1975. Visual pigments of the American eel *Anguilla rostrata*. *Vision Research*, **15**, 771-776. [Ch. 10]

Beatty, D.D., 1984. Visual pigments and the labile scotopic visual system of fish. *Vision Research*, **24**, 1563-1574. [Ch. 10]

Beebe, W.M., 1934. Deep-sea fishes of the Bermuda oceanographic expeditions. IV. Family Idiacanthidae. *Zoologica, New York*, **16**, 149-241. [Ch. 16]

Beebe, W.M., 1935. *Half Mile Down*. London: Bodley Head. [Ch. 16]

Beebe, W.M. & Vander Pyl, M., 1944. Eastern Pacific expeditions of the New York Zoological Society. XXXIII. Pacific Myctophidae. *Zoologica, New York*, **29**, 59-95. [Ch. 16]

Beer, S. & Levy, I., 1983. Effects of photon fluence rate and light spectrum composition on growth, photosynthesis and pigments relations in *Gracilaria* sp. *Journal of Phycology*, **19**, 516-522. [Ch. 5]

Beer, Th., 1897. Die Akkomodation des Kephalopodenauges. *Archiv für die Gesammte Physiologie*, **67**, 541-586. [Ch. 8]

Beijerinck, M.W., 1902. Photobacteria as a reactive in the investigation of the chlorophyll-function. *Proceedings of the Academy of Science, Amsterdam*, **4**, 45-49. [Ch. 18]

Ben Yami, M., 1976. *Fishing with Light*. Farnham: Fishing News Books. [FAO Fishing Manuals.] [Ch. 15]

Bentrup, F.-W., 1963. Vergleichende untersuchungen zur polaritatsinduktion durch das licht an der *Equisiteum* spore und der *Fucus* zygote. *Planta*, **59**, 472-491. [Ch. 7]

Berkner, L.V. & Marshall, L.C., 1965. On the origin and rise of oxygen concentration in the Earth's atmosphere. *Journal of Atmospheric Science*, **22**, 225-261. [Ch. 4]

Berner, T., Dubinsky, Z., Wyman, K. & Falkowski, P.G., 1989. Photoadaptation and the 'package' effect in *Dunaliella tertiolecta* (Chlorophyceae). *Journal of Phycology*, **25**, 70-78. [Ch. 5]

Berridge, M.J., 1987. Inositol trisphosphate and diacylglycerol: two interacting second messengers. *Annual Review of Biochemistry*, **56**, 159-193. [Ch. 7]

Berridge M.J. & Irvine, R.F., 1984. Inositol trisphosphate, a novel second messenger in cellular signal transduction. *Nature, London*, **312**, 315-321. [Ch. 7]

Bertsch, H., 1968. Effect of feeding by *Armina californica* on the bioluminescence of *Renilla koellikeri*. *Veliger*, **10**, 440-441. [Ch. 16]

Bessen, M., Fay, R.B. & Whitman, G.B., 1980. Calcium control of waveform in isolated flagellar axonemes of *Chlamydomonas*. *Journal of Cell Biology*, **86**, 446-455. [Ch. 7]

Bienfang, P.K. & Szyper, J.P., 1981. Phytoplankton dynamics in the subtropical Pacific Ocean off Hawaii. *Deep-Sea Research*, **28**A, 981-1000. [Ch. 6]

Bilbaut, A., 1975. Etude de la bioluminescence chez l'octocoralliaire *Veretillum cynomorium* Pall. II. Les réponses lumineuses de la colonie. *Archives de Zoologie Expérimentale et Générale*, **116**, 321-341. [Ch. 16]

Bilbaut, A. & Bassot, J.-M., 1977. Bioluminescence des élytres d'*Acholoë*. II. Données photometriques. *Biologie Cellulaire*, **28**, 145-154. [Ch. 16]

Bird, R.E., 1984. A simple, solar spectral model for direct-normal and diffuse-horizontal irradiance. *Solar Energy*, **32**, 461-471. [Ch. 1]

Blaxter, J.H.S., 1964. Spectral sensitivity of the herring, *Clupea harengus* L. *Journal of Experimental Biology*, **41**, 155-162. [Ch. 13]

Blaxter, J.H.S., 1968. Visual thresholds and spectral sensitivity of herring larvae. *Journal of Experimental Biology*, **48**, 39-53. [Ch. 13]

Blaxter, J.H.S., 1970. Light: animals: fishes. In *Marine Ecology*, vol. 1 (ed. O. Kinne), pp. 213-285. London: Wiley Interscience. [Ch. 13, 15]

Blaxter, J.H.S., 1973. Monitoring the vertical movements and light responses of herring and plaice larvae. *Journal of the Marine Biological Association of the United Kingdom*, **53**, 635-647. [Ch. 13]

Blaxter, J.H.S., 1975. Fish vision and applied reseach. In *Vision in Fishes* (ed. M.A. Ali), pp. 757-773. New York: Plenum Publishing Corporation. [Ch. 15]

Blaxter, J.H.S., 1976. The role of light in the vertical migration of fish - a review. In *Light as an Ecological Factor* (ed. G.C. Evans *et al.*), pp. 189-210. Oxford: Blackwell Scientific Publications. [Ch. 15]

Blaxter, J.H.S. & Batty, R.S., 1985. Herring behaviour in the dark: responses to stationary and continuously vibrating obstacles. *Journal of the Marine Biological Association of the United Kingdom*, **65**, 1031-1049. [Ch. 13]

Blaxter, J.H.S. & Batty, R.S., 1987. Comparisons of herring behaviour in the light and dark: changes in activity and responses to sound. *Journal of the Marine Biological Association of the United Kingdom*, **67**, 849-860. [Ch. 13]

Blaxter, J.H.S. & Currie, R.J., 1967. The effect of artificial lights on acoustic scattering layers in the ocean. *Symposia of the Zoological Society of London*, no. 19, 1-14. [Ch. 15]

Blaxter, J.H.S., Gray, J.A.B. & Denton, E.J., 1981. Sound and startle responses in herring shoals. *Journal of the Marine Biological Association of the United Kingdom*, **61**, 851-869. [Ch. 13]

Blaxter, J.H.S. & Hunter, J.R., 1982. The biology of the clupeoid fishes. *Advances in Marine Biology*, **20**, 1-223. [Ch. 13]

Blaxter, J.H.S. & Jones, M.P., 1967. The development of the retina and retinomotor responses in the herring. *Journal of the Marine Biological Association of the United Kingdom*, **47**, 677-697. [Ch. 13]

Blaxter, J.H.S. & Parrish, B.B., 1958. The effect of artifical lights on fish and other marine organisms at sea. *Marine Research*, no. 2, 24 pp. [Ch. 13, 15]

Blaxter, J.H.S. & Parrish, B.B., 1964. The importance of vision in the reaction of fish to drift nets and trawls. In *Modern Fishing Gear of the World*, vol. 2 (ed. H. Kristjonsson), pp. 529-536. London: Fishing News Books. [Ch. 15]

Blaxter, J.H.S. & Parrish, B.B., 1965. The importance of light in shoaling, avoidance of nets and vertical migration by herring. *Journal du Conseil*, **30**, 40-57. [Ch. 13, 15]

Blaxter, J.H.S. & Parrish, B.B., 1966. The reaction of marine fish to moving netting and other devices in tanks. *Marine Research*, no. 1, 15 pp. [Ch. 13]

Bloomquist, B.T., Shortridge, R.D., Schneuwly, S., Perdew, M., Montell, C., Steller, H., Rubin, G. & Pak, W.L., 1988. Isolation of a putative phospholipase C gene of Drosophila, *norpA*, and its role in phototransduction. *Cell*, **54**, 723-733. [Ch. 12]

Blowers, D.P. & Trewavas, A.J., 1988. Phosphatidyl inositol kinase activity of a plasma-membrane associated calcium-activated protein kinase from pea. *FEBS Letters*, **88**, 87-89. [Ch. 7]

Blumenfeld, A., Erusalimsky, J., Heichal, O., Selinger, Z. & Minke, B., 1985. Light-activated guanosinetriphosphatase in *Musca* eye membranes resembles the prolonged depolarizing afterpotential in photoreceptor cells. *Proceedings of the National Academy of Sciences of the United States of America*, **82**, 7116-7120. [Ch. 12]

Boczar, B.A. & Prezelin, B.B., 1987. Chlorophyll-protein complexes from the red-tide dinoflagellate, *Gonyaulax polyedra* Stein. Isolation, characterization, and the effect of growth irradiance on chlorophyll distribution. *Plant Physiology*, **83**, 805-812. [Ch. 5]

Boden, B.P., Kampa, E.M. & Snodgrass, J.M., 1961. Photoreception of a planktonic crustacean in relation to light penetration in the sea. In *Progress in Photobiology* (ed. B.C. Christianen and B. Buchmann), pp. 189-196. Amsterdam: Elsevier. [Ch. 16]

Booty, B., 1989. *Measurement of the Reflectance Ratio of Natural Light in the Sea*. Wormley, Surrey: Institute of Oceanographic Sciences Deacon Laboratory. [Report no. 265.] [Ch. 2]

Bowmaker, J.K., Dartnall, H.J.A. & Herring, P.J., 1988. Longwave-sensitive visual pigments in some deep-sea fishes: segregation of 'paired' rhodopsins and porphyropsins. *Journal of Comparative Physiology*, **163**A, 685-698. [Ch. 8, 16]

Bowmaker, J.K. & Kunz, Y.W., 1987. Ultraviolet receptors, tetrachromatic colour vision and retinal mosaics in the brown trout (*Salmo trutta*): age dependent changes. *Vision Research*, **27**, 2101-2108. [Ch. 10]

Brandt, A. von, 1984. *Fish Catching Methods of the World*, 3rd ed. Farnham: Fishing News Books. [Ch. 15]

Brauer, A., 1908. Die Tiefseefische, 2. Anatomische Teil. *Wissenschaftliche Ergebnisse der Deutschen Tiefsee-Expedition auf dem Dampfer 'Valdivia'*, **15**, 266 pp. [Ch. 8]

Brehm, P., 1977. Electrophysiology and luminescence of an ophiuroid radial nerve. *Journal of Experimental Biology*, **71**, 213-227. [Ch. 16]

Bricaud, A., Bedhomme, A.-L. & Morel, A., 1988. Optical properties of diverse phytoplankton species: experimental results and theoretical interpretations. *Journal of Plankton Research*, **10**, 851-873. [Ch. 4]

Bricaud, A., Morel, A. & Prieur, L., 1983. Optical efficiency factors of some phytoplankters. *Limnology and Oceanography*, **28**, 816-832. [Ch. 4]

Briggs, W.R. & Iino, M., 1983. Blue light absorbing photoreceptors in plants. *Philosophical Transactions of the Royal Society* (B), **303**, 347-359. [Ch. 7]

Brine, D.T. & Iqbal, M., 1983. Diffuse and global spectral irradiance under cloudless skies. *Solar Energy*, **30**, 447-453. [Ch. 1]

Bristow, M., Nielsen, D., Bundy, D. & Furtek, R., 1981. Use of water Raman emission to correct airborne laser fluorosensor data for effects of water optical attenuation. *Applied Optics*, **20**, 2889-2906. [Ch. 2]

Brown, D.R.E., 1952. *Natural Illumination Charts*. [Report No 374-1.] Washington, DC: Department of the Navy, Bureau of Ships. [Ch. 8]

Brown, J.E. Coles, J.A. & Pinto, L.H., 1977. Effects of injections of calcium and EGTA into the outer segments of retinal rods of *Bufo marinus. Journal of Physiology*, **269**, 707-722. [Ch. 11]

Brown, J.E., Watkins, D.C. & Malbon, C.C., 1987. Light-induced changes in the content of inositol phosphates in squid, *Loligo pealei*, retina. *Biochemical Journal*, **247**, 293-297. [Ch. 12]

Brown, J.S., 1985. Three photosynthetic antenna porphyrins in a primitive green alga. *Biochimica et Biophysica Acta*, **807**, 143-146. [Ch. 5]

Brown, J.S., 1988. Photosynthetic pigment organization in diatoms (Bacillariophyceae). *Journal of Phycology*, **24**, 96-102. [Ch. 5]

Brownlee, C., 1989. Visualizing cytoplasmic calcium in polarized and polarizing *Fucus* zygotes. *Biological Bulletin. Marine Biological Laboratory, Woods Hole, Mass.*, in press. [Ch. 7]

Brownlee, C. & Pulsford, A.L., 1988. Visualization of the cytoplasmic calcium gradient in *Fucus serratus* rhizoids: correlation with cell ultrastructure and polarity. *Journal of Cell Science*, **91**, 249-256. [Ch. 7]

Brownlee, C. & Wood, J.W., 1986. A gradient of cytoplasmic free calcium in growing rhizoid cells of *Fucus serratus. Nature, London*, **320**, 624-626. [Ch. 7]

Bryan, J.R., Riley, J.P. & Williams, P.J.le B., 1976. A Winkler procedure for making precise measurements of oxygen concentration for productivity and related studies. *Journal of Experimental Marine Biology and Ecology*, **21**, 191-197. [Ch. 4]

Buck, J., 1973. Bioluminescent behaviour in *Renilla*. 1. Colonial responses. *Biological Bulletin. Marine Biological Laboratory, Woods Hole, Mass.*, **144**, 19-42. [Ch. 16]

Buck, J., 1978. Functions and evolutions of bioluminescence. In *Bioluminescence in Action* (ed. P.J. Herring), pp. 419-460. London: Academic Press. [Ch. 15, 16, 17]

Bullard, R.K., 1983. Detection of marine contours from Landsat film and tape. In *Remote Sensing Applications in Marine Science and Technology* (ed. A.P. Cracknell), pp. 373-381. Dordrecht: D. Reidel Publishing Co. [Ch. 2]

Burkenroad, M.D., 1943. A possible function of bioluminescence. *Journal of Marine Research*, **5**, 161-164. [Ch. 16]

Buschbaum, G., 1981. The retina as a two-dimensional detector array in the context of color vision theories and signal detection theory. *Proceedings of the IEEE*, **69**, 772-785. [Ch. 10]

Buskey, E.J. & Swift, E., 1983. Behavioral responses of the coastal copepod *Acartia hudsonica* (Pinhey) to simulated dinoflagellate bioluminescence. *Journal of Experimental Marine Biology and Ecology*, **72**, 43-58. [Ch. 16]

Buskey, E.J. & Swift, E., 1985. Behavioral responses of oceanic zooplankton to simulated bioluminescence. *Biological Bulletin. Marine Biological Laboratory, Woods Hole, Mass.*, **168**, 263-275. [Ch. 16]

Buskey, E.J., Mann, C.G. & Swift, E., 1987. Photophobic responses of calanoid copepods: possible adaptive value. *Journal of Plankton Research*, **9**, 857-870. [Ch. 16]

Buskey, E.J., Mills, L. & Swift, E., 1983. The effects of dinoflagellate bioluminescence on the swimming behaviour of a marine copepod. *Limnology and Oceanography*, **28**, 575-579. [Ch. 16]

Buskey, E.J., Reynolds, G.T., Swift, E. & Walton, A.J., 1985. Interactions between copepods and bioluminescent dinoflagellates: direct observations using image intensification. *Biological Bulletin. Marine Biological Laboratory, Woods Hole, Mass.*, **169**, 530. [Ch. 16]

Calhoon, R., Tsuda, M. & Ebrey, T.G., 1980. A light-activated GTPase from *Octopus* photoreceptors. *Biochemical and Biophysical Research Communications*, **94**, 1452-1457. [Ch. 12]

Campbell, A.K., 1974. Extraction, partial purification and properties of the Ca^{2+}-activated photoprotein obelin from the hydroid *Obelia geniculata*. *Biochemical Journal* **143**, 411-418. [Ch. 18]

Campbell, A.K., 1983. *Intracellular Calcium: its Universal Role as Regulator*. Chichester: Wiley. [Ch. 18]

Campbell, A.K., 1988. *Chemiluminescence: Principles and Application in Biology and Medicine*. Chichester: Horwood/VCH. [Ch. 17, 18]

Campbell, A.K., 1989. Living light - biochemistry, function and biomedical applications. *Essays in Biochemistry*, **24**, 41-81. [Ch. 17]

Campbell, A.K. & Hallett, M.B., 1983. Measurement of intracellular calcium ions and oxygen radicals in polymorphonuclear leucocyte-erythrocyte 'ghost' hybrids. *Journal of Physiology*, **338**, 537-550. [Ch. 18]

Campbell, A.K. & Herring, P.J., 1987. A novel red fluorescent protein from the deep sea luminous fish *Malacosteus niger*. *Comparative Biochemistry and Physiology*, **86**B, 411-417. [Ch. 18]

Campbell, A.K. & Herring, P.J., 1989. Imidazolopyrazine chemiluminescence in copepods and other marine animals. *Marine Biology*, in press. [Ch. 18]

Campbell, A.K. & Luzio, J.P., 1981. Intracellular free calcium as a pathogen in cell damage initiated by the immune system. *Experientia*, **37**, 1110-1112. [Ch. 18]

Campbell, A.K. & Patel, A., 1983. A homogeneous immunoassay for cyclic nucleotides based on chemiluminescence energy transfer. *Biochemical Journal*, **216**, 185-194. [Ch. 18]

Campbell, A.K., Patel, A., Razavi, Z.S. & McCapra, F., 1988. Formation of the Ca^{2+}-activated photoprotein obelin from apoobelin and mRNA in human neutrophils. *Biochemical Journal*, **252**, 143-149. [Ch. 18]

Campbell, A.K., Roberts, P.A. & Patel, A., 1985. Chemiluminescence energy transfer: a technique for homogeneous immunoassay. In *Alternative Immunoassays* (ed. W.P. Collings), pp. 152-183. Chichester: Wiley. [Ch. 18]

Campbell, L. & Iturriaga, R., 1988. Identification of *Synechococcus* spp. in the Sargasso Sea by immunofluorescence and fluorescence excitation spectroscopy performed on individual cells. *Limnology and Oceanography*, **33**, 1196-1201. [Ch. 5]

Canuto, V.M., Levine, J.S., Augustsson, T.R. & Imhoff, C.L., 1982. UV radiation from the young sun and oxygen and ozone levels in the prebiological palaeoatmosphere. *Nature, London*, **296**, 816-820. [Ch. 4]

Capovilla, M., Caretta, A., Cervetto, L. & Torre, V., 1983. Ionic movements through light-sensitive channels of toad rods. *Journal of Physiology*, **343**, 295-310. [Ch. 11]

Carlisle, D.B. & Denton, E.J., 1959. On the metamorphosis of the visual pigments of *Anguilla anguilla* (L.). *Journal of the Marine Biological Association of the United Kingdom*, **38**, 97-102. [Ch. 10]

Caron, L. & Brown, J., 1987. Chlorophyll-carotenoid protein complexes from the diatom *Phaeodactylum tricornutum*: spectrophotometric, pigment and polypeptide analyses. *Plant and Cell Physiology*, **28**, 775-785. [Ch. 5]

Caron, L., Dubacq, J.P., Berkaloff, C. & Jupin, H., 1985. Subchloroplast fractions from the brown alga *Fucus serratus*: phosphatidylglycerol contents. *Plant and Cell Physiology*, **26**, 131-139. [Ch. 5]

Caron, L., Remy, R. & Berkaloff, C., 1988. Polypeptide composition of light-harvesting complexes from some brown algae and diatoms. *FEBS Letters*, **229**, 11-15. [Ch. 5]

Case, J.F., 1984. Vision in mating behaviour of fireflies. In *Insect Communication* (ed. T. Lewis), pp. 195-222. London: Academic Press. [Ch. 16]

Case, J.F., Warner, J., Barnes, A.T. & Lowenstine, M., 1977. Bioluminescence of lantern fish (Myctophidae) in response to changes in light intensity. *Nature, London*, **265**, 179-181. [Ch. 16]

Catalani, L.H. & Wilson, T., 1989. Electron transfer and chemiluminescence. Two inefficient systems: 1,4-dimethoxy-9,10-diphenylanthracene peroxide and diphenoyl peroxide. *Journal of the American Chemical Society*, **111**, 2633-2639. [Ch. 17]

Chang, J.J., 1954. Analysis of the luminescent response of the ctenophore *Mnemiopsis leidyi* to stimulation. *Journal of Cellular and Comparative Physiology*, **44**, 365-394. [Ch. 16]

Chisholm, S.W., Olson, R.J., Zettler, E.R., Goericke, R., Waterbury, J.B. & Welschmeyer, N.A., 1988. A novel free-living prochlorophyte abundant in the oceanic euphotic zone. *Nature, London*, **334**, 340-343. [Ch. 6]

Chu, Z.-X. & Anderson, J.M., 1985. Isolation and characterisation of a siphonaxanthin-chlorophyll *a/b*-protein complex of Photosystem I from a *Codium* species (Siphonales). *Biochimica et Biophysica Acta*, **806**, 154-160. [Ch. 5]

Clark, D.K., 1981. Phytoplankton pigment algorithms for the Nimbus-7 CZCS. In *Oceanography from Space* (ed. J.F.R. Gower), pp. 227-237. New York: Plenum Press. [Ch. 2]

Clarke, G.L., 1936. On the depth at which fishes can see. *Ecology*, **17**, 452-456. [Ch. 10]

Clarke, G.L. & Denton, E.J., 1962. Light and animal life. In *The Sea*, vol. 1 (ed. M.N. Hill), pp. 456-468. New York: Interscience Publishers. [Ch. 8, 15]

Clarke, G.L. & Hubbard, C.J., 1959. Quantitative records of the luminescent flashing of oceanic animals at great depths. *Limnology and Oceanography*, **4**, 163-180. [Ch. 8]

Clarke, M.R. & Pascoe, P.L., 1985. The influence of an electric light on the capture of deep-sea animals by a midwater trawl. *Journal of the Marine Biological Association of the United Kingdom*, **65**, 373-393. [Ch. 15]

Clarke, M.R., Pascoe, P.L. & Maddock, L., 1986. Influence of 70 watt electric lights on the capture of fish by otter trawl off Plymouth. *Journal of the Marine Biological Association of the United Kingdom*, **66**, 711-720. [Ch. 15]

Clarke, T., 1986. *A Two Layer Model of the Seasonal Thermocline and its Application to the North-west European Continental Shelf*. Report U86-2. Menai Bridge, Gwynedd: University College of North Wales Unit for Coastal and Estuarine Studies. [Ch. 4]

Cobbs, W.H., Barkdoll, A.E. III & Pugh, E.N., Jr, 1985. Cyclic GMP increases photocurrent and light sensitivity of retinal cones. *Nature, London*, **317**, 64-66. [Ch. 11]

Cobbs, W.H. & Pugh, E.N., Jr, 1985. Cyclic GMP can increase rod outer-segment light-sensitive current 10-fold without delay of excitation. *Nature, London*, **313**, 585-587. [Ch. 11]

Cockcroft, S. & Gomperts, B.D., 1985. Role of guanine nucleotide binding protein in the activation of phosphoinositide phosphodiesterase. *Nature, London*, **314**, 534-536. [Ch. 7]

Cohen, A.C. & Morin, J.G., 1986. Three new luminescent ostracodes of the genus *Vargula* (Myodocopida, Cypridinidae) from the San Blas region of Panama. *Contributions in Science of the Natural History Museum of Los Angeles County*, no. 373, 23 pp. [Ch. 16]

Cohen, A.C. & Morin, J.G., 1989. Six new luminescent ostracodes of the genus *Vargula* (Myodocopida: Cypridinidae) from the San Blas region of Panama. *Journal of Crustacean Biology*, **9**, 297-340. [Ch. 16]

Cohen, A.I., 1973. An ultrastructural analysis of the photoreceptors of the squid and their synaptic connections. I. Photoreceptive and non-synaptic regions of the retina. *Journal of Comparative Neurology*, **147**, 351-378. [Ch. 12]

Cohen, R.J. & Atkinson, M.M., 1978. Activation of *Phycomyces* adenosine 3', 5'-monophosphate phosphodiesterase by blue light. *Biochemical and Biophysical Research Communications*, **83**, 616-621. [Ch. 7]

Cohen-Bazire, G. & Bryant, D.A., 1982. Phycobilisomes: composition and structure. In *The Biology of Cyanobacteria* (ed. N.G. Carr and B.A. Whitton), pp. 143-190. Oxford: Blackwell. [Ch. 6]

Cohn, D.A., Ogden, R.C., Abelson, J.M., Baldwin, T.O., Nealson, K.H., Simon, M.I. & Mileham, A.J., 1983. Cloning of the *Vibrio harveyi* luciferase genes: use of a synthetic oligonucleotide probe. *Proceedings of the National Academy of Sciences of the United States of America*, **80**, 120-123. [Ch. 18]

Colbert, J.T., 1988. Molecular biology of phytochrome. *Plant Cell Environment*, **11**, 305-318. [Ch. 7]

Colijn, F., Gieskes, W.W.C. & Zevenboom, W., 1983. The measurement of primary production: problems and recommendations. *Hydrobiological Bulletin*, **17**, 29-51. [Ch. 4]

Collette, B. B. & Talbot, F. H., 1972. Activity patterns of coral reef fishes with emphasis on nocturnal-diurnal change-over. *Bulletin of the Natural History Museum of Los Angeles County*, **14**, 98-124. [Ch. 14]

Collin, S.P. & Pettigrew, J.D., 1988*a*. Retinal topography in reef fishes. I. Some species with well-developed areae but poorly developed streaks. *Brain Behaviour and Evolution*, **31**, 269-282. [Ch. 9, 10]

Collin, S.P. & Pettigrew, J.D., 1988*b*. Retinal topography in reef fishes. II. Some species with prominent horizontal streaks and high-density areae. *Brain Behaviour and Evolution*, **31**, 283-295. [Ch. 9, 10]

Cormier, M.J., 1978. Comparative biochemistry of animal systems. In *Bioluminescence in Action* (ed. P.J. Herring), pp. 75-108. London: Academic Press. [Ch. 17]

Cormier, M.J. Ward, W.W. & Charbonneau, H., 1977. Role of oxygen in coelenterate bioluminescence. In *Superoxide and Superoxide Dismutases* (ed. A.M. Michelson *et al.*), pp. 451-458. New York: Academic Press. [Ch. 17]

Corrachano, L.M., Galland, P., Lipson, E.D. & Cerdo-Olmedo, E., 1988. Photomorphogenesis in Phycomyces: fluence response curves and action spectra. *Planta*, **174**, 315-320. [Ch. 7]

Cousteau, J.-Y., Jausseran, C., Laban, A. & Lieberman, M., 1964. Mesure de l'absorption dans l'eau de mer, de la lumière émise par une source artificielle, à diverses profondeurs, verticalement et horizontalement. *Bulletin de l' Institut Océanographique*, **63**, 17 pp. [Ch. 8]

Craig, R.E. & Priestley, R., 1960. Photographic studies of fish populations. *Nature, London*, **188**, 333-334. [Ch. 13]

Cram, D.L., 1973. Pilchard stocks surveyed by remote sensors. *Fishing News International*, **12** (3), 35-39. [Ch. 15]

Cram, D.L. & Hampton, I., 1976. A proposed aerial/acoustic strategy for pelagic fish stock assessment. *Journal du Conseil*, **37**, 91-97.

Crane, J.M., 1965. Bioluminescent courtship display in the teleost *Porichthys notatus. Copeia*, **1965**, 239-241. [Ch. 16]

Cutter, M.A., Lobb, D.R., Ramon, Y., Ratier, G. & Bezy, J., 1989. A medium resolution imaging spectrometer for the European polar orbiting platform. *Proceedings of the Society of Photo-optic Instrumentation Engineers*, **1129**, in press. [Ch. 2]

Daniel, A. & Jothinayagam, J.T., 1979. Observations on nocturnal swarming of the planktonic ostracod *Cypridina dentata* (Müller) for mating in the northern Arabian Sea. *Bulletin of the Zoological Survey of India*, **2**, 25-28. [Ch. 16]

Dartnall, H.J.A. & Lythgoe J.N., 1965. The spectral clustering of visual pigments. *Vision Research*, **5**, 81-100. [Ch. 10]

Davenport, D. & Nicol, J.A.C., 1955. Luminescence in Hydromedusae. *Proceedings of the Royal Society* (B), **144**, 399-412. [Ch. 16]

David, C.N. & Conover, R.J., 1961. Preliminary investigation on the physiology and ecology of luminescence in the copepod *Metridia lucens. Biological Bulletin. Marine Biological Laboratory, Woods Hole, Mass.*, **121**, 92-107. [Ch. 16]

Davies, W. P. & Birdsong, R., 1973. Coral Reef fishes which forage in the water column. *Helgoländer Wissenschaftliche Meeresuntersuchungen*, **24**, 292-306. [Ch. 14]

Deluca, M. A. (ed.), 1978. Bioluminescence and chemiluminescence. *Methods in Enzymology*, **57**, 1-653. [Ch. 18]

Deluca, M.A. & McElroy, W.D. (ed.) 1987. Bioluminescence and chemiluminescence, part B. *Methods in Enzymology*, **133**, 1-649. [Ch. 18]

Dempsey, C.H., 1978. Chemical stimuli as a factor in feeding and intraspecific behaviour of herring larvae. *Journal of the Marine Biological Association of the United Kingdom*, **58**, 739-747. [Ch. 13]

Denton, E.J., 1970. On the organization of the reflecting surfaces in some marine animals. *Philosophical Transactions of the Royal Society* (B) **258**, 285-313. [Ch. 8, 9]

Denton, E.J. & Blaxter, J.H.S., 1976. The mechanical relationships between the clupeoid swimbladder, inner ear and lateral line. *Journal of the Marine Biological Association of the United Kingdom*, **56**, 787-807. [Ch. 13]

Denton, E.J., Gilpin-Brown, J.B. & Wright, P.G., 1970. On the 'filters' in the photophores of mesopelagic fish and on a fish emitting red light and especially sensitive to red light. *Journal of Physiology*, **208**, 72-73P. [Ch. 8]

Denton, E.J., Gilpin-Brown, J.B. & Wright, P.G., 1972. The angular distribution of the light produced by some mesopelagic fish in relation to their camouflage. *Proceedings of the Royal Society* (B), **182**, 145-158. [Ch. 16]

Denton, E.J., Herring, P.J., Widder, E.A., Latz, M.I. & Case, J.F., 1985. The roles of filters in the photophores of oceanic animals and their relation to vision in the oceanic environment. *Proceedings of the Royal Society* (B), **225**, 63-97. [Ch. 8, 16]

Denton, E.J. & Locket, N.A., 1989. Possible wavelength discrimination by multibank retinae in deep-sea fishes. *Journal of the Marine Biological Association of the United Kingdom*, **69**, 409-435. [Ch. 8, 10]

Denton, E.J. & Nicol, J.A.C., 1965. Studies on reflexion of light from silvery surfaces of fishes, with special reference to the bleak, *Alburnus alburnus. Journal of the Marine Biological Association of the United Kingdom*, **45**, 683-703. [Ch. 8]

Denton, E.J. & Nicol, J.A.C., 1966. A survey of reflectivity in silvery teleosts. *Journal of the Marine Biological Association of the United Kingdom*, **46**, 685-722. [Ch. 10]

Denton, E.J. & Pirenne, M.H., 1954. The absolute sensitivity and functional stability of the human eye. *Journal of Physiology*, **123**, 417-442. [Ch. 8]

Denton, E.J. & Shaw, T.I., 1963. The visual pigments of some deep-sea elasmobranchs. *Journal of the Marine Biological Association of the United Kingdom*, **43**, 65-70. [Ch. 8]

Denton, E.J. & Warren, F.J., 1957. The photosensitive pigments in the retinae of deep-sea fish. *Journal of the Marine Biological Association of the United Kingdom*, **36**, 651-662. [Ch. 8]

De Silva, S.S., 1973. Food and feeding habits of the herring. *Clupea harengus* and sprat *C. sprattus* in inshore waters off the west coast of Scotland. *Marine Biology*, **20**, 282-290. [Ch. 13]

DeWet, J.R., Wood, K.V., Helinski, D.R. & Deluca, M., 1985. Firefly luciferase cDNA and the expression of active luciferase in *Escherichia coli. Proceedings of the National Academy of Sciences of the United States of America*, **82**, 7870-7873. [Ch. 18]

DeWet, J.R., Wood, K.W., Deluca, M., Helinski, D.R. & Subramami, S., 1987. Firefly luciferase gene: structure and expression in mammalian cells. *Molecular and Cellular Biology*, **7**, 725-737. [Ch. 18]

Dilella, A.G., Hope, D.A., Chen, H., Trumbauer, M., Schwartz, R.J. & Smith, R.G., 1988. Utility of firefly luciferase as a reporter gene for promotor activity in transgenic mice. *Nucleic Acids Research*, **16**, 4159. [Ch. 18]

Dirks, R.W.J. & Spitzer, D., 1987. On the radiative transfer in the sea, including fluorescence and stratification effects. *Limnology and Oceanography*, **32**, 942-953. [Ch. 1]

Doherty, P.J., 1987. Light-traps: selective but useful devices for quantifying the distributions and abundances of larval fishes. *Bulletin of Marine Science*, **41**, 423-431. [Ch. 15]

Domm, S. B. & Domm, A. J., 1973. The sequence of appearance at dawn and disappearance at dusk of some coral reef fishes. *Pacific Science*, **27**, 128-135. [Ch. 14]

Donner, K.O. & Hemilä, S., 1978. Excitation and adaptation in the vertebrate rod photoreceptor. *Medical Biology*, **56**, 52-63. [Ch. 11]

Doughty, M.J., Grieser, R. & Diehn, B., 1980. *Biochimica Biophysica Acta*, **602**, 10-23. [Ch. 7]

Douglas, R.H., 1986. Photopic spectral sensitivity of a teleost fish, the roach (*Rutilus rutilus*), with special reference to its ultraviolet sensitivity. *Journal of Comparative Physiology*, **159**(A), 415-421. [Ch. 10]

Douglas, R.H., 1989. The spectral transmission of the lens and cornea of the brown trout (*Salmo trutta*) and goldfish (*Carasius auratus*) - effects of age and implications for ultraviolet vision. *Vision Research*, **29**, 861-869. [Ch. 10]

Douglas, R.H. & McGuigan, C.M., 1989. The spectral transmissions of freshwater teleost ocular media - an interspecific comparison and a guide to potential ultraviolet sensitivity. *Vision Research*, **29**, 871-879. [Ch. 10]

Dowling, J.E. & Ripps, H., 1972. Adaptation in skate photoreceptors. *Journal of General Physiology*, **60**, 698-719. [Ch. 11]

Dratz, E. & Hargrave, P., 1983. The structure of rhodopsin in the rod outer segment disk membrane. *Trends in Biochemical Sciences*, **8**, 128-133. [Ch. 12]

Dring, M.J., 1967. Phytochrome in the red alga *Porphyra tenera*. *Nature, London*, **215**, 1411-1412. [Ch. 7]

Dring, M.J., 1981. Chromatic adaptation of photosynthesis in benthic marine algae: an examination of its ecological significance using a theoretical model. *Limnology and Oceanography*, **26**, 271-284. [Ch. 5]

Dring, M.J., 1982. *The Biology of Marine Plants*. London: Edward Arnold. [Ch. 5]

Dring, M.J., 1986. Pigment composition and photosynthetic action spectra of sporophytes of *Laminaria* (Phaeophyta) grown in different light qualities and irradiances. *British Phycological Journal*, **21**, 199-207. [Ch. 5]

Dring, M.J., 1987. Marine plants and blue light. In *Blue Light Responses: Phenomena and Occurrence in Plants and Microorganisms* (ed. H. Senger), pp. 121-140. Florida: CRC Press Inc. [Ch. 7]

Dring, M.J., 1988. Photocontrol of development in algae. *Annual Review of Plant Physiology*, **39**, 157-174. [Ch. 7]

Dring, M.J. & Lünning, K., 1975. Induction of two dimensional growth and hair formation by blue light in the brown alga *Scitosyphon lomentaria*. *Zeitschrift für Pflanzenphysiologie*, **75**, 107-117. [Ch. 7]

Dring, M.J. & Lüning, K., 1985. Emerson enhancement effect and quantum yield of photosynthesis for marine macroalgae in simulated underwater light fields. *Marine Biology*, **87**, 109-117. [Ch. 5]

Droop, M.R., 1983. Twenty-five years of algal growth kinetics, a personal view. *Botanica Marina*, **26**, 99-112. [Ch. 4]

Droop, M.R., Mickelson, M.J., Scott, J.M. & Turner, M.F., 1982. Light and nutrient status of algal cells. *Journal of the Marine Biological Association of the United Kingdom*, **62**, 403-434. [Ch. 4]

Dubin, R. E. & Baker, J. E., 1982. Two types of cover-seeking at sunset by the Priacen parrotfish, *Scarus taeniopterus*, at Barbados, West Indies. *Bulletin of Marine Science*, **32**, 572-583. [Ch. 14]

Dubois, R., 1887. Function photogenique chez le *Pholas dactylus*. *Compte Rendu des Séances de la Société de Biologie*, **3**, 564-565. [Ch. 18]

Dugdale, R.C. & Goering, J.J., 1967. Uptake of new and regenerated nitrogen in primary production. *Limnology and Oceanography*, **12**, 196-206. [Ch. 4]

Duncan, G. & Pynsent, P.B., 1979. An analysis of the wave forms of photoreceptor potentials in the retina of the cephalopod *Sepiola atlantica*. *Journal of Physiology*, **288**, 171-188. [Ch. 12]

Duncan, M.J. & Foreman, R.E., 1980. Phytochrome-mediated stipe elongation in the kelp *Nereocystis* (Phaeophyceae). *Journal of Phycology*, **16**, 138-142. [Ch. 7]

Dunlap, J.C., Shimomura, O. & Hasting, J.W., 1981. Dinoflagellate luciferin is structurally related to chlorophyll. *FEBS Letters*, **135**, 273-276. [Ch. 17]

Duntley, S.Q., 1962. Underwater visibility. In *The Sea*, vol. 1, *Physical Oceanography* (ed. M.N. Hill), pp. 452-455. New York: Interscience. [Ch. 10]

Duntley, S.Q., 1963. Light in the sea. *Journal of the Optical Society of America*, **53**, 214-233. [Ch. 10]

Ebeling, A. W. & Bray, R. N., 1976. Day versus night activity of reef fishes in a kelp forest off Santa Barbara, California. *Fisheries Bulletin. National Oceanic and Atmospheric Administration of the United States*, **74**, 703-717. [Ch. 14]

Eguchi, E., Nemoto, A., Meyer-Rochow, V.B. & Ohba, N., 1984. A comparative study of spectral sensitivity curves in three diurnal and eight nocturnal species of Japanese fireflies. *Journal of Insect Physiology*, **30**, 607-612. [Ch. 16]

Ehrlich, P. R., Talbot, F. H., Russell, B. C. & Anderson, G. R. V., 1977. The behaviour of chaetodontid fishes with special reference to Lorenz's "poster colouration" hypothesis. *Journal of Zoology*, **183**, 213-228. [Ch. 14]

Eisen, A. & Reynolds, G.J., 1985. Source and sinks for the calcium release released during fertilisation of single sea urchin eggs. *Journal of Cell Biology*, **100**, 1522-1527. [Ch. 18]

Emery, A.R., 1978. The basis of fish community structure: marine and freshwater comparisons. *Environmental Biology of Fishes*, **3**, 33-47. [Ch. 14]

Emery, W.J., Thomas, A.C., Collins, M.J., Crawford, W.R. & Maklas, D.L., 1986. An objective method for computing advective surface velocities from sequential infrared satellite images. *Journal of Geophysical Research*, **91**, 12865-12878. [Ch. 2]

Engebrecht, J.B. & Silverman, M., 1987. Nucleotide sequence of the regulatory locus controlling expression of bacterial genes for bioluminescence. *Nucleic Acids Research*, **15**, 10455-10467. [Ch. 18]

Engelmann, T.W., 1883. Farbe und Assimilation. *Botanische Zeitung*, **41**, 1-29. [Ch. 5]

Eppley, R.W. & Peterson, B.J., 1979. Particulate organic matter flux and planktonic new production in the deep ocean. *Nature, London*, **282**, 677-680. [Ch. 4]

Eppley, R.W., Stewart, E., Abbott, M.R. & Heyman, U., 1985. Estimating ocean primary production from satellite chlorophyll. Introduction to regional differences and statistics for the Southern California Bight. *Journal of Plankton Research*, **7**, 57-70. [Ch. 1, 2]

ESA, 1987. *Ocean Colour*. Paris: European Space Agency. [Report by the ESA Ocean Colour Working Group. ESA SP-1083.] [Ch. 2]

Esaias, W.E. & Curl, H.C., 1972. Effect of dinoflagellate bioluminescence on copepod ingestion rates. *Limnology and Oceanography*, **17**, 901-906. [Ch. 16]

Esaias, W.E., Feldman, G.C., McClain, C.R. & Elrod, J.A., 1986. Monthly satellite-derived phytoplankton pigment distribution for the North Atlantic Ocean basin, *Eos*, **67**, 835-837. [Ch. 2]

Evans, L.V., Callow, J. A. & Callow, M. E., 1982. The biology and biochemistry of reproduction and early development in *Fucus*. *Progress in Phycological Research*, **1**, 68-110. [Ch. 7]

Exner, S., 1891. *Die Physiologie der facettirten Augen von Krebsen und Insecten*. Leipzig & Wien: Deuticke. [Translated, 1989. Berlin: Springer.] [Ch. 9]

Fain, G.L., 1976. Sensitivity of toad rods: dependence on wave-length and background illumination. *Journal of Physiology*, **261**, 71-101. [Ch. 11]

Fain, G.L., Lamb, T.D., Matthews, H.R. & Murphy, R.L.W., 1989. Cytoplasmic calcium as the messenger for light adaptation in salamander rods. *Journal of Physiology*, **416**, 215-243. [Ch. 11]

Falkowski, P.G., 1980. Light-shade adaptations in marine phytoplankton. In *Primary Productivity in the Sea* (ed. P.G. Falkowski), pp. 99-199. New York: Plenum Press. [Ch. 4]

Falkowski, P.G., 1983. Light-shade adaptation and vertical mixing of marine phytoplankton: a comparative field study. *Journal of Marine Research*, **41**, 215-237. [Ch. 4]

Falkowski, P.G. & Owens, T.G., 1980. Light-shade adaptation. Two strategies in marine phytoplankton. *Plant Physiology*, **66**, 592-595. [Ch. 4]

Fasham, M.R., Platt, T., Irwin, B. & Jones, K., 1985. Factors affecting the spatial pattern of the deep chlorophyll maximum in the region of the Azores Front. *Progress in Oceanography*, **14**, 129-165. [Ch. 4]

Faust, M.A., Sager, J.C. & Meeson, B.W., 1982. Response of *Prorocentrum mariae-lebouriae* (Dinophyceae) to light of different spectral qualities and irradiances: growth and pigmentation. *Journal of Phycology*, **18**, 349-356. [Ch. 5]

Fawley, M.W. & Grossman, A.R., 1986. Polypeptides of a light-harvesting complex of the diatom *Phaeodactylum tricornutum* are synthesized in the cytoplasm of the cell as precursors. *Plant Physiology*, **81**, 149-155. [Ch. 5]

Fawley, M.W., Morton, S.J., Stewart, K.D. & Mattox, K.R., 1987. Evidence for a common evolutionary origin of light-harvesting fucoxanthin chlorophyll *a/c*-protein complexes of *Pavlova gyrans* (Prymnesiophyceae) and *Phaeodactylum tricornutum* (Bacillariophyceae). *Journal of Phycology*, **23**, 377-381. [Ch. 5]

Feiler, R., Harris, W.A., Kirschfeld, K., Wehrhahn, C.A. & Zuker, C.S., 1988. Targeted misexpression of a *Drosophila* opsin gene leads to altered visual function. *Nature, London*, **333**, 737-741. [Ch. 10]

Fernald, R.D., 1988. Aquatic adaptations in fish eyes. In *Sensory Biology of Aquatic Animals* (ed. J. Atema, *et al.*), pp. 435-466. New York: Springer-Verlag. [Ch. 10]

Fernandez, R. & Tsuji, F.I., 1976. Photopigment and spectral sensitivity in the bioluminescent fish *Porichthys notatus*. *Marine Biology*, **34**, 101-107. [Ch. 16]

Fesenko, E.E., Kolesnikov, S.S. & Lyubarsky, A.L., 1985. Induction by cyclic GMP of cationic conductance in plasma membrane of retinal rod outer segment. *Nature, London*, **313**, 310-313. [Ch. 11]

Findlay, J.B.C. & Pappin, D.J.C., 1986. The opsin family of proteins. *Biochemical Journal*, **238**, 625-642. [Ch. 12]

Fineran, B.A. & Nicol, J.A.C., 1976. Novel cones in the retina of the anchovy (*Anchoa*). *Journal of Ultrastructure Research*, **54**, 296-303. [Ch. 10]

Fletcher, A., Murphy, T. & Young, A., 1954. Solutions of two optical problems. *Proceedings of the Royal Society* (A), **223**, 216-225. [Ch. 9]

Flores, E.E.C., 1982. Light attraction techniques in squid fishing. In *Proceedings of the International Squid Symposium, 1981, Boston, Massachusetts*, pp. 55-67. New York: Unipub. [Ch. 15]

Fogg, G.E., 1986. Picoplankton. *Proceedings of the Royal Society* (B), **228**, 1-30. [Ch. 6]

Foote, K.G. & Ona, E., 1985. Tilt angles of schooling penned saithe. *Journal du Conseil*, **43**, 118-121. [Ch. 13]

Frank, T. & Case, J.F., 1988*a*. Visual spectral sensitivities of bioluminescent deep-sea crustaceans. *Biological Bulletin. Marine Biological Laboratory, Woods Hole, Mass.*, **175**, 261-273. [Ch. 16]

Frank, T. & Case, J.F., 1988*b*. Visual spectral sensitivity of the bioluminescent deep-sea mysid, *Gnathophausia ingens*. *Biological Bulletin. Marine Biological Laboratory, Woods Hole, Mass.*, **175**, 274-283. [Ch. 16]

Freeman, G., 1987. Localization of bioluminescence in the siphonophore *Nanomia cara*. *Marine Biology*, **93**, 535-541. [Ch. 16]

Freeman, G. & Reynolds, G.T., 1973. The development of bioluminescence in the ctenophore *Mnemiopsis leidyi*. *Developmental Biology*, **31**, 61-100. [Ch. 16]

Friedman, A.L. & Alberte, R.S., 1986. Biogenesis and light regulation of the major light-harvesting chlorophyll-proteins of diatoms. *Plant Physiology*, **80**, 43-51. [Ch. 5]

Friedman, A.L. & Alberte, R.S., 1987. Phylogenetic distribution of the major light-harvesting pigment-protein determined by immunological methods. *Journal of Phycology*, **23**, 427-433. [Ch. 5]

Fuiman, L.A., 1989. Vulnerability of Atlantic herring larvae to predation by yearling herring. *Marine Ecology - Progress Series*, **51**, 291-299. [Ch. 13]

Gaarder, T. & Gran, H.H., 1927. Investigations of the production of plankton in the Oslo Fjord. *Rapport et Procès-Verbaux des Réunions. Conseil Permanent International pour l'Exploration de la Mer*, **42**, 1-48. [Ch. 4]

Galland, P.A., 1987. Action Spectroscopy. In *Blue Light Responses: Phenomena and Occurrence in Plants and Microorganisms* (ed. H. Senger), pp. 37-52. Florida: CRC Press Inc. [Ch. 7]

Gallegos, C.L. & Platt, T., 1982. Phytoplankton production and water motion in surface mixed layers. *Deep-Sea Research*, **29**, 65-76. [Ch. 4]

Galt, C.P., 1978. Bioluminescence: dual mechanism in a planktonic tunicate produces brilliant coastal display. *Science, New York*, **200**, 70-72. [Ch. 16]

Galt, C.P., Grober, M.S. & Sykes, P.F., 1985. Taxonomic correlates of bioluminescence among appendicularians (Urochordata: Larvacea). Biological Bulletin. Marine Biological Laboratory, Woods Hole, Mass., **168**, 125-134. [Ch. 16]

Galt, C.P. & Sykes, P.F., 1983. Sites of bioluminescence in the appendicularians *Oikopleura dioica* and *O. labradoriensis* (Urochordata: Larvacea). *Marine Biology*, **77**, 155-159. [Ch. 16]

Garcia, C.A.E., 1989. *Dynamics in Shallow Seas Inferred from Ocean Colour Imagery*. PhD Thesis. University of Southampton. [Ch. 2]

Garcia, C.A.E. & Robinson, I.S., 1989. Sea surface velocities in shallow seas extracted from sequential CZCS satellite data. *Journal of Geophysical Research*, **94**C, 12681-12692. [Ch. 2]

Geider, R.J. & Osborne, B.A., 1986. Light absorption, photosynthesis and growth of *Nannochloris atomus* in nutrient-saturated cultures. *Marine Biology*, **93**, 351-360. [Ch. 4]

Geider, R.J., Osborne, B.A. & Raven, J.A., 1986. Growth, photosynthesis and maintenance metabolic costs in the diatom *Phaeodactylum tricornutum* at very low light levels. *Journal of Phycology*, **22**, 39-48. [Ch. 4]

Geiger, R., Hauber, R. & Miska, W., 1989. New, bioluminescence-enhanced detection systems for use in enzyme activity tests, in enzyme immunoassays, in protein blotting and nucleic acid hybridization. *Molecular and Cellular Probes*, in press. [Ch. 18]

Gibson, R.N., 1986. Intertidal teleosts: life in a fluctuating environment. In *The Behaviour of Teleost Fishes* (ed. T. J. Pitcher), pp. 388-408. London: Croom Helm. [Ch. 14]

Gibson, R.N., 1988. Development, morphometry and particle retention capability of the gill rakers in the herring, *Clupea harengus* L. *Journal of Fish Biology*, **32**, 949-962. [Ch. 13]

Gibson, R.N. & Ezzi, I.A., 1985. Effect of particle concentration on filter - and particulate - feeding in the herring, *Clupea harengus*. *Marine Biology*, **88**, 109-116. [Ch. 13]

Gieskes, W.W. & Kraay, G.W., 1983. Dominance of Cryptophyceae during the phytoplankton spring bloom in the central North Sea detected by HPLC analysis of pigments. *Marine Biology*, **75**, 179-185. [Ch. 5]

Gieskes, W.W. & Kraay, G.W., 1986. Analysis of phytoplankton pigments by HPLC before, during and after mass occurrence of the microflagellate *Corymbellus aureus* during the spring bloom in the open northern North Sea in 1983. *Marine Biology*, **92**, 45-52. [Ch. 5]

Gilman, A.G., 1987. G proteins: transducers of receptor-generated signals. *Annual Review of Biochemistry*, **56**, 615-650. [Ch. 7]

Gladfelter,W. B., 1979. Twilight migration and foraging activities of the copper sweeper, *Pempheris schomburgki* (Teleostei, Pempheridae). *Marine Biology*, **51**, 109-119. [Ch. 14]

Glass, C.W. & Wardle, C.S., 1989. Comparison of the reactions of fish to a trawl gear, at high and low light intensities. *Fisheries Research*, in press. [Ch. 15]

Glazer, A.N., 1981. Photosynthetic accessory pigments with bilin prosthetic groups. In *The Biochemistry of Plants*, vol. 8 (ed. M.D. Hatch and N.K. Boardman), pp. 51-96. New York: Academic Press. [Ch. 5]

Glover, H.E., Keller, M.D. & Guillard, R.L., 1986. Light quality and oceanic ultraphytoplankters. *Nature, London*, **319**, 142-143. [Ch. 5]

Glover, H.E., Keller, M.D. & Spinrad, R.W., 1987. The effects of light quality and intensity on photosynthesis and growth of marine eukaryotic and prokaryotic phytoplankton clones. *Journal of Experimental Marine Biology and Ecology*, **105**, 137-159. [Ch. 5]

Glover, H.E., Phinney, D.A. & Yentsch, C.S., 1985. Photosynthetic characteristics of picoplankton compared with those of larger phytoplankton populations in various water masses in the Gulf of Maine. *Biological Oceanography*, **3**, 223-248. [Ch. 6]

Glover, H.E., Prezelin, B.B., Campbell, L. & Wyman, M., 1988. Pico- and ultraplankton Sargasso Sea communities: variability and comparative distributions of *Synechococcus* spp. and algae. *Marine Ecology - Progress Series* , **49**, 127-139. [Ch. 6]

Glover, R.S., 1967. The Continuous Plankton Recorder Survey of the North Atlantic. *Symposia of the Zoological Society of London*, no. 19, 189-210. [Ch. 3]

Goedheer, J.C., 1964. Fluorescence bands and chlorophyll *a* forms. *Biochimica et Biophysica Acta*, **88**, 304-317. [Ch. 1]

Goodyear, R.H. & Gibbs, R.H., 1969. Ergebnisse der Forschungsreisen des F.F.S. 'Walter Herwig' nach Südamerika. X. Systematics and zoogeography of stomiatoid fishes of the *Astronesthes cyaneus* species group (Family Astronesthidae), with descriptions of three new species. *Archiv für Fischereiwissenschaft*, **20**, 107-131. [Ch. 16]

Gordon, H.R., 1974. Spectral variation in volume scattering function at large angles in natural waters. *Journal of the Optical Society of America*, **64**, 773-775. [Ch. 1]

Gordon, H.R., 1978. Removal of atmospheric effects from satellite imagery of the oceans. *Applied Optics*, **17**, 1631-1636. [Ch. 2]

Gordon, H.R., 1979. Diffuse reflectance of the ocean: the theory of its augmentation by chlorophyll *a* fluorescence at 685 nm. *Applied Optics*, **18**, 1161-1166. [Ch. 1]

Gordon, H.R., 1981. A preliminary assessment of the Nimbus-7 CZCS atmospheric correction algorithm in a horizontally inhomogeneous atmosphere. In *Oceanography from Space* (ed. J.F.R. Gower), pp. 257-265. New York: Plenum Press. [Ch. 2]

Gordon, H.R., Austin, R.W., Clark, D.K., Hovis, W.A. & Yentsch, C.S., 1985. Ocean color measurements. *Advances in Geophysics*, **27**, 297-333. [Ch. 2]

Gordon, H.R., Brown, J.W. & Evans, R.H., 1988. Exact Rayleigh scattering calculations for use with the Nimbus-7 Coastal Zone Color Scanner. *Applied Optics*, **27**, 862-871. [Ch. 2]

Gordon, H.R. & Brown, O.B., 1973. Irradiance reflectivity of a flat ocean as a function of its optical properties. *Applied Optics*, **12**, 1549-1551. [Ch. 2]

Gordon, H.R. & Brown, O.B., 1974. Influence of bottom depth and albedo on the diffuse reflectance of a flat homogeneous ocean. *Applied Optics*, **13**, 2153-2159. [Ch. 2]

Gordon, H.R. & Brown, O.B., 1975. Diffuse reflectance of the ocean: some effects of vertical structure. *Applied Optics*, **14**, 417-427. [Ch. 2]

Gordon, H.R. & Castano, D.J., 1987. Coastal Zone Color Scanner atmospheric correction algorithm: multiple scattering effects. *Applied Optics*, **26**, 2111-2122. [Ch. 2]

Gordon, H.R. & Clark, D., 1980. Atmospheric effects in the remote sensing of phytoplankton pigments. *Boundary-layer Meteorology*, **18**, 299-313. [Ch. 2]

Gordon, H.R. & Clark, D.K., 1981. Clear water radiances for atmospheric correction of coastal zone color scanner imagery. *Applied Optics*, **20**, 4175-4180. [Ch. 3]

Gordon, H.R., Clark, D., Brown, J.W., Brown, O.B., Evans, R.H. & Broenkow, W., 1983. Phytoplankton pigment concentrations in the Middle Atlantic Bight: comparison of ship determinations and CZCS estimates. *Applied Optics*, **22**, 20-36. [Ch. 2]

Gordon, H.R. & Morel, A.Y., 1983. *Remote Assessment of Ocean Color for Interpretation of Satellite Visible Imagery* 4. New York: Springer Verlag. [Ch. 2]

Gower, J.F.R, 1980. Observations of *in situ* fluorescence of chlorophyll-*a* in Saanich Inlet. *Boundary-Layer Meteorology*, **18**, 235-245. [Ch. 1]

Gower, J.F.R. & Borstad, G., 1981. Use of the *in vivo* fluorescence line at 685 nm for remote sensing surveys of surface chlorophyll-*a*. In *Oceanography from Space* (ed. J.F.R. Gower), pp. 329-339. New York: Plenum Press. [Ch. 2]

Gower, J.F.R., Denman, K.L. & Holyer, R.J., 1980. Phytoplankton patchiness indicates the fluctuation spectrum of mesoscale oceanic structure. *Nature, London*, **288**, 157-159. [Ch. 2]

Graham, J.B. & Rosenblatt, R.H., 1970. Aerial vision: unique adaptation in an intertidal fish. *Science, New York*, **168**, 586-588. [Ch. 9]

Gran, H.H. & Braarud, T., 1935. A quantitative study of the phytoplankton in the Bay of Fundy and the Gulf of Maine. *Journal of the Biological Board of Canada*, **1**, 279-467. [Ch. 4]

Grassl, H., 1987. Use of chlorophyll fluorescence measurements from space for separating constituents of sea water. In *Ocean Colour: Report by the ESA Ocean Colour Working Group*. Paris: European Space Agency. [ESA SP-1083, 103] [Ch. 2]

Green, B.R., Camm, E.L. & Van Houten, J., 1982. The chlorophyll-protein complexes of *Acetabularia*. A novel chlorophyll *a/b* complex which forms oligomers. *Biochimica et Biophysica Acta*, **681**, 248-255. [Ch. 5]

Green, D.G. & Seigel, I.M., 1975. Double branched flicker fusion curves from the all-rod skate retina. *Science, New York*, **188**, 1120-1122. [Ch. 10]

Griendling, K.K., Rittenhouse, S.E., Brock, T.A., Ekstein, L.S., Gimbrone, L.A., Jr & Alexander, R.W., 1986. Sustained diacylglycerol formation from inositol phospholipids in angiotensin II-stimulated vascular smooth muscle cells. *Journal of Biological Chemistry*, **261**, 5901-5906. [Ch. 7]

Grinnell, A.D., Narins, P.M., Awbrey, F.T., Hamner, W.M. & Hamner, P.P., 1988. Eye/photophore co-ordination and light following in krill, *Euphausia superba*. *Journal of Experimental Biology*, **134**, 61-77. [Ch. 16]

Grober, M.S, 1988a. Brittle-star bioluminescence functions as an aposematic signal to deter crustacean predators. *Animal Behaviour*, **36**, 493-501. [Ch. 16]

Grober, M.S., 1988b. Responses of tropical reef fauna to brittle-star luminescence (Echinodermata: Ophiuroidea). *Journal of Experimental Marine Biology and Ecology*, **115**, 157-168. [Ch. 16]

Groom, S.B. & Holligan, P.M., 1987. Remote sensing of coccolithophore blooms. *Advances in Space Research*, **7**, 73-78. [Ch. 3]

Guillard, R.R.L., Murphy, L.S., Foss, P. & Liaaen-Jensen, S., 1985. *Synechococcus* spp. as likely zeaxanthin-dominant ultraphytoplankton in the North Atlantic. *Limnology and Oceanography*, **30**, 412-414. [Ch. 5]

Gundermann, K-D. & McCapra, F., 1987. *Chemiluminescence in Organic Chemistry*. Berlin: Springer Verlag. [Ch. 17]

Hagins, F.M., 1973. Purification and partial characterization of the protein component of squid rhodopsin. *Journal of Biological Chemistry*, **248**, 3298-3304. [Ch. 12]

Hagins, W.A., 1965. Electrical signs of information flow in photoreceptors. *Cold Spring Harbor Symposia on Quantitative Biology*, **30**, 403-415. [Ch. 12]

Hagins, W.A., 1972. The visual process: excitatory mechanisms in the primary receptor cells. *Annual Review of Biophysics and Bioengineering*, **1**, 131-158. [Ch. 11]

Hagins, W.A., Penn, R.D. & Yoshikami, S., 1970. Dark current and photocurrent in retinal rods. *Biophysical Journal*, **10**, 380-412. [Ch. 11]

Hailman, J.P., 1977. *Optical Signals*. Bloomington: Indiana University Press. [Ch. 16]

Hallett, M.B. & Campbell, A.K., 1982. Measurement of changes in cytoplasmic free Ca^{2+} in fused cell hybrids. *Nature, London*, **295**, 155-158. [Ch. 18]

Hallett, M.B., Luzio, J.P. & Campbell, A.K., 1981. Stimulation of Ca-dependent chemiluminescence in rat polymorphonuclear leucocytes by polystyrene beads and the non-lytic action of complement. *Immunology*, **44**, 569-576. [Ch. 18]

Hamabe, M., Hamuro, C. & Ogura, M., 1982. *Squid Jigging from Small Boats*. Farnham: Fishing News Books. [FAO Fishing Manuals.] [Ch. 15]

Hamabe, M. & Shimizu, T., 1957. The copulation behaviour of Yari-ika *Loligo bleekeri* K. *Report of the Japan Sea and Regional Fisheries Research Laboratory*, no. 3, 131-136. [Ch. 16]

Hamner, W.M., Madin, L.P., Alldredge, A.L., Gilmer, R.W. & Hamner, P.P., 1975. Underwater observations of gelatinous zooplankton: sampling problems, feeding biology and behavior. *Limnology and Oceanography*, **20**, 907-917. [Ch. 16]

Haneda, Y., 1956. Squid producing an abundant luminous secretion found in Suruga Bay, Japan. *Science Reports of the Yokosuka City Museum*, **1**, 27-31. [Ch. 16]

Haneda, Y. & Tsuji, F.I., 1976. The luminescent systems of pony fishes. *Journal of Morphology*, **150**, 539-552. [Ch. 16]

Harden Jones, F.R., 1962. Further observations on the movements of herring (*Clupea harengus* L.) shoals in relation to tidal currents. *Journal du Conseil*, **21**, 52-76. [Ch. 13]

Harding, L.W., 1988. The time-course of photoadaptation to low-light in *Prorocentrum marielebouriae* (Dinophyceae). *Journal of Phycology*, **24**, 274-281. [Ch. 4]

Hardy, A.C., 1939. Ecological investigations with the Continuous Plankton Recorder: object, plan and methods. *Hull Bulletins of Marine Ecology*, **1**, 1-57. [Ch. 3]

Hardy, A.C., 1954. *The Open Sea*. Part I. London: Collins. [Ch. 9]

Harkness, L. & Bennet-Clark, H.C., 1978. The deep fovea as a focus indicator. *Nature, London*, **272**, 814-816. [Ch. 8]

Harosi, F.I., 1975. Microspectrophotometry: the technique and some of its pitfalls. In *Vision in Fishes* (ed. M.A. Ali), pp. 43-54. New York: Plenum. [Ch. 10]

Harosi, F.I., 1985. Ultraviolet- and violet-absorbing vertebrate visual pigments: dichroic and bleaching properties. In *The Visual System* (ed. A. Fein and J.S. Levine), pp. 41-55. New York: Liss. [Ch. 10]

Harosi, F.I. & Hashimoto, Y., 1983. Ultraviolet visual pigment in a vertebrate: a tetrachromatic cone system in the dace. *Science, New York*, **222**, 1021-1023. [Ch. 10]

Harris, G.P., 1978. Photosynthesis, productivity and growth: the physiological ecology of phytoplankton. *Archiv für Hydrobiologie* (Beihefte Ergebnisse der Limnologie), **10**, 1-171. [Ch. 4]

Hartline, H.K., 1938. The discharge of impulses in the optic nerve of *Pecten* in response to illumination of the eye. *Journal of Cellular and Comparative Physiology*, **11**, 465-477. [Ch. 9]

Hartline, P.H. & Lange, G.D., 1984. Visual systems of cephalopods. In *Comparative Physiology of Sensory Systems* (ed. L. Bolis), pp. 335-355. Cambridge: Cambridge University Press. [Ch. 9]

Harvey, E.N., 1916. On the production of light by certain substances in the presence of oxidases. *American Journal of Physiology*, **41**, 454-464. [Ch. 18]

Harvey, E.N., 1952. *Bioluminescence*. New York: Academic Press. [Ch. 16]

Hastings, J.W., 1971. Light to hide by: ventral luminescence to camouflage the silhouette. *Science, New York*, **173**, 1016-1017. [Ch. 16]

Hastings, J.W., 1983. Biological diversity, chemical mechanisms and the evolutionary origins of bioluminescent systems. *Journal of Molecular Evolution*, **19**, 309-321. [Ch. 17]

Hawryshyn, C.W., Arnold, M.G., Chaisson, D.J. & Martin, P.C., 1989. The ontogeny of ultraviolet photosensitivity in rainbow trout (*Salmo gairdneri*). *Visual Neuroscience*, **2**, 247-254. [Ch. 10]

Hawryshyn, C.W. & Beauchamp, R.D., 1985. Ultraviolet photosensitivity in the goldfish: an independent UV retinal mechanism. *Vision Research*, **25**, 11-20. [Ch. 10]

Hawryshyn, C.W. & McFarland, W.N., 1987. Cone photoreceptor mechanisms and the detection of polarized light in fish. *Journal of Comparative Physiology*, **160**(A), 459-466. [Ch. 10]

Haxo, F.T., 1985. Photosynthetic action spectrum of the coccolithophorid, *Emiliania huxleyi* (Haptophyceae): 19' hexanoyloxyfucoxanthin as antenna pigment. *Journal of Phycology*, **21**, 282-287. [Ch. 5]

Haxo, F.T. & Blinks, L.R., 1950. Photosynthetic action spectra of marine algae. *Journal of General Physiology*, **33**, 389-422. [Ch. 5]

Haynes, L. & Yau, K.-W., 1985. Cyclic GMP-sensitive conductance in outer segment membrane of catfish cones. *Nature, London*, **317**, 61-64. [Ch. 11]

Heath, M.R., Henderson, E.W. & Baird, D.L., 1988. Vertical distribution of herring larvae in relation to physical mixing and illumination. *Marine Ecology - Progress Series*, **47**, 211-228. [Ch. 13]

Hecht, S. & Pirenne, M., 1940. The sensitivity of the long-eared owl in the spectrum. *Journal of General Physiology*, **23**, 709-717. [Ch. 8]

Hecht, S. & Verjip, C.D., 1933. Intermittent stimulation by light. *Journal of General Physiology*, **17**, 237-282. [Ch. 10]

Helfman, G. S., 1978. Patterns of community structure in fishes: summary and overview. *Environmental Biology of Fishes*, **3**, 129-148. [Ch. 14]

Helfman, G. S., 1986. Fish behaviour by day, night and twilight. In *The Behaviour of Teleost Fishes* (ed. T. J. Pitcher), pp. 366-87. London: Croom Helm. [Ch. 14]

Helfman, G. S., Meyer, J. L. & McFarland, W. N., 1982. The ontogeny of twilight migration patterns in grunts (Pisces: Haemulidae). *Animal Behaviour*, **30**, 317-326. [Ch. 14]

Helfman, G. S. & Schultz, E. T., 1984. Social transmission of behaviour traditions in a coral reef fish. *Animal Behaviour*, **32**, 379-384. [Ch. 14]

Hemmings, C.C. & Lythgoe, J.N., 1965. The visibility of underwater objects. In *Symposium of the Underwater Association for Malta* (ed. J.N. Lythgoe and J.D. Woods), pp. 23-30. London: Underwater Association. [Ch. 10]

Hernando, A.M. & Flores, E.C., 1981. The Philippine squid fishery; a review. *Marine Fisheries Review*, **43**, 13-20. [Ch. 15]

Herring, P.J., 1974. New observations on the bioluminescence of echinoderms. *Journal of Zoology, London*, **172**, 401-418. [Ch. 16]

Herring, P.J., 1977. Luminescence in cephalopods and fish. *Symposia of the Zoological Society of London*, no. 38, 127-159. [Ch. 15]

Herring, P.J., 1978. Bioluminescence in invertebrates other than insects. In *Bioluminescence in Action* (ed. P.J. Herring), pp. 199-240. London: Academic Press. [Ch. 15, 16]

Herring, P.J., 1981. Studies on bioluminescent marine amphipods. *Journal of the Marine Biological Association of the United Kingdom*, **61**, 161-176. [Ch. 16]

Herring, P.J., 1982. Aspects of the bioluminescence of fishes. *Oceanography and Marine Biology, an Annual Review*, **20**, 415-470. [Ch. 16]

Herring, P.J., 1983. The spectral characteristics of luminous marine organisms. *Proceedings of the Royal Society* (B), **220**, 183-217. [Ch. 8, 10, 16]

Herring, P.J., 1985. Bioluminescence in the Crustacea. *Journal of Crustacean Biology*, **5**, 557-573. [Ch. 16]

Herring, P.J., 1986. How to survive in the dark: bioluminescence in the deep sea. *Symposia of the Society for Experimental Biology*, no. 39, 323-350. [Ch. 16]

Herring, P.J., 1987. Systematic distribution of bioluminescence in living organisms. *Journal of Bioluminescence and Chemiluminescence*, **1**, 147-167. [Ch. 18]

Herring, P.J., 1988a. Luminescent organs. In *The Mollusca* (ed. E.R. Trueman and M.R. Clarke), **11**, 449-489. [Ch. 16]

Herring, P.J., 1988*b*. Copepod luminescence. In *Biology of Copepods* (ed. G.A. Boxshall and H.R. Schminke). *Hydrobiologia*, **167/168**, 183-195. [Ch. 16]

Herring, P.J. & Locket, N.A., 1978. The luminescence and photophores of euphausiid crustaceans. *Journal of Zoology, London*, **186**, 431-462. [Ch. 16]

Herring, P.J. & Morin, J.G., 1978. Bioluminescence in fishes. In *Bioluminescence in Action* (ed. P.J. Herring), pp. 273-329. London: Academic Press. [Ch. 15, 16]

Herring, P.J. & Roe, H.S.J., 1988. The photoecology of pelagic oceanic decapods. *Symposia of the Zoological Society of London*, no. 59, 263-290. [Ch. 16]

Hiller, R.G. & Larkum, A.W.D., 1984. The chlorophyll-protein complexes of *Prochloron* sp. (Prochlorophyta). *Biochimica et Biophysica Acta*, **806**, 184-193. [Ch. 5]

Hiller, R.G., Larkum, A.W.D. & Wrench, P.M., 1988. Chlorophyll proteins of the prymnesiophyte *Pavlova lutherii* (Droop) comb. nov.; identification of the major light-harvesting complex. *Biochimica et Biophysica Acta*, **932**, 223-231. [Ch. 5]

Hiller-Adams, P. & Case, J.F., 1988. Eye size of pelagic crustaceans as a function of habitat depth and possession of photophores. *Vision Research*, **28**, 667-680. [Ch. 16]

Hobson, E. S., 1965. Nocturnal-diurnal activity of some inshore fishes in the Gulf of California. *Copeia*, **3**, 291-302. [Ch. 14]

Hobson, E. S., 1968. Predatory behaviour of some shore fishes in the Gulf of California. *Research Report. United States Fish and Wildlife Service*, no. 73, 1-92. [Ch. 14]

Hobson, E. S., 1972. Activity of Hawaiian reef fishes during the evening and morning transitions between daylight and darkness. *Fisheries Bulletin. National Oceanic and Atmospheric Administration of the United States*, **70**, 715-740. [Ch. 14]

Hobson, E. S., 1973. Diel feeding migrations in tropical reef fishes. *Helgoländer Wißenschaftliche Meersuntersuchungen*, **24**, 361-370. [Ch. 14]

Hobson, E. S., 1974. Feeding relationships of teleost fishes on coral reefs in Kona, Hawaii. *Fisheries Bulletin. National Oceanic and Atmospheric Administration of the United States*, **72**, 915-1031. [Ch. 14]

Hobson, E. S., 1986. Predation on the Pacific sand lance, *Ammodytes hexapterus* (Pisces: Ammodytidae), during the transition between day and night on south-eastern Alaska. *Copeia*, **1986**, 223-226. [Ch. 14]

Hobson, E. S. & Chess, J. R., 1978. Tropical relationships among fishes and plankton in the Lagoon at Enewetak Atoll, Marshall Islands. *Fisheries Bulletin. National Oceanic and Atmospheric Administration of the United States*, **76**, 133-153. [Ch. 14]

Hobson, E. S. & Chess, J. R., 1986. Diel movements of residents and transient zooplankters above lagoon reefs at Enewetak Atoll, Marshall Islands. *Pacific Science*, **40**, 1-4. [Ch. 14]

Hobson, E.S., McFarland, W.N. & Chess, J.R., 1981. Crepuscular and nocturnal activities of Californian nearshore fishes, with consideration of their scotopic visual pigments and the photic environment. *Fishery Bulletin. National Oceanic and Atmospheric Administration of the United States*, **79**, 1-30. [Ch. 14, 16]

Hodgkin, A.L., McNaughton, P.A. & Nunn, B.J., 1985. The ionic selectivity and calcium dependence of the light-sensitive pathway in toad rods. *Journal of Physiology*, **358**, 447-468. [Ch. 11]

Hodgkin, A.L., McNaughton, P.A. & Nunn, B.J., 1986. Effects of changing Ca before and after light flashes in salamander rods. *Journal of Physiology*, **372**, 54*P*. [Ch. 11]

Hodgkin, A.L., McNaughton, P.A. & Nunn, B.J., 1987. Measurement of sodium-calcium exchange in salamander rods. *Journal of Physiology*, **391**, 347-370. [Ch. 11]

Hodgkin, A.L., McNaughton, P.A., Nunn, B.J. & Yau, K.-W., 1984. Effects of ions on retinal rods from *Bufo marinus*. *Journal of Physiology*, **350**, 649-680. [Ch. 11]

Hodgkin, A.L. & Nunn, B.J., 1988. Control of light-sensitive current in salamander rods. *Journal of Physiology*, **403**, 439-471. [Ch. 11]

Hoge, F.E. & Swift, R.N., 1981. Airborne simultaneous spectroscopic detection of laser-induced water Raman backscatter and fluorescence from chlorophyll *a* and other naturally occurring pigments. *Applied Optics*, **20**, 3197-3205. [Ch. 1, 2]

Holligan, P.M., 1981. Biological implications of fronts on the Northwest European continental shelf. *Philosophical Transactions of the Royal Society* (A), **302**, 547-562. [Ch. 2]

Holligan, P.M., Viollier, M., Harbour, D.S., Camus, P. & Champagne-Phillippe, M., 1983. Satellite and ship studies of coccolithophore production along a continental shelf edge, *Nature, London*, **304**, 339-342. [Ch. 2]

Hooks, C.E., Bidigare, R.R., Keller, M.D. & Guillard, R.R.L., 1988. Coccoid eukaryotic marine ultraplankters with four different HPLC pigment signatures. *Journal of Phycology*, **24**, 571-580. [Ch. 5]

Horwitz, B.A. & Gressel, J.B., 1987. First measureable effects following photoinduction of morphogenesis. In *Blue Light Responses: Phenomena and Occurrence in Plants and Microorganisms* (ed. H. Senger), pp. 53-70. Florida: CRC Press Inc. [Ch. 7]

Hovis, W.A. *et al.*, 1980. Nimbus-7 Coastal Zone Colour Scanner: system description and initial imagery. *Science, New York*, **210**, 60-63. [Ch. 2]

Howard, K.M. & Joint, I., 1989. Physiological ecology of picoplankton in the North Sea. *Marine Biology*, **102**, 275-281. [Ch. 6]

Hubbell, W.L. & Bownds, M.D., 1979. Visual transduction in vertebrate photoreceptors. *Annual Review of Neuroscience*, **2**, 17-34. [Ch. 11]

Hughes, A., 1977. The topography of vision in mammals of contrasting life style: comparative optics and retinal organization. In *Handbook of Sensory Physiology*, vol. VII/5 (ed. F. Crescitelli), pp. 613-716. Berlin: Springer-Verlag. [Ch. 9]

Hummelen, J.C. , Luider, T.M. & Wynberg, A., 1987. Stable 1,2-dioxetanes as labels for thermochemiluminescent immunoassay. *Methods in Enzymology*, **133**, 531-557. [Ch. 18]

Hurley, J.B., 1987. Molecular properties of the cGMP cascade of vertebrate photoreceptors. *Annual Review of Physiology*, **49**, 793-812. [Ch. 11]

Højerslev, N., 1975. A spectral light absorption meter for measurements in the sea. *Limnology and Oceanography*, **20**, 1024-1034. [Ch. 3]

Højerslev, N., 1980. Water colour and its relation to primary production. *Boundary-layer Meteorology*, **18**, 203-220. [Ch. 2]

Højerslev, N., 1981. Daylight measurements appropriate for photosynthetic studies in natural waters. *Journal du Conseil*, **38**, 131-146. [Ch. 4]

Iino, M., 1987. Kinetic modelling of phototropism in maize coleoptiles. *Planta*, **171**, 110-126. [Ch. 7]

Inoue, S. Okada, K., Tanino, H. & Kakoi, H., 1980. A new synthesis of *Watasenia* preluciferin. *Chemical Letters*, 299-300. [Ch. 17]

Inouye, S., Noguchi, M., Sakaki, Y., Takagi, I., Miyata, T., Iwanaga, S. & Tsuji, F.I., 1985. Cloning and sequence analysis of cDNA for the luminescent protein aequorin. *Proceedings of the National Acadamy of Sciences of the United States of America*, **82**, 3154-3158. [Ch. 18]

Itagaki, T., Nakayama, K. & Okada, M., 1986. Chlorophyll-protein complexes associated with Photosystem I isolated from the green alga *Bryopsis maxima*. *Plant and Cell Physiology*, **27**, 1241-1247. [Ch. 5]

Ivanoff, A. & Waterman, T.H., 1958. Factors, mainly depth and wavelength, affecting the degree of underwater light polarization. *Journal of Marine Research*, **16**, 283-307. [Ch. 10]

Jacobs, G.H., 1981. *Comparative Color Vision*. New York: Academic Press. [Ch. 10]

Jacobs, G.H. & Neitz, J., 1985. Spectral positioning of mammalian cone pigments. *Journal of the Optical Society of America*, **2**(A), **P23**. [Ch. 10]

Jaffe, L.F., 1958. Tropistic responses of zygotes of the Fucaceae to polarized light. *Experimental Cell Research*, **15**, 282-299. [Ch. 7]

Jassby, A.D. & Platt, T., 1976. Mathematical formulation of the relationship between photosynthesis and light for phytoplankton. *Limnology and Oceanography*, **21**, 540-547. [Ch. 4]

Jeffrey, S.W., 1976. The occurrence of chlorophyll c_1 and c_2 in algae. *Journal of Phycology*, **12**, 349-354. [Ch. 5]

Jeffrey, S.W. & Wright, S.W., 1987. A new spectrally distinct component in preparations of chlorophyll c from the micro-alga *Emiliania huxleyi* (Prymnesiophyceae). *Biochimica et Biophysica Acta*, **894**, 180-188. [Ch. 5]

Jenkin, P.M., 1937. Oxygen production by the diatom *Coscinodiscus excentricus* Ehr. in relation to submarine illumination in the English Channel. *Journal of the Marine Biological Association of the United Kingdom*, **22**, 301-343. [Ch. 4]

Jerlov, N.G., 1968. *Optical Oceanography*. Amsterdam: Elsevier. [Ch. 9]

Jerlov, N.G., 1974. Significant relationships between optical properties of the sea. In *Optical Aspects of Oceanography* (ed. N.G. Jerlov and E.S. Nielsen), pp. 77-94. London: Academic Press. [Ch. 2]

Jerlov, N.G., 1976. *Marine Optics*. Amsterdam: Elsevier. [Ch. 1, 4, 8, 10]

Jitts, H.R., Morel, A. & Saijo, Y., 1976. The relation of oceanic primary production to available photosynthetic irradiance. *New Phytologist*, **75**, 11-20. [Ch. 4]

Johannes, R. E., 1978. Reproductive strategies of coastal marine fishes in the tropics. *Environmental Biology of Fishes*, **3**, 65-84. [Ch. 14]

Johnson, E., Robinson, P. & Lisman, J., 1986. Involvement of cGMP in the excitation of invertebrate photoreceptors. *Nature, London*, **324**, 468-470. [Ch. 12]

Johnson, P. & Sieburth, J.McN., 1979. Chroococcoid cyanobacteria in the sea: a ubiquitous and diverse phototrophic biomass. *Limnology and Oceanography*, **24**, 928-935. [Ch. 6]

Johnson, P.W. & Sieburth, J.McN., 1982. *In situ* morphology and occurrence of eucaryotic phototrophs of bacterial size in the plankton of estuarine and oceanic waters. *Journal of Phycology*, **18**, 318-327. [Ch. 6]

Joint, I.R., 1986. Physiological ecology of picoplankton in various oceanographic provinces. *Canadian Bulletin of Fisheries and Aquatic Sciences*, no. 214, 287-309. [Ch. 6]

Joint, I.R. & Pipe, R.K., 1984. An electron microscope study of a natural population of picoplankton from the Celtic Sea. *Marine Ecology - Progress Series*, **20**, 113-118. [Ch. 6]

Joint, I.R. & Pomroy, A.J., 1983. Production of picoplankton and small nanoplankton in the Celtic Sea. *Marine Biology*, **77**, 19-27. [Ch. 6]

Joint, I.R. & Pomroy, A.J., 1986. Photosynthetic characteristics of nanoplankton and picoplankton from the surface mixed layer. *Marine Biology*, **92**, 465-474. [Ch. 6]

Joint, I.R., Owens, N.J.P. & Pomroy, A.J., 1986. Seasonal production of photosynthetic picoplankton and nanoplankton in the Celtic Sea. *Marine Ecology - Progress Series*, **28**, 51-258. [Ch. 6]

Kamiya, A. & Miyachi, S., 1984. Blue-green and green light adaptations on photosynthetic activity in some algae collected from subsurface chlorophyll layer in the western Pacific Ocean. In *Blue Light Effects in Biological Systems* (ed. H. Senger), pp. 517-523. Berlin: Springer-Verlag. [Ch. 5]

Kampa, M., 1970. Underwater daylight and moonlight measurements in the eastern North Atlantic. *Journal of the Marine Biological Association of the United Kingdom*, **50**, 397-420. [Ch. 8]

Kana, T.M. & Glibert, P.M., 1987. Effect of irradiances up to 2000 μE m^{-2} s^{-1} on marine *Synechococcus* WH7803 - I. Growth, pigmentation, and cell composition. *Deep-Sea Research*, **34**, 479-495. [Ch. 5, 6]

Kana, T.M., Glibert, P.M., Goericke, R. & Welschmeyer, N.A., 1988. Zeaxanthin and β–carotene in *Synechococcus* WH7803 respond differently to irradiance. *Limnology and Oceanography*, **33**, 1623-1627. [Ch. 6]

Kastendiek, J., 1976. Behavior of the sea pansy *Renilla kollikeri* Pfeffer (Coelenterata: Pennatulacea) and its influence on the distribution and biological interactions of the species. *Biological Bulletin. Marine Biological Laboratory, Woods Hole, Mass.*, **151**, 518-537. [Ch. 16]

Kataoka, H. & Weisenseel, M.H., 1988. Blue light promotes ionic current influx at the growing apex of *Vaucheria terrestris*. *Planta*, **173**, 490-499. [Ch. 7]

Kato, S. & Hardwick, J.E., 1976. The Californian squid fishery. *FAO Fisheries Reports*, no. 170, 107-127. [Ch. 15]

Kawamura, S. & Bownds, M.D., 1981. Light adaptation of the cyclic GMP phosphodiesterase of frog photoreceptor membranes mediated by ATP and calcium ions. *Journal of General Physiology*, **77**, 571-591. [Ch. 11]

Kaye, G.W.C. & Laby, T.H., 1966. *Table of Physical Constants*. 13th ed. London: Longmans. [Ch. 10]

Keller, G.-A., Gould., S., Deluca, M.B. & Subramani, S., 1987. Firefly luciferase is targetted to peroxisomes in mammalian cells. *Proceedings of the National Academy of Science of the United States of America*, **84**, 3264-3268. [Ch. 18]

Kerneis, A., 1975. Etude comparée d'organes photorécepteurs de Sabellidae (Annélides Polychètes). *Journal of Ultrastructure Research*, **53**, 164-179. [Ch. 9]

Kiefer, D.A., 1973. Chlorophyll *a* fluorescence in marine centric diatoms: responses of chloroplasts to light and nutrient stress. *Marine Biology*, **23**, 39-46. [Ch. 1]

Kikkawa, U., Kitano, T., Saito, N., Kishimoto, A., Taniyama, K., Tanaka, C. & Nishizuka, Y., 1986. Role of protein kinase C in calcium-mediated signal transduction. In *Calcium and the Cell* (ed. D. Evered and J. Whelan), pp. 197-207. Chichester: John Wiley. [Ch. 7]

Kirby, R., 1986. *Suspended Fine Cohesive Sediment in the Severn Estuary and Inner Bristol Channel, UK.* ETSU Report no. ETSU-STB-4042. [Report to the UKAEA, Ravensrod Consultants Ltd.] [Ch. 2]

Kirk, J.T.O., 1975. A theoretical analysis of the contribution of algal cells to the attenuation of light within natural waters. I. General treatment of suspensions of pigmented cells. *New Phytologist,* **75**, 11-20. [Ch. 5]

Kirk, J.T.O., 1976. A theoretical analysis of the contribution of algal cells to the attenuation of light within natural waters. III. Cylindrical and spheroidal cells. *New Phytologist,* **77**, 341-358. [Ch. 5]

Kirk, J.T.O., 1977. Thermal dissociation of fucoxanthin-protein binding in pigment complexes from chloroplasts of *Hormosira* (Phaeophyta). *Plant Science Letters,* **9**, 373-380. [Ch. 5]

Kirk, J.T.O., 1981. A Monte Carlo study of the nature of the underwater light field in, and the relationships between optical properties of turbid waters. *Australian Journal of Marine and Freshwater Research,* **32**, 517-532. [Ch. 2, 3]

Kirk, J.T.O., 1983. *Light and Photosynthesis in Aquatic Ecosystems.* Cambridge: Cambridge University Press. [Ch. 1, 2, 3, 5, 6, 8]

Klein, B. & Sournia, A., 1987. A daily study of the diatom spring bloom at Roscoff (France) in 1985. II. Phytoplankton pigment composition studied by HPLC analysis. *Marine Ecology - Progress Series,* **37**, 265-275. [Ch. 5]

Knowles, A. & Dartnall, H.J.A., 1977. The photobiology of vision. In *The Eye* (ed. H. Davson), pp. 1-689. London: Academic Press. [Ch. 10]

Kobayashi, H., 1962. A comparative study on electroretinogram in fish with special reference to ecological aspects. *Journal of the Shimonoseki College of Fisheries,* **11**, 17-148. [Ch. 10]

Koch, K.-W. & Stryer, L., 1988. Highly cooperative feedback control of retinal rod guanylate cyclase by calcium ions. *Nature, London,* **334**, 64-66. [Ch. 11]

Kohata, K. & Watanabe, M., 1988. Diel changes in the composition of photosynthetic pigments and cellular carbon and nitrogen in *Chattonella antiqua* (Raphidophyceae). *Journal of Phycology,* **24**, 58-66. [Ch. 5]

Koslow, J.A., 1979. Vertical migrators see the light? *Limnology and Oceanography,* **24**, 783-784. [Ch. 16]

Kricka, L.J. & Thorpe, G.H.G., 1987. Photographic detection of chemiluminescent and bioluminescent reactions. *Methods in Enzymology* **133**, 404-420. [Ch. 18]

Kropf, D.L., Berge, S.K. & Quatrano, R.S., 1989a. Actin localization during *Fucus* embryogenesis. *Plant Cell,* **1**, 191-200. [Ch. 7]

Kropf, D.L., Hopkins, R. & Quatrano, R.S., 1989b. Protein synthesis during embryogenesis in *Fucus. Plant Cell,* in press. [Ch. 7]

Kropf, D.L., Kloareg, B. & Quatrano, R.S., 1988. Cell wall is required for fixation of the embryonic axis in *Fucus* zygotes. *Science, New York,* **239**, 187-190. [Ch. 7]

Kropf, D.L. & Quatrano, R.S., 1987. Localization of membrane-associated calcium during development of fucoid algae using chlorotetracycline. *Planta,* **171**, 158-170. [Ch. 7]

Kurc, G., 1969. L'application à la peche des réactions phototropiques des poissons. *FAO Fisheries Reports,* no. 62, vol. 2, 283-296. [Ch. 15]

Labas, Y. A., 1977. Triggering and regulatory mechanisms of ciliary beating in Ctenophora. 1. Coordination of ciliary beating with intracellular bioluminescence and muscle contractions. *Tsitologiya,* **19**, 514-521. [Ch. 16]

Lagnado, L., Cervetto, L. & McNaughton, P.A., 1988. Ion transport by the Na-Ca exchange in isolated rod outer segments. *Proceedings of the National Academy of Sciences of the United States of America,* **85**, 4548-4552. [Ch. 11]

Lamb, T.D., McNaughton, P.A. & Yau, K.-W., 1981. Spatial spread of activation and background desensitization in toad rod outer segments. *Journal of Physiology,* **319**, 463-496. [Ch. 11]

Lamb, T.D. & Matthews, H.R., 1988. External and internal actions in the response of salamander rods to altered external calcium concentration. *Journal of Physiology,* **403**, 473-494. [Ch. 11]

Lamb, T.D., Matthews, H.R. & Torre, V., 1986. Incorporation of calcium buffer into salamander retinal rods: a rejection of the calcium hypothesis of phototransduction. *Journal of Physiology*, **372**, 315-349. [Ch. 11]

Lampitt, R.S., 1985. Evidence for the seasonal deposition of detritus to the deep-sea floor and its susequent resuspension. *Deep-Sea Research*, **32**, 885-897. [Ch. 4]

Land, M.F., 1965. Image formation by a concave reflector in the eye of the scallop, *Pecten maximus. Journal of Physiology*, **179**, 138-153. [Ch. 9]

Land, M.F., 1972. The physics and biology of animal reflectors. *Progress in Biophysics and Molecular Biology*, **24**, 75-106. [Ch. 9]

Land, M.F., 1978. Animal eyes with mirror optics. *Scientific American*, **239**(6), 126-134. [Ch. 9]

Land, M.F., 1979. The optical mechanism of the eye of *Limulus. Nature, London*, **280**, 396-397. [Ch. 9]

Land, M.F., 1980. Eye movements and the mechanism of vertical steering in euphausiid crustacea. *Journal of Comparative Physiology*, **137**(A), 255-265. [Ch. 16]

Land, M.F., 1981a. Optics and vision in invertebrates. In *Handbook of Sensory Physiology*, vol. VII/6B. *Comparative Physiology and Evolution of Vision in Invertebrates* (ed. H. Autrum), pp. 471-592. Berlin: Springer-Verlag. [Ch. 9, 10, 12]

Land, M.F., 1981b. Optics of the eyes of *Phronima*, and other deep-sea amphipods. *Journal of Comparative Physiology* (A), **145**, 209-226. [Ch. 9]

Land, M.F., 1982. Scanning eye movements in a heteropod mollusc. *Journal of Experimental Biology*, **96**, 427-430. [Ch. 9]

Land, M.F., 1984. Crustacea. In *Photoreception and Vision in Invertebrates* (ed. M.A. Ali), pp. 401-438. New York: Plenum. [Ch. 9]

Land, M.F., 1987. Vision in air and water. In *Comparative Physiology: Life in Water and on Land* (ed. P. Dejours), pp. 289-302. New York: Liviana/Springer. [Ch. 9]

Land, M.F., 1988. The functions of eye and body movements in *Labidocera* and other copepods. *Journal of Experimental Biology*, **140**, 381-391. [Ch. 9]

Land, M.F., 1989. The eyes of hyperiid amphipods: relations of optical structure to depth. *Journal of Comparative Physiology* (A), **164**, 751-762. [Ch. 9]

Land, M.F., Burton, F.A. & Meyer-Rochow, V.B., 1979. The optical geometry of euphausiid eyes. *Journal of Comparative Physiology* (A), **130**, 49-62. [Ch. 9]

Lane, E.D., 1967. A study of the Atlantic midshipman, *Porichthys porosissimus*, in the vicinity of Port Aransas, Texas. *Contributions in Marine Science, University of Texas*, **12**, 1-53. [Ch. 16]

Larkum, A.W.D. & Barrett, J., 1983. Light-harvesting processes in algae. *Advances in Botanical Research*, **10**, 3-222. [Ch. 5]

Larkum, A.W.D. & Weyrauch, S.K., 1977. Photosynthetic action spectra and light-harvesting in *Griffithsia monilis* (Rhodophyta). *Photochemistry and Photobiology*, **25**, 65-72. [Ch. 5]

Latham, J.A. & Walsh C., 1987. Mechanism-based inactivation of the flavoenzyme cyclohexanone oxygenase during oxygenation of cyclic thiol ester substrates. *Journal of the American Chemical Society*, **109**, 3421-3427. [Ch. 17]

Latz, M.I. & Case, J.F., 1982. Light organ and eyestalk compensation to body tilt in the luminescent midwater shrimp, *Sergestes similis. Journal of Experimental Biology*, **98**, 83-104. [Ch. 16]

Lavett Smith, C. & Tyler, J. C., 1972. Space resource sharing in a coral reef fish community. *Bulletin of the Natural History Museum of Los Angeles County*, **14**, 125-178. [Ch. 14]

Laws, E.A. & Bannister, T.T., 1980. Nutrient- and light- limited growth of *Thalassiosira fluviatilis* in continuous culture, with implications for phytoplankton growth in the ocean. *Limnology and Oceanography*, **25**, 457-473. [Ch. 4]

Lederman, T.C. & Tett, P., 1981. Problems in modelling the photosynthesis-light relationship for phytoplankton. *Botanica Marina*, **24**, 125-134. [Ch. 4]

Lee, J., 1985. Mechanism of bacterial bioluminescence. In *Chemi- and Bioluminescence* (ed. J. Burr), pp. 401-437. New York: Marcel Dekker. [Ch. 17]

Lees, B.J., 1989. *The Use of Remote Sensing as a Tool to Further the Understanding of Suspended Sediment Dynamics in the Bristol Channel*. PhD Thesis. University of Southampton. [Ch. 2]

Le Grand, Y., 1952. *Optique Physiologique*, T2. *Lumière et Couleurs*. Paris: Editions de la Revue d'Optique. [Ch. 8]

Leong, T.-Y. & Briggs, W.R., 1981. Partial purification and characterization of a blue light-sensitive cytochrome-flavin complex from corn membranes. *Plant Physiology*, **67**, 1042-1046. [Ch. 7]

Leong, T.-Y., Vierstra, R.D. & Briggs, W.R., 1981. A blue light-sensitive cytochrome-flavin complex from corn root coleoptiles. Further characterization. *Photochemistry and Photobiology*, **34**, 697-703. [Ch. 7]

Levavasseur, G., 1989. Analyse comparée des complexes pigment-proteines de chlorophyco-phytes marines benthiques. *Phycologia*, **28**, 1-14. [Ch. 5]

Levi, B. & Sivak, J.G., 1980. Mechanisms of accommodation in the bird eye. *Journal of Comparative Physiology*, **137**(A), 267-272. [Ch. 9]

Levine, J.S. & MacNichol, E.F., Jr, 1979. Visual pigments in teleost fishes: effects of habitat, microhabitat, and behavior on visual system evolution. *Sensory Processes*, **3**, 95-131. [Ch. 10]

Levine, J.S. & MacNichol, E.F., Jr, 1982. Color vision in fishes. *Scientific American*, **246**, 140-149. [Ch. 10]

Levitus, S., 1982. *Climatological Atlas of the World Ocean*. [NOAA Professional Paper no. 13.] Rockville, Maryland: US Department of Commerce. [Ch. 4]

Lewis, M.R., Cullen, J.J. & Platt, T., 1982. Relationships between vertical mixing and photoadap-tation of phytoplankton: similarity criteria. *Marine Ecology - Progress Series*, **15**, 141-149. [Ch. 4]

Lewis, M.R., Horne, E.P.W., Cullen, J.J., Oakey, N.S. & Platt, T., 1984. Turbulent motions may control phytoplankton photosynthesis in the upper ocean. *Nature, London*, **311**, 49-50. [Ch. 4]

Lewis, M.R., Warnock, R.E., Irwin, B. & Platt, T., 1985. Measuring photosynthetic action spectra of natural phytoplankton populations. *Journal of Phycology*, **21**, 310-315. [Ch. 1]

Lewis, M.R., Warnock, R.E. & Platt, T., 1986. Photosynthetic response of marine picoplankton at low photon flux. *Canadian Bulletin of Fisheries and Aquatic Sciences*, no. 214, 235-250. [Ch. 6]

Li, W.K.W., 1986. Experimental approaches to field measurements: methods and interpretation. *Canadian Bulletin of Fisheries and Aquatic Sciences*, no. 214, 251-286. [Ch. 1]

Li, W.K.W., Subba Rao, D.V., Harrison, W.G., Smith, J.C., Cullen, J.J., Irwin, B. & Platt, T., 1983. Autotrophic picoplankton in the tropical ocean. *Science, New York*, **219**, 292-295. [Ch. 6]

Lichtlé, C., Duval, J.C. & Lemoine, Y., 1987. Comparative biochemical, functional and ultrastruc-tural studies of photosystem particles from a cryptophyceae: *Cryptomonas rufescens*; isolation of an active phycoerythrin particle. *Biochimica et Biophysica Acta*, **894**, 76-90. [Ch. 5]

Liebman, P.A., 1972. Microspectrophotometry of photoreceptors. In *Handbook of Sensory Physiology*, vol. VII/1. *Photochemistry of Vision* (ed. H.J.A. Dartnall), pp. 481-528. Berlin: Springer-Verlag. [Ch. 10]

Liebman, P.A., Parker, K.R. & Dratz, E.A., 1987. The molecular mechanism of visual excitation and its relation to the structure and composition of the rod outer segment. *Annual Review of Physiology*, **49**, 765-791. [Ch. 11]

Liebman, P.A. & Pugh, E.N., Jr, 1979. The control of phosphodiesterase in rod disk membranes: kinetics, possible mechanisms and significance for vision. *Vision Research*, **19**, 375-380. [Ch. 11]

Lipetz, E.R. & Cronin, T.W., 1988. Application of an invariant spectral form to the visual pigments of crustaceans: implications regarding the binding of the chromophore. *Vision Research*, **28**, 1083-1093. [Ch. 10]

Lipps, J.H., 1970. Plankton evolution. *Evolution*, **24**, 1-22. [Ch. 4]

Littler, M.M., Littler, D.S., Blair, S.M. & Norris, J.N., 1985. Deepest known plant life discovered on an uncharted seamount. *Science, New York*, **227**, 57-69. [Ch. 5]

Litvin, F.F., Sineshchekov, O.A. & Sineshchekov, V.A., 1978. Photoreceptor electric potential in the phototaxis of the alga *Haematococcus pluvialis*. *Nature, London*, **271**, 476-478. [Ch. 7]

Lloyd, J.E., 1977. Bioluminescence and communication. In *How Animals Communicate* (ed. T.A. Sebeok), pp. 164-183. Bloomington: Indiana University Press. [Ch. 16]

Lloyd, J.E., 1983. Bioluminescence and communication in insects. *Annual Review of Entomology*, **28**, 131-160. [Ch. 16]

Lloyd, J.E., 1984. On deception, a way of all flesh, and firefly signalling and systematics. *Oxford Surveys of Evolutionary Biology*, **1**, 48-84. [Ch. 16]

Lobel, P. S., 1978. Diel, lunar, and seasonal periodicity in the reproductive behaviour of the pomacanthid fish, *Centropyge potteri*, and some other reef fishes in Hawaii. *Pacific Science*, **32**, 193-207. [Ch. 14]

Locket, N.A., 1977. Adaptations to the deep-sea environment. In *Handbook of Sensory Physiology*, vol. VII/5. *The Visual System in Vertebrates* (ed. F. Crescitelli), pp. 67-192. Berlin: Springer-Verlag. [Ch. 8, 9]

Locket, N.A., 1985. The multiple bank fovea of *Bajacalifornia drakei*, an alepocephalid deep-sea teleost. *Proceedings of the Royal Society* (B), **224**, 7-22. [Ch. 8]

Loeblich, A.R., Jr, 1974. Protistan phylogeny as indicated by the fossil record. *Taxon*, **23**, 277-290. [Ch. 4]

Loew, E.R. & Dartnall, H.J.A., 1976. Vitamin A1/A2-based visual pigment mixtures in cones of the rudd. *Vision Research*, **16**, 891-896. [Ch. 10]

Loew, E.R. & Lythgoe, J.N., 1978. The ecology of cone pigments in teleost fishes. *Vision Research*, **18**, 715-722. [Ch. 10]

Loew, E.R. & Lythgoe, J.N., 1985. The ecology of colour vision. *Endeavour*, **14**, 170-174. [Ch. 10]

Lolley, R.N. & Racz, E., 1982. Calcium modulation of cyclic GMP synthesis in rat visual cells. *Vision Research*, **22**, 1481-1486. [Ch. 11]

Lorenzen, C.J., 1966. A method for the continuous measurement of *in vivo* chlorophyll concentration. *Deep-Sea Research*, **13**, 223-227. [Ch. 1]

Lovelock, J.E., 1987. *Gaia. A New Look at Life on Earth*. Oxford: Oxford University Press. [Ch. 4]

Lüning, K. & Dring, M.J., 1985. Action spectra and spectral quantum yield of photosynthesis in marine macroalgae with thin and thick thalli. *Marine Biology*, **87**, 119-129. [Ch. 5]

Lythgoe J.N., 1968. Visual pigments and visual range underwater. *Vision Research*, **8**, 997-11012. [Ch. 10]

Lythgoe, J.N., 1972. List of vertebrate visual pigments. In *Handbook of Sensory Physiology*, vol. VII/1. *Photochemistry of Vision* (ed. H.J. Dartnall), pp. 604-624. Berlin: Springer-Verlag. [Ch. 8, 10]

Lythgoe, J.N., 1979. *The Ecology of Vision*. Oxford: Clarendon Press. [Ch. 9, 10, 14]

Lythgoe, J.N., 1988. Light and vision in the aquatic environment. In *Sensory Biology of Aquatic Animals* (ed. J. Atema *et al.*), pp. 57-82. New York: Springer-Verlag. [Ch. 10]

Lythgoe, J.N. & Hemmings. C.C., 1967. Polarized light and underwater vision. *Nature, London*, **213**, 893-894. [Ch. 10]

Lythgoe, J.N. & Partridge, J.C., 1989. Visual pigments and the acquisition of visual information. *Journal of Experimental Biology*, **146**, 1-20. [Ch. 10]

Lythgoe, J.N. & Shand, J., 1983. Diel changes in the neon tetra *Paracheirodon innesi*. *Environmental Biology of Fishes*, **8**, 249-254. [Ch. 10]

McCapra, F. & Manning, M.J., 1973. Bioluminescence of coelenterates: chemiluminescent model compounds. *Journal of the Chemical Society of London. Chemical Communications*, 467-468. [Ch. 17]

McCapra, F. & Perring, K., 1985. General organic chemiluminescence. In *Chemi- and Bioluminescence* (ed. J. Burr), pp. 259-320. New York: Marcel Dekker. [Ch. 17]

McCapra, F., Perring, K., Hart, R.J. & Hann, R.A., 1981. Photochemistry without light: reaction of oxalate esters with anthracenophanes. *Tetrahedron Letters*, 5087-5090. [Ch. 17]

McCapra, F. & Razavi, Z., 1975. A model for luciferin biosynthesis. *Journal of the Chemical Society of London. Chemical Communications*, 42-43. [Ch. 17]

McCapra, F. & Razavi, Z., 1976. Biosynthesis of luciferin in *Pyrophorus pellucens*. *Journal of the Chemical Society of London. Chemical Communications*, 153-154. [Ch. 17]

McCapra, F. & Richardson, D.G., 1964. The mechanism of chemiluminescence: a new chemiluminescent reaction. *Tetrahedron Letters*, 3147-3150. [Ch. 17]

McCapra, F. & Roth, M., 1972. Cyclisation of a dehydropeptide derivative: a model for *Cypridina* luciferin biosynthesis. *Journal of the Chemical Society of London. Chemical Communications*, 894-895. [Ch. 17]

McCosker, J.E., 1977. Flashlight fishes. *Scientific American*, **236**, 106-114. [Ch. 16]

McElroy, M.B., 1983. Marine biological controls on atmospheric CO_2 and climate. *Nature, London*, **302**, 328-329. [Ch. 4]

McElroy, W.D., 1947. The energy source for bioluminescence in an isolated system. *Proceedings of the National Academy of Sciences of the United States of America*, **33**, 342-345. [Ch. 18]

McElroy, W.D. & Seliger, H.H., 1962. Origin and evolution of bioluminescence. In *Horizons in Biochemistry* (ed. M. Kasha and B. Pullman), pp. 901-102. New York: Academic Press. [Ch. 17]

McFall-Ngai, M.J. & Dunlap, P.V., 1983. Three new modes of luminescence in the leiognathid fish *Gazza minuta*: discrete projected luminescence, ventral body flash and buccal luminescence. *Marine Biology*, **73**, 227-237. [Ch. 16]

McFall-Ngai, M.J. & Dunlap, P.V., 1984. External and internal sexual dimorphism in leiognathid fishes: morphological evidence for sex-specific bioluminescent signaling. *Journal of Morphology*, **182**, 71-83. [Ch. 16]

McFarland, W.N., 1986. Light in the sea - correlations with behaviors of fishes and invertebrates. *American Zoologist*, **26**, 389-401. [Ch. 10]

McFarland, W.N. & Loew, E.R., 1983. Wave induced changes in underwater light and their relation to vision. *Environmental Biology of Fishes*, **8**, 173-184. [Ch. 10]

McFarland, W.N. & Munz, F.W., 1975. Part III: The evolution of photopic visual pigments in fishes. *Vision Research*, **15**, 1071-1080. [Ch. 10]

McFarland, W. N. & Munz, F. W., 1976. The visible spectrum during twilight and its implications to vision. In *Light as an Ecological Factor*, vol. 2. (ed. G. C. Evans *et al.*), pp. 249-270. Oxford: Blackwell. [Ch. 14]

McFarland, W. N., Ogden, J. C. & Lythgoe, J. N., 1979. The influence of light on the twilight migrations of grunts. *Environmental Biology of Fishes*, **4**, 9-22. [Ch. 14]

MacFarlane, N. & Robinson, I.S., 1984. Atmospheric correction of Landsat MSS data for a multidate suspended sediment algorithm. *International Journal of Remote Sensing*, **5**, 561-576. [Ch. 2]

Mackie, G.O. & Bone, Q., 1978. Luminescence and associated effector activity in *Pyrosoma* (Tunicata: Pyrosomida). *Proceedings of the Royal Society* (B), **202**, 483-495. [Ch. 16]

Mackie, G.O., Pugh, P.R. & Purcell, J.E., 1987. Siphonophore biology. *Advances in Marine Biology*, **24**, 97-262. [Ch. 16]

MacLeish, P.R., Schwartz, E.A. & Tachibana, M., 1984. Control of the generator current in solitary rods of the *Ambystoma tigrinum* retina. *Journal of Physiology*, **348**, 645-664. [Ch. 11]

McNaughton, P.A. Cervetto, L. & Nunn, B.J., 1986. Measurement of the intracellular free calcium concentration in salamander rods. *Nature, London*, **322**, 261-263. [Ch. 11]

MacNichol, F.E., Jr, Feinberg, R. & Harosi, F.I., 1973. Colour discrimination processes in the retina. In *Colour 73*, pp. 191-251. London: Adam Hilger. [Ch. 10]

Major, P. F., 1977. Predator-prey interactions in schooling fishes during periods of twilight: a study of the silverside *Pranesus* in Hawaii. *Fisheries Bulletin. National Oceanic and Atmospheric Administration of the United States*, **75**, 415-426. [Ch. 14]

Makiguchi, N., Arita, M. & Asai, Y., 1980. Application of a luminous bacterium to fish attracting purpose. *Bulletin of the Japanese Society of Scientific Fisheries*, **46**, 1307-1312. [Ch. 15]

Mann, J.E. & Myers, J.E., 1968. Photosynthetic enhancement in the diatom *Phaeodactylum tricornutum*. *Plant Physiology*, **43**, 1991-1995. [Ch. 5]

Manodori, A. & Melis, A., 1984. Photochemical apparatus organisation in *Anacystis nidulans* (Cyanophyceae). Effect of CO_2 during cell growth. *Plant Physiology*, **74** 67-71. [Ch. 6]

Markert, R.E., Markert, B.J. & Vertrees, N.J., 1961. Lunar periodicity in spawning and luminescence in *Odontosyllis enopla*. *Ecology*, **42**, 414-415. [Ch. 16]

Marler, P., 1968. Visual systems. In *Animal Communication* (ed. T.A. Sebeok), pp. 103-126. Bloomington: Indiana University Press. [Ch. 16]

Marra, J., 1978. Phytoplankton photosynthetic response to vertical movement in a mixed layer. *Marine Biology*, **46**, 203-208. [Ch. 4]

Marsac, N.T. de, 1977. Occurrence and nature of chromatic adaptation in cyanobacteria. *Journal of Bacteriology*, **130**, 82-91. [Ch. 5]

Marshall, N.B., 1954. *Aspects of Deep-Sea Biology*. London: Hutchinson. [Ch. 8, 9]

Marshall, N.B., 1979. *Developments in Deep-Sea Biology*. Poole: Blandford. [Ch. 9]

Marshall, S.M. & Orr, A.P., 1927. The relation of the plankton to some physical and chemical factors in the Clyde Sea area. *Journal of the Marine Biological Association of the United Kingdom*, **14**, 837-868. [Ch. 4]

Marshall, S.M. & Orr, A.P., 1928. The photosynthesis of diatom cultures in the sea. *Journal of the Marine Biological Association of the United Kingdom*, **15**, 321-360. [Ch. 4]

Marshall, S.M. & Orr, A.P., 1930. A study of the spring diatom increase in Loch Striven. *Journal of the Marine Biological Association of the United Kingdom*, **16**, 853-878. [Ch. 4]

Martin, G.R., 1982. The owl's eye: schematic and visual performance. *Journal of Comparative Physiology*, **145**, 341-349. [Ch. 8]

Martin, P.J., 1985, Simulation of the mixed layer at OWS November and Papa with several models. *Journal of Geophysical Research*, **90**, 903-916. [Ch. 2]

Matsui, S., Seidou, M., Horiuchi, S., Uchigama, I. & Kito, Y., 1988a. Adaptation of a deep-sea cephalopod to the photic environment. *Journal of General Physiology*, **92**, 55-66. [Ch. 8, 12, 16]

Matsui, S., Seidou, M., Uchiyama, I., Sekiya, N., Hiraki, K., Yoshihara, K. & Kito, Y., 1988b. 4-hydroxyretinal, a new visual pigment chromophore found in the bioluminescent squid, *Watasenia scintillans Biochimica et Biophysica Acta*, **966**, 370-374. [Ch. 8]

Matthews, G., 1986. Comparison of the light-sensitive and cyclic GMP-sensitive conductances of the rod photoreceptor: noise characteristics. *Journal of Neuroscience*, **6**, 2521-2526. [Ch. 11]

Matthews, G., 1987. Single-channel recordings demonstrate that cGMP opens the light-sensitive ion channel of the rod photoreceptor. *Proceedings of the National Academy of Sciences of the United States of America*, **84**, 299-302. [Ch. 11]

Matthews, G. & Watanabe, S.-I., 1987. Properties of ion channels closed by light and opened by guanosine 3',5'-cyclic monophosphate in toad retinal rods. *Journal of Physiology*, **389**, 691-715. [Ch. 11]

Matthews, H.R., Fain, G.L., Murphy, R.L.W. & Lamb, T.D., 1990. Light adaptation in cone photoreceptors of the salamander: a role for cytoplasmic calcium. *Journal of Physiology*, in press. [Ch. 11]

Matthews, H.R., Murphy, R.L.W., Fain, G.L. & Lamb, T.D., 1988. Photoreceptor light adaptation is mediated by cytoplasmic calcium concentration. *Nature, London*, **334**, 67-69. [Ch. 11]

Matthews, H.R., Torre, V. & Lamb, T.D., 1985. Effects on the photoresponse of calcium buffers and cyclic GMP incorporated into the cytoplasm of retinal rods. *Nature, London*, **313**, 582-585. [Ch. 11]

Matthews, J.A., Batki, A., Hynds, C.C. & Kricka, L.J., 1985. Enhanced chemiluminescent method for the detection of DNA dot-hybridisation assays. *Analytical Biochemistry*, **151**, 205-209. [Ch. 18]

Mauchline, J., 1960. The biology of the euphausiid crustacean *Meganyctiphanes norvegica* (M. Sars). *Proceedings of the Royal Society of Edinburgh* (B), **67**, 141-179. [Ch. 16]

Maull, G, 1985. *Introduction to Satellite Oceanography*. Dordrecht: D. Reidel Publishing Co. [Ch. 2]

Messenger, J.B., 1981. Comparative physiology of vision in molluscs. In *Handbook of Sensory Physiology*, vol. VII/6C (ed. H. Autrum), pp. 93-200. Berlin: Springer-Verlag. [Ch. 12]

Meyer, T.L., Cooper, R.A. & Langton, R.W., 1979. Relative abundance, behaviour, and food habits of the American sand lance, *Ammodytes americanus*, from the Gulf of Maine. *Fisheries Bulletin. National Oceanic and Atmospheric Administration of the United States*, **77**, 243-253. [Ch. 14]

Middleton, W.E.K., 1952. *Vision Through the Atmosphere*. Toronto: University of Toronto Press. [Ch. 10]

Miki, N., Baraban, J.M., Keirns, J.J., Boyce, J.J. & Bitensky, M.W., 1975. Purification and properties of the light-activated cyclic nucleotide phosphodiesterase of rod outer segments. *Journal of Biological Chemistry*, **250**, 6320-6327. [Ch. 11]

Miller, W.H. (ed.), 1981. *Current Topics in Membranes and Transport. 15. Molecular Mechanisms of Photoreceptor Transduction*. New York: Academic Press. [Ch. 11]

Miska, W. & Geiger, R., 1987. Modified luciferin sbstrates jfor enzyme analysis. *Journal of Chemical and Clinical Biochemistry*, **25**, 23-30. [Ch. 18]

Mitchelson, E.G., Jacob, N.J. & Simpson, J.H., 1986. Ocean colour algorithms from the Case 2 waters of the Irish Sea in comparison to algorithms from Case 1 waters. *Continental Shelf Research*, **5**, 403-418. [Ch. 2, 4]

Miyamoto, C., Boyland, M., Graham, A. & Meighen, E., 1987. Cloning and expression of the genes from the bioluminescence system of marine bacteria. *Methods in Enzymology*, **133**, 70-83. [Ch. 18]

Moeller, H.W., Bennett, B., Coughlin, S. & Getz, D., 1972. Predator prey relationships under luminous conditions. *Marine Behaviour and Physiology*, **1**, 257-260. [Ch. 16]

Mohr, H., 1987. Mode of coaction between blue/UV light absorbed by phytochrome in higher plants. In *Blue Light Responses: Phenomena and Occurrence in Microorganisms* (ed. H. Senger), pp. 145-162. Florida: CRC Press Inc. [Ch. 7]

Montell, C. & Rubin, G., 1988. The *Drosophila nina C* locus encodes two photoreceptor cell specific proteins with domains homologous to protein kinases and the myosin heavy chain head. *Cell*, **52**, 757-772. [Ch. 12]

Morel, A., 1974. Optical properties of pure water and pure sea water. In *Optical Aspects of Oceanography* (ed. N.G. Jerlov and E. Steeman Nielsen), pp. 1-24. London: Academic Press. [Ch. 10]

Morel, A., 1978. Available, usable and stored radiant energy in relation to marine photosynthesis. *Deep-Sea Research*, **25**, 673-688. [Ch. 4]

Morel, A.Y., 1980. In-water and remote measurements of ocean colour. *Boundary-layer Meteorology*, **18**, 177-201. [Ch. 2]

Morel, A., 1988. Optical modeling of the upper ocean in relation to its biogenous matter content (case I waters). *Journal of Geophysical Research*, **93**, 10749-10768. [Ch. 1, 3, 4]

Morel, A.Y. & Prieur, L., 1977. Analysis of variations in ocean colour. *Limnology and Oceanography*, **22**, 709-722. [Ch. 1, 2, 3]

Morel, A. & Smith, R.C., 1974. Relation between total quanta and total energy for aquatic photosynthesis. *Limnology and Oceanography*, **19**, 591-600. [Ch. 4]

Morin, J.G., 1974. Coelenterate bioluminescence. In *Coelenterate Biology: Reviews and New Perspectives* (ed. L. Muscatine and H.M. Lenhoff), pp. 397-438. New York: Academic Press. [Ch. 16]

Morin, J.G., 1976. Probable functions of bioluminescence in the Pennatulacea (Cnidaria, Anthozoa). In *Coelenterate Ecology and Behavior* (ed. G.O. Mackie), pp. 629-638. New York: Plenum. [Ch. 16]

Morin, J.G., 1981. Bioluminescent patterns in shallow tropical marine fishes. *Proceedings of the Fourth International Coral Reef Symposium, Manila, 1981*, **2**, 569-574. [Ch. 16]

Morin, J.G., 1983. Coastal bioluminescence: patterns and functions. *Bulletin of Marine Science*, **33**, 787-817. [Ch. 16]

Morin, J.G., 1986. 'Firefleas' of the sea: luminescent signaling in marine ostracode crustaceans. *Florida Entomologist*, **69**, 105-121. [Ch. 16]

Morin, J.G. & Cohen, A.C., 1988. Two new luminescent ostracodes of the genus *Vargula* (Myodocopida: Cypridinidae) from the San Blas region of Panama. *Journal of Crustacean Biology*, **8**, 620-638. [Ch. 16]

Morin, J.G. & Cooke, I.M., 1971a. Behavioural physiology of the colonial hydroid *Obelia*. II. Stimulus-initiated electrical activity and bioluminescence. *Journal of Experimental Biology*, **54**, 707-721. [Ch. 16]

Morin, J.G. & Cooke, I.M., 1971b. Behavioural physiology of the colonial hydroid *Obelia*. III. Characteristics of the bioluminescent system. *Journal of Experimental Biology*, **54**, 723-735. [Ch. 16]

Morin, J.G., Harrington, A., Nealson, K., Krieger, N., Baldwin, T.O. & Hastings, J.W., 1975. Light for all reasons: versatility in the behavioral repertoire of the flashlight fish. *Science, New York*, **190**, 74-76. [Ch. 16]

Morris, I. & Glover, H.E., 1981. Physiology of photosynthesis by marine coccoid cyanobacteria - some ecological implications. *Limnology and Oceanography*, **26**, 957-961. [Ch. 6]

Morse, M.J., Crain, R.C., Cote, G.G. & Satter, R.L., 1989. Light-stimulated inositol phospholipid turnover in *Samanea saman* Pulvini. *Plant Physiology*, **89**, 724-727. [Ch. 7]

Morse, M.J., Crain, R.C. & Satter, R.L., 1987. Light-stimulated phospholipid turnover in *Samanea saman*. *Proceedings of the National Academy of Sciences of the United States of America*, **84**, 7075-7078. [Ch. 7]

Mueller, J.L., 1973. *The Influence of Phytoplankton on Ocean Color Spectra*. PhD Thesis. School of Oceanography, Oregon State University. [Ch. 1]

Munk, O., 1966a. Ocular anatomy of some deep-sea teleosts. *Dana Reports*, no. 70, 62 pp. [Ch. 8, 10]

Munk, O., 1966b. On the retina of *Diretmus argenteus* Johnson, 1863 (Diretmidae, Pisces). *Videnskabelige Meddelelser fra Dansk Naturhistorik Forening i Kjøbenhavn*, **129**, 73-80. [Ch. 8]

Munk, O., 1970. On the occurrence and significance of horizontal band-shaped retinal areae in teleosts. *Videnskabelige Meddeleleser fra Dansk Naturhistorisk Forening i Kjøbenhavn*, **133**, 85-120. [Ch. 9, 10]

Muñoz, V., Brody, S. & Butler, W.L., 1974. Photoreceptor pigment for blue light responses in *Neurospora crassa*. *Biochemical and Biophysical Research Communications*, **58**, 421. [Ch. 7]

Muñoz, V. & Butler, W.L., 1975. Photoreceptor pigment for blue light in *Neurospora crassa*. *Plant Physiology*, **55**, 421-426. [Ch. 7]

Muntz, W.R.A., 1976. On yellow lenses in mesopelagic animals. *Journal of the Marine Biological Association of the United Kingdom*, **56**, 963-976. [Ch. 8]

Muntz, W.R.A., 1978. A penetração de luz águas de rias amazônicos. *Acta Amazonica* **8**, 613-619. [Ch. 10]

Muntz, W.R.A. & Raj, U., 1984. On the visual system of *Nautilus pompilius*. *Journal of Experimental Biology*, **109**, 253-263. [Ch. 9]

Munz, F.W., 1958. Photosensitive pigments from retinae of certain deep-sea fishes. *Journal of Physiology*, **140**, 220-235. [Ch. 8]

Munz, F. W. & McFarland, W. N., 1973. The significance of spectral position in the rhodopsins of tropical marine fishes. *Vision Research*, **13**, 1829-1874. [Ch. 14]

Munz, F.W. & McFarland, W.N., 1977. Evolutionary adaptations of fishes to the photic environment. In *Handbook of Sensory Physiology*, vol. VIII/5. *The Visual System in Vertebrates* (ed. F. Crescitelli), pp. 193-274. New York: Springer Verlag. [Ch. 10]

Murata, M., 1989. Population assessment, management and fishery forecasting for the Japanese common squid, *Todarodes pacificus*. In *Marine Invertebrate Fisheries: their Assessment and Management* (ed. J.F. Caddy), pp. 613-636. New York: Wiley Interscience. [Ch. 15]

Murphy, L.S. & Haugen, E.M., 1985. The distribution and abundance of phototrophic ultraplankton in the North Atlantic. *Limnology and Oceanography*, **30**, 47-58. [Ch. 5]

Murray, J. & Hjort, J., 1912. *The Depths of the Ocean*. London: Macmillan. [Ch. 8]

Musicki, B., Kishi, Y. & Shimomura, O., 1986. Structure of the functional part of photoprotein aequorin. *Journal of the Chemical Society of London. Chemical Communications*, 1566-1568. [Ch. 17]

Nakatani, K. & Yau, K.-W., 1988a. Calcium and magnesium fluxes across the plasma membrane of the toad rod outer segment. *Journal of Physiology*, **395**, 695-729. [Ch. 11]

Nakatani, K. & Yau, K.-W., 1988b. Guanosine 3',5'-cyclic monophosphate-activated conductance studied in a truncated rod outer segment of the toad. *Journal of Physiology*, **395**, 731-753. [Ch. 11]

Nakatani, K. & Yau, K.-W., 1988c. Calcium and light adaptation in retinal rods and cones. *Nature, London*, **334**, 69-71. [Ch. 11]

Nakayama, K., Itagaki T. & Okada, M., 1986. Pigment composition of chlorophyll-protein complexes isolated from the green alga *Bryopsis maxima*. *Plant and Cell Physiology*, **27**, 311-317. [Ch. 5]

NASA, 1987. *MODIS, Moderate-Resolution Imaging Spectrometer Earth Observing System Instrument Panel Report*, vol. IIb. Washington, DC: National Aeronautics and Space Administration. [Ch. 2]

Nathans, J., Thomas, D. & Hogness, D.S., 1986. Molecular genetics of human colour vision: the genes encoding blue, green, and red pigments. *Science, New York*, **232**, 193-202. [Ch. 10]

National Academy of Sciences, 1982. Two special issues in satellite oceanography. *Ocean Dynamics and Biological Oceanography*. Washington, DC: National Academy Press. [Ch. 3]

Nealson, K.H., Arneson, A.C. & Huber, M.E., 1986. Identification of marine organisms using kinetic and spectral properties of their bioluminescence. *Marine Biology*, **91**, 77-83. [Ch. 16]

Neori, A., Vernet, M., Holm-Hansen, O. & Haxo, F.T., 1988. Comparison of chlorophyll far-red and red fluorescence excitation action spectra with photosynthetic oxygen action spectra for photosystem II in algae. *Marine Ecology - Progress Series*, **44**, 297-302. [Ch. 5]

Neshyba, S., 1967. Pulsed light stimulation of marine bioluminescence *in situ*. *Limnology and Oceanography*, **12**, 22-235. [Ch. 16]

Neville, R.A. & Gower, J.F.R., 1977. Passive remote sensing of phytoplankton via chlorophyll *a* fluorescence. *Journal of Geophysical Research*, **82**, 3487-3493. [Ch. 1]

Newton, I., 1730. *Opticks*. Book 1, part II. Prop. X. Prob. V. Fourth edition. London. [Dover Publications, 1952.] [Ch. 8]

Nicol, J.A.C., 1952. Studies on *Chaetopterus variopedatus* (Renier). II. Nervous control of light production. *Journal of the Marine Biological Association of the United Kingdom*, **30**, 433-452. [Ch. 16]

Nicol, J.A.C., 1953. Luminescence in polynoid worms. *Journal of the Marine Biological Association of the United Kingdom*, **32**, 65-84. [Ch. 16]

Nicol, J.A.C., 1954. Fatigue of the luminescent responses of *Chaetopterus*. *Journal of the Marine Biological Association of the United Kingdom*, **33**, 177-186. [Ch. 16]

Nicol, J.A.C., 1955. Nervous regulation of luminescence in the sea pansy *Renilla köllikeri*. *Journal of Experimental Biology*, **32**, 619-635. [Ch. 16]

Nicol, J.A.C., 1958a. Observations on luminescence in pelagic animals. *Journal of the Marine Biological Association of the United Kingdom*, **37**, 705-752. [Ch. 8]

Nicol, J.A.C., 1958b. Observations on the luminescence of *Pennatula phosphorea*, with a note on the luminescence of *Virgularia mirabilis*. *Journal of the Marine Biological Association of the United Kingdom*, **37**, 551-563. [Ch. 16]

Nicol, J.A.C., 1959. Studies on luminescence. Attraction of animals to a weak light. *Journal of the Marine Biological Association of the United Kingdom*, **38**, 477-479. [Ch. 16]

Nicol, J.A.C., 1960. *The Biology of Marine Animals*. London: Pitman. [Ch. 8]

Nicol, J.A.C., 1989. *The Eyes of Fishes*. Oxford: Clarendon Press. [Ch. 10, 14]

Nikonorov, I.V., 1959. The basic principles of fishing for the Caspian kilka by underwater light. In *Modern Fishing Gear of the World*, vol. 1 (ed. H. Kristjonsson), pp. 559-566. London: Fishing News Books. [Ch. 15]

Nikonorov, I.V., 1968. Methods of continuous fishing. In *Trawling and New Methods of Continuous Fishing*. (ed. Yu. N. Kostyunin *et al.*). Jerusalem: Israel Program for Scientific Translations. [Translated from Russian, 1971.]

Nilsson, D.-E., 1982. The transparent compound eye of *Hyperia* (Crustacea): examination with a new method for analysis of refractive index gradients. *Journal of Comparative Physology* (A), **147**, 339-349. [Ch. 9]

Nilsson, D.-E., 1988. A new type of imaging optics in compound eyes. *Nature, London*, **332**, 76-78. [Ch. 9]

Nilsson, D.-E. & Nilsson, H.L., 1981. A crustacean compound eye adapted for low light intensities (Isopoda). *Journal of Comparative Physiology*, **143**(A), 503-510. [Ch. 9]

Noma, Y., Sideras, P., Naito, T., Bergstedt-Lindquist, S., Azyma, C., Severinson, E., Tanabe, T., Kinashi, T., Matsuda, F., Yaoita, Y. & Honjo, T., 1986. Cloning of cDNA encoding the marine IgG1 induction factor by a novel strategy using SP6 promoter. *Nature, London*, **319**, 640-646. [Ch. 18]

O'Day, W.T., 1973. Luminescent silhouetting in stomiatoid fishes. *Contributions in Science from the Natural History Museum of Los Angeles County*, no. 246, 8 pp. [Ch. 16]

O'Day, W.T. & Fernandez, H.R., 1974. *Aristostomias scintillans* (Malacosteidae): a deep-sea fish with visual pigments apparently adapted to its own bioluminescence. *Vision Research*, **14**, 545-550. [Ch. 8]

Ogden, J. C. & Ehrlich, P. R., 1977. The behaviour of heterotypic resting schools of juvenile grunts (Pomadasyidae). *Marine Biology*, **42**, 273-280. [Ch. 14]

Okada, K., Iio, H. & Goto, T., 1976. Biosynthesis of firefly luciferin. Probable formation from benzothiazole and cysteine. *Journal of the Chemical Society of London. Chemical Communications*, 32-33. [Ch. 17]

Okada, K., Iio, H., Kubota, I. & Goto, T., 1974. Firefly bioluminescence: conversion of oxyluciferin to luciferin in the firefly. *Tetrahedron Letters*, 2771-2774. [Ch. 17]

Olson, R.J., Chisholm, S.W., Zettler, E.R. & Armbrust, E.V., 1988. Analysis of *Synechococcus* pigment types in the sea using single and dual beam flow cytometry. *Deep-Sea Research*, **35**, 425-440. [Ch. 5]

Otte, D., 1974. Effects and functions in the evolution of signaling systems. *Annual Review of Ecology and Systematics*, **5**, 385-417. [Ch. 16]

Ovchinnikov, Yu. A., Abdulaev, N.G., Zolotarev, A.S., Artamonov, I.D., Bespalov, I.A., Dergachev, A.E. & Tsuda, M., 1988. *Octopus* rhodopsin. Amino acid sequence deduced from cDNA. *FEBS Letters*, **232**, 69-72. [Ch. 12]

Ow, D.W., Wood, K.V., DeLuca, M.D. DeWet, J.R., Helinski, D.R. & Howells, S.A., 1986. Transient and stable expression of the firefly luciferase gene in plant cells and transgenic plants. *Science, New York*, **234**, 856-859. [Ch. 18]

Owen, W.G., 1987. Ionic conductances in rod photoreceptors. *Annual Review of Physiology*, **49**, 743-764. [Ch. 11]

Owens, T.G., Gallagher, J.C. & Alberte, R.S., 1987. Photosynthetic light-harvesting function of violaxanthin in *Nannochloropsis* spp. (Eustigmatophyceae). *Journal of Phycology*, **23**, 79-85. [Ch. 5]

Owens, T.G. & Wold, E.R., 1986*a*. Light-harvesting function in the diatom *Phaeodactylum tricornutum*. I. Isolation and characterization of pigment-protein complexes. *Plant Physiology*, **80**, 732-738. [Ch. 5]

Owens, T.G. & Wold, E.R., 1986*b*. Light-harvesting function in the diatom *Phaeodactylum tricornutum*. II. Distribution of excitation energy between the photosystems. *Plant Physiology*, **80**, 739-746. [Ch. 5]

Paltridge, G.W. & Platt, C.M.R., 1976. *Radiative Processes in Meteorology and Climatology*. New York: Elsevier. [Ch. 1]

Parrish, B.B., 1969. A review of some experimental studies of fish reactions to stationary and moving objects of relevance to fish capture processes. *FAO Fisheries Reports*, no. 62 (vol. 2), 233-245. [Ch. 15]

Parsons, T.R., 1979. Some ecological, experimental and evolutionary aspects of the upwelling ecosystem. *South African Journal of Science*, **75**, 536-540. [Ch. 4]

Partridge, J.C., 1986. *Microspectrophotometry of Vertebrate Photoreceptors*. PhD Thesis. University of Bristol. [Ch. 10]

Partridge, J.C., 1989. The visual pigments of deep-sea fishes: ecophysiology and molecular biology. *Progress in Underwater Science*, **14**, 17-31. [Ch. 10]

Partridge, J.C., Archer, S.N. & Lythgoe, J.N., 1988. Visual pigments in the individual rods of deep-sea fishes. *Journal of Comparative Physiology*, **162**(A), 543-550. [Ch. 8, 10]

Partridge, J.C., Shand, J., Archer, S.N., Lythgoe, J.N. & van Groningen-Luyben, W.A.H.M., 1989. Interspecific variation in the visual pigments of deep-sea fishes. *Journal of Comparative Physiology*, **164**(A), 513-529. [Ch. 10]

Patrick, P.H., Christie, A.E., Sager, D.R., Hocutt, C.H. & Stauffer, J.R., Jr, 1985. Responses of fish to a strobe light/air bubble barrier. *Fisheries Research*, **3**, 157-172. [Ch. 15]

Patterson, K., 1987. Fishy events in the Falklands. *New Scientist*, **114**, no. 1562, 44-48. [Ch. 15]

Paulsen, R., Zinkler, D. & Delmelle, M., 1983. Architecture and dynamics of microvillar photoreceptor membranes of a cephalopod. *Experimental Eye Research* **36**, 47-56. [Ch. 12]

Payne, R., 1986. Phototransduction by microvillar photoreceptors of invertebrates: mediation of a visual cascade by inositol trisphosphate. *Photobiochemistry and Photobiophysics*, **13**, 373-397. [Ch. 12]

Payne, R., Walz, B., Levy, S. & Fein, A., 1988. The localization of calcium release by inositol trisphosphate in *Limulus* photoreceptors and its control by negative feedback. *Philosophical Transactions of the Royal Society* (B), **320**, 359-379. [Ch. 12]

Pepe, I.M., Panfoli, I. & Cugnoli, C., 1986. Guanylate-cyclase in rod outer segments of the toad retina. *FEBS Letters*, **203**, 73-76. [Ch. 11]

Peterson, B.J. & Melillo, J.M., 1985. The potential storage of carbon caused by eutrophication in the biosphere. *Tellus*, **37**B, 117-127. [Ch. 4]

Piantanida, T., 1988. The molecular genetics of color vision and color blindness. *Trends in Genetics*, **4**, 319-323. [Ch. 10]

Pietsch, T.W. & Grobacker, D.B., 1978. The compleat angler: aggressive mimicry in an antennariid anglerfish. *Science, New York*, **201**, 369-370. [Ch. 16]

Piggins, D.J., 1970. Refraction of the harp seal, *Pagophilus groenlandicus* (Erxleben 1777). *Nature, London*, **227**, 78-79. [Ch. 9]

Pilgrim, D.A., 1988. *Optical Attenuation Coefficients in Oceanic and Estuarine Waters*. PhD Thesis. Plymouth Polytechnic. [Ch. 3]

Pilgrim, D.A., Redfern, T.A., MacLachlan, G.S. & Marsh, R.I., 1989. Estimation of optical coefficients from diver observations of visibility. *Progress in Underwater Science*, **14**, 33-52. [Ch. 10]

Pingree, R.D., Pugh, P.R., Holligan, P.M. & Fister, G.R., 1975. Summer phytoplankton blooms and red tides along tidal fronts in the approaches to the English Channel. *Nature, London*, **258**, 672-677. [Ch. 4]

Pinto, L.H. & Brown, J.E., 1977. Intracellular recordings from photoreceptors of the squid (*Loligo pealii*). *Journal of Comparative Physiology*, **122**, 241-250. [Ch. 12]

Pirenne, M.H. & Denton, E.J., 1952. Accuracy and sensitivity of the human eye. *Nature, London*, **170**, 1039-1042. [Ch. 8]

Platt, T., 1986. Primary production of the ocean water column as a function of surface light intensity: algorithms for remote sensing. *Deep-Sea Research*, **33**, 149-163. [Ch. 2, 4]

Platt, T., Denman, K.L. & Jassby, A.D., 1977. Modelling the productivity of phytoplankton. In *The Sea*, vol. 6 (ed. E.D. Goldberg *et al.*), pp. 807-856. New York: Wiley-Interscience. [Ch. 1]

Platt, T., Gallegos, C.L. & Harrison, W.G., 1980. Photoinhibition of photosynthesis in natural assemblages of marine phytoplankton. *Journal of Marine Research*, **38**, 687-701. [Ch. 4]

Platt, T., Harrison, W.G., Irwin, B., Horne, E.P. & Gallegos, C.L., 1982. Photosynthesis and photoadaptation of marine phytoplankton in the Arctic. *Deep-Sea Research*, **29**, 1159-1170. [Ch. 1]

Platt, T. & Herman, A.W., 1983. Remote sensing of phytoplankton in the sea: surface-layer chlorophyll as an estimate of water-column chlorophyll and primary production. *International Journal of Remote Sensing*, **4**, 343-351. [Ch. 1, 2]

Platt, T. & Jassby, A.D., 1976. The relationship between photosynthesis and light for natural assemblages of coastal marine phytoplankton. *Journal of Phycology*, **12**, 421-430. [Ch. 4]

Platt, T. & Sathyendranath, S., 1988. Oceanic primary production: estimation by remote sensing at local and regional scales. *Science, New York*, **241**, 1613-1620. [Ch. 1]

Platt, T., Subba Rao, D.V. & Irwin, B., 1983. Photosynthesis of picoplankton in the oligotrophic ocean. *Nature, London*, **24**, 702-704. [Ch. 6]

Poff, K.L., 1983. Perception of a unilateral light stimulus. *Philosophical Transactions of the Royal Society* (B), **303**, 479-488. [Ch. 7]

Poff, K.L. & Butler, W.L., 1974. Absorbance changes induced by blue light in *Phycomyces blakesleeanus* and *Dicteostelium discoidium*. *Nature, London*, **248**, 799-800. [Ch. 7]

Pooviah, B.W. & Reedy, A.S.N., 1987. Calcium messenger systems in plants. *CRC Critical Reviews in Plant Science*, **6**, 47-103. [Ch. 7]

Porter, K.G. & Porter, J.W., 1979. Bioluminescence in marine plankton: a coevolved antipredation strategy. *American Naturalist*, **114**, 458-461. [Ch. 16]

Postuma, K.H., 1957. The vertical migration of feeding herring in relation to light and the vertical temperature gradient. *International Council for the Exploration of the Sea, Herring Committee*, mimeographed report. [Ch. 13]

Potts, G.W., 1980. The predatory behaviour of *Caranx melampygus* (Pisces) in the Channel environment of Aldabra Atoll (Indian Ocean), *Journal of Zoology*, **192**, 323-50. [Ch. 14]

Potts, G.W., 1981. Behavioural interactions between Carangidae (Pisces) and their prey on the fore-reef slope of Aldabra, with notes on other predators. *Journal of Zoology*, **195**, 385-404. [Ch. 14]

Potts, G.W., 1983. The predatory tactics of *Caranx melampygus* and the response of its prey. In *Predators and Prey in Fishes* (ed. D. L. G. Noakes *et al.*), pp. 181-191. The Hague: Dr W. Junk Publishers. [Ch. 14]

Potts, G.W., 1986. Predator-prey interactions between pollack and sandeels. *Progress in Underwater Science*, **11**, 69-79. [Ch. 10, 14]

Prasher, D., McCann, R.O. & Cormier, M.J., 1985. Cloning and expression of the cDNA coding for aequorin, a bioluminescent calcium-binding protein. *Biochemical and Biophysical Research Communications*, **112**, 1259-1268. [Ch. 18]

Preisendorfer, R.W., 1959. Theoretical proof of the existence of characteristic diffuse light in natural waters. *Journal of Marine Research*, **18**, 1-9. [Ch. 1]

Preisendorfer, R.W., 1964. A model for radiance distribution in natural hydrosols. In *Physical Aspects of Light in the Sea* (ed. J.E. Tyler), pp. 51-59. University of Hawaii Press. [Ch. 10]

Preisendorfer, R.W., 1976a. *Hydrologic Optics*, vol. I. *Introduction*. Honolulu, Hawaii: US Department of Commerce, NOAA Environmental Research Laboratories.

Preisendorfer, R.W., 1976b. *Hydrologic Optics*, vol. V. *Properties*. Honolulu, Hawaii: US Department of Commerce, NOAA Environmental Research Laboratories. [Ch. 1]

Preisendorfer, R.W. & Mobley, C.D., 1986. Albedos and glitter patterns of a wind-roughened sea surface. *Journal of Physical Oceanography*, **16**, 1293-1316. [Ch. 1]

Prezelin, B.B., 1976. The role of peridinin-chlorophyll *a*-proteins in the photosynthetic light adaptation of the marine dinoflagellate, *Glenodinium* sp. *Planta*, **130**, 225-233. [Ch. 5]

Prezelin, B.B. & Boczar, B.A., 1986. Molecular bases of cell absorption and fluorescence in phytoplankton: potential applications to studies in optical oceanography. *Progress in Phycological Research*, **4**, 359-464. [Ch. 5]

Prezelin, B.B. & Haxo, F.T., 1976. Purification and characterization of peridinin-chlorophyll *a*-proteins from the marine dinoflagellate *Glenodinium* sp. and *Gonyaulax polyedra*. *Planta*, **128**, 133-141. [Ch. 5]

Prezelin, B.B., Ley, A.C. & Haxo, F.T., 1976. Effect of growth irradiance on the photosynthetic action spectra of the marine dinoflagellate *Glenodinium* sp. *Planta*, **130**, 251-256. [Ch. 5]

Prezelin, B.B. & Sweeney, B.M., 1978. Photoadaptation of photosynthesis in *Gonyaulax polyedra*. *Marine Biology*, **48**, 27-35. [Ch. 4]

Prieur, L., 1976. *Transferts Radiatifs dans les Eaux de Mer. Application à la Détermination des Paramètres Optiques Caractérisant leur Teneur en Substances Dissoutes et leur Contenu en Particules*. Thèse, Doctorat d'Etat, Université P. & M. Curie, Paris. [Ch. 1]

Prieur, L. & Morel, A., 1975. Relations théoriques entrée le facteur de reflexion diffuse de l'eau de mer à diverses profondeurs et leurs caracteristiques optiques. (Abstract.) In *Abstracts from the International Union of Geophysics and Geodesy, 16th General Assembly, Grenoble, 1975*, pp. 250-251. Paris: IUGG. [Ch. 1]

Prieur, L. & Sathyendranath, S., 1981. An optical classification of coastal and oceanic waters based on the specific spectral absorption curves of phytoplankton pigments, dissolved organic matter and other particulate materials. *Limnology and Oceanography*, **26**, 671-688. [Ch. 10]

Protasov, V.R., 1968. *Vision and Near Orientation of Fish*. Jerusalem: Israel Program for Scientific Translations. [Translated from Russian, 1970.] [Ch. 15]

Pugh, E.N., Jr, 1987. The nature and identity of the internal excitational transmitter of vertebrate phototransduction. *Annual Review of Physiology*, **49**, 715-741. [Ch. 11]

Pugh, E.N., Jr & Cobbs, W.H., 1986a. Properties of cytoplasmic transmitters of excitation in vertebrate rods and evaluation of candidate intermediary transmitters. In *The Molecular Mechanism of Photoreception* (ed. H. Stieve), pp. 127-158. Berlin: Springer-Verlag. [Ch. 11]

Pugh, E.N., Jr & Cobbs, W.H., 1986b. Visual transduction in vertebrate rods and cones: a tale of two transmitters, calcium and cyclic GMP. *Vision Research*, **26**, 1613-1643. [Ch. 11]

Pumphrey, R.J., 1948. The theory of the fovea. *Journal of Experimental Biology*, **25**, 299-312. [Ch. 8]

Pumphrey, R.J., 1961. Concerning vision. In *The Cell and the Organism* (ed. J.A. Ramsay and V.B. Wigglesworth), pp. 193-208. Cambridge: Cambridge University Press. [Ch. 9]

Quail, P.H., 1983. Rapid action of phytochrome in photomorphogenesis. In *Encyclopedia of Plant Physiology (NS): Photomorphogenesis* (ed. W. Shropshire and H. Mohr), pp. 178-212. Berlin: Springer-Verlag. [Ch. 7]

Radakov, D.V., 1960. Observations on herring during a voyage of research submarine 'Severyanka'. [In Russian.] *Bulletin of the Oceanographic Commission of the Academy of Sciences of the USSR*, **6**, 39-40. [Ch. 13]

Radmer, R.J. & Kok, B., 1977. Light conversion efficiency in photosynthesis. In *Encyclopaedia of Plant Physiology*, vol 5, *Photosynthesis I: Photosynthetic Electron Transport and Photophosphorylation* (ed. A. Trebst and M. Avron), pp. 125-135. Berlin: Springer-Verlag. [Ch. 4]

Ramus, J., 1983. A physiological test of the theory of complementary chromatic adaptation. II. Brown, green and red seaweeds. *Journal of Phycology*, **19**, 173-178. [Ch. 5]

Ranjeva, R. & Boudet, A.M., 1987. Phosphorylation of proteins in plants: regulatory effects and potential involvement in stimulus/response coupling. *Annual Review of Plant Physiology*, **38**, 73-93. [Ch. 7]

Ratto, G.M., Payne, R., Owen, W.G. & Tsien, R.Y., 1988. The concentration of cytosolic free calcium in vertebrate rod outer segments measured with fura-2. *Journal of Neuroscience*, **8**, 3240-3246. [Ch. 11]

Raven, J.A., 1981. Light quality and solute transport. In *Plants and the Daylight Spectrum* (ed. H. Smith), pp. 375-390. London: Academic Press. [Ch. 7]

Raven, J.A., 1983. Do plant photoreceptors act at the membrane level? *Philosophical Transactions of the Royal Society* (B), **303**, 403-418. [Ch. 7]

Raven, J.A. & Beardall, J., 1981. Respiration and photorespiration. *Canadian Bulletin of Fisheries and Aquatic Sciences*, no. 210, 55-82. [Ch. 4]

Reuter, H., 1986. Voltage-dependent mechanisms for raising intracellular free calcium concentration: calcium channels. In *Calcium and the Cell* (ed. D. Evered and J. Whelan), pp. 5-22. Chichester: Wiley. [Ch. 7]

Reynolds, G.T., 1978. Application of photosensitive devices to bioluminescence studies. *Photochemistry and Photobiology*, **27**, 405-442. [Ch. 18]

Richardson, K., Beardall, J. & Raven, J.A., 1983. Adaptation of unicellular algae to irradiance: an analysis of strategies. *New Phytologist*, **93**, 157-191. [Ch. 4, 5, 6]

Ricketts, T.R., 1966. Magnesium 2,4-divinylphaeoporphyrin a_5 monomethyl ester, a protochlorophyll-like pigment present in some unicellular flagellates. *Phytochemistry*, **5**, 223-229. [Ch. 5]

Ridgway, E.B. & Ashley, C.C., 1967. Calcium transients in single muscle fibres. *Biochemical and Biophysical Research Communications*, **29**, 229-234. [Ch. 18]

Riggs, C.D. & Chrispeels, M.J., 1987. Luciferase reporter gene cassettes for plant gene expression studies. *Nucleic Acids Research*, **15**, 8115. [Ch. 18]

Roberts, P.A., Knight, J. & Campbell, A.K., 1987. Pholasin- a bioluminescent indicator for detecting activation of single neutrophils. *Analytical Biochemistry*, **160**, 139-148. [Ch. 18]

Robertson, D. R. & Sheldon, J. M., 1979. Competitive interactions and the availability of sleeping sites for a diurnal coral reef fish. *Journal of Experimental Marine Biology and Ecology*, **40**, 285-98. [Ch. 14]

Robinson, I.S., 1983. Satellite observations of ocean colour. *Philosophical Transactions of the Royal Society* (A), **309**, 415-432. [Ch. 2]

Robinson, I.S., 1985. *Satellite Oceanography*. Chichester: Ellis Horwood. [Ch. 2]

Robinson, I.S., 1988. Ocean applications of colour and thermal imagery from space. *Hydrographic Journal*, no. 49, 9-21. [Ch. 2]

Robinson, K.R. & Jaffe, L.F., 1975. Polarizing fucoid eggs drive a calcium current through themselves, *Science, New York*, **187**, 70-73. [Ch. 7]

Robinson, M.K., Bauer, R.A. & Schroeder, E.H., 1979. *Atlas of North Atlantic-Indian Ocean Monthly Mean Temperatures and Mean Salinities of the Surface Layer*. St Louis, Mississippi: United States Naval Oceanographic Office. [RP-18.] [Ch. 4]

Robinson, P.R., Kawamura, S., Abramson, B. & Bownds, M.D., 1980. Control of the cyclic GMP phosphodiesterase of frog photoreceptor membranes. *Journal of General Physiology*, **76**, 631-645. [Ch. 11]

Rodhouse, P.G., 1988. Squid fisheries in the South Atlantic. *NERC News*, no. 5, 20-21. [Ch. 15]

Roe, H.S.J., Baker, A. de C., Carson, R.M., Wild, R. & Shale, D.M., 1980. Behaviour of the Institute of Oceanographic Science's rectangular midwater trawls: theoretical aspects and experimental observations. *Marine Biology*, **56**, 247-259. [Ch. 15]

Roe, H.S.J. & Shale, D.M., 1979. A new multiple rectangular midwater trawl (RMT 1+8M) and some modifications to the Institute of Oceanographic Sciences' RMT 1+8. *Marine Biology*, **50**, 283-288. [Ch. 15]

Roithmayr, C.M. & Wittmann, F.P., 1972. Low light level sensors for marine resource assessment. In *Proceedings of the Eighth Annual Conference of the Marine Technology Society, Washington DC.* Preprints pp. 277-288. [Ch. 15]

Rose, A., 1973. *Vision Human and Electronic.* New York: Plenum. [Ch. 10]

Roux, S.J., Wayne, R.O. & Datta, N., 1986. Role of calcium ions in phytochrome responses: an update. *Plant Physiology*, **66**, 344-348. [Ch. 7]

Ruby, E.G. & Morin, J.G., 1979. Luminous enteric bacteria of marine fishes: a study of their distribution, densities and dispersion. *Applied and Environmental Microbiology*, **38**, 406-411. [Ch. 16]

Rushton, W.A.H., 1956. The rhodopsin density in human rods. *Journal of Physiology*, **134**, 673-680. [Ch. 8]

Russell, G.A., Bemis, A.G., Geels, E.J., Janzen, E.G. & Moye, A.J., 1967. Oxidation of carbanions. *Advances in Chemical Series*, **75**, 174-202. [Ch. 17]

Ruttner, M.G., 1971. *The Origin of Life.* Amsterdam: Elsevier. [Ch. 4]

Ryther, J.H., 1956. Photosynthesis in the ocean as a function of light intensity. *Limnology and Oceanography*, **1**, 61-70. [Ch. 4]

Ryther, J.H. & Guillard, R.R.L., 1962. Studies of marine planktonic diatoms. III. Some effects of temperature on respiration of five species. *Canadian Journal of Microbiology*, **8**, 447-453. [Ch. 4]

Sager, D.R., Hocutt, C.H. & Stauffer, J.R., Jr, 1987. Estuarine fish responses to strobe light, bubble curtains and strobe light/bubble curtain combinations as influenced by water flow rate and flash frequencies. *Fisheries Research*, **5**, 383-399. [Ch. 15]

Saibil, H.R., 1982. An ordered membrane-cytoskeleton network in squid photoreceptor microvilli. *Journal of Molecular Biology*, **158**, 435-456. [Ch. 12]

Saibil, H., 1984. A light-stimulated increase of cyclic GMP in squid photoreceptors. *FEBS Letters*, **168**, 213-216. [Ch. 12]

Saibil, H., 1990. Structure and function of the squid eye. In *Squid as Experimental Animals* (ed. D. Gilbert *et al.*), in press. Plenum Publishing Corporation. [Ch. 12]

Saibil, H. & Hewat, E., 1987. Ordered transmembrane and extracellular structure in squid photoreceptor microvilli. *Journal of Cell Biology*, **105**, 19-28. [Ch. 12]

Saibil, H. & Michel-Villaz, M., 1984. Cross-reactions between rhodopsin and GTP-binding protein in bovine and squid photoreceptors. *Proceedings of the National Academy of Sciences of the United States of America*, **81**, 5111-5115. [Ch. 12]

Saibil, H. & White, N., 1989. Recent advances in biological imaging. *Bioscience Reports*, **9**, 437-449. [Ch. 12]

Sathyendranath, S. & Morel, A., 1983. Light emerging from the sea - interpretation and uses in remote sensing. In *Remote Sensing Applications in Marine Science and Technology* (ed. A.P. Cracknell), pp. 323-357. Dordrecht: D. Reidel Publishing Co. [Ch. 1, 2]

Sathyendranath, S. & Platt, T., 1988. The spectral irradiance field at the surface and in the interior of the ocean: a model for applications in oceanography and remote sensing. *Journal of Geophysical Research*, **93**, 9270-9280. [Ch. 1]

Sathyendranath, S. & Platt, T., 1989. Computation of aquatic primary production: extended formalism to include effect of angular and spectral distribution of light. *Limnology and Oceanography*, **34**, 188-198. [Ch. 1]

Satter, R.L., Guggino, S.E., Lonergan, T.A. & Galston, A.W., 1981. The effects of blue and far red light on rhythmic leaflet movements in *Samanea* and *Albizia*. *Plant Physiology*, **67**, 965-968. [Ch. 7]

Scharfe, J., 1953. Uber die Verwendung kunstlichen Lichtes in der Fischerei. [The application of artificial light in fishing.] *Protokolle zur Fischereitechnik*, **8**(15), 2-29. [Ch. 15]

Schenk, H., 1957. On the focusing of sunlight by ocean waves. *Journal of the Optical Society of America*, **47**, 53-657. [Ch. 10]

Schmidt, W., 1983. The physiology of blue light systems. In *The Biology of Photoreception* (ed. D.J. Cosens and D. Vince-Prue), pp. 305-330. Cambridge: Cambridge University Press. [Ch. 7]

Schmidt, W., 1987. Primary reactions and optical spectroscopy of blue light photoreceptors. In *Blue Light Responses: Phenomena and Occurrence in Plants and Microorganisms* (ed. H. Senger), pp. 19-36. Florida: CRC Press Inc. [Ch. 7]

Schmidt, W., Tunnermann, M. & Idziak, E.-M., 1988. Transient reduction of responsiveness of blue light-mediated hair-whorl morphogenesis in *Acetabularia mediterranea* induced by blue light. *Planta*, **174**, 373-379. [Ch. 7]

Schnepf, E., 1986. Cellular polarity. *Annual Review of Plant Physiology*, **37**, 23-72. [Ch. 7]

Schuster, G.B., 1979. Chemiluminescence of organic peroxides. Conversion of ground state reagents to excited state products by the CIEEL mechanism. *Accounts of Chemical Research*, **12**, 366-373. [Ch. 17]

Scolding, N.J., Morgan, B.P., Houston, W.A.J., Linington, C., Campbell, A.K. & Compston, D.A.S., 1989. Oligodendrocytes activate complement but resist lysis by vesicular removal of membrane attack complexes. *Nature, London*, **339**, 620-622. [Ch. 18]

Segal, E., 1970. Light: animals: invertebrates. In *Marine Ecology*, vol. 1 (ed. O. Kinne), pp. 159-211. London: Wiley Interscience. [Ch. 15]

Seliger, H.H., 1975. The origin of bioluminescence. *Photochemistry and Photobiology*, **21**, 355-361. [Ch. 16, 17]

Seliger, H.H., 1987. The evolution of bioluminescence in bacteria. *Photochemistry and Photobiology*, **45**, 291-297. [Ch. 16]

Seliger, H.H., Lall, A.B., Lloyd, J.E. & Biggley, W.H., 1982. The colors of firefly bioluminescence. II. Experimental evidence for the optimization model. *Photochemistry and Photobiology*, **36**, 681-688. [Ch. 16]

Selivertsov, A.S., 1974. Vertical migrations of larvae of Atlanto-Scandian herring. In *The Early Life History of Fish* (ed. J.H.S. Blaxter), pp. 253-262. Berlin: Springer Verlag. [Ch. 13]

Senger, H. (ed.), 1987*a*. *Blue Light Responses: Phenomena and Occurrence in Plants and Microorganisms*. Florida: CRC Press Inc. [Ch. 7]

Senger, H., 1987*b*. Sun and shade effects of blue light on plants. In *Blue Light Responses: Phenomena and Occurrence in Plants and Microorganisms*, vol. 2 (ed. H. Senger), pp. 141-149. Boca Raton, Florida: CRC Press. [Ch. 5]

Shand, J., Partridge, J. C., Archer, S. N., Potts, G. W. & Lythgoe, J. N., 1988. Spectral absorbance shifts in the violet/blue sensitive cone of the pollack, *Pollachius pollachius*. *Journal of Comparative Physiology*, **163**(A), 699-703. [Ch. 10, 14]

Shefner, J.M. & Levine, M.O., 1976. A psychophysical demonstration of goldfish trichromacy. *Vision Research*, **16**, 671-673. [Ch. 10]

Shimomura, O., 1980. Chlorophyll derived bile pigment in bioluminescent euphausiids. *FEBS Letters*, **116**, 203-206. [Ch. 17]

Shimomura, O., 1987. Presence of coelenterazine in non-bioluminescent marine organisms. *Comparative Biochemistry and Physiology*, **86**B, 361-363. [Ch. 17]

Shimomura, O., Inoue, S., Johnson F.H. & Haneda Y., 1980. Widespread occurrence of coelenterazine in marine bioluminescence. *Comparative Biochemistry and Physiology*, **65**B, 435-437. [Ch. 17, 18]

Shimomura, O. & Johnson, F.H., 1968. The structure of *Latia* luciferin. *Biochemistry*, **7**, 1734-1738. [Ch. 17]

Shimomura, O. & Johnson, F.H., 1978. Peroxidised coelenterazine, the active group in bioluminescence of the photoprotein aequorin. *Proceedings of the National Academy of Sciences of the United States of America*, **75**, 2611-2615. [Ch. 17]

Shimomura, O., Musicki, B. & Kishi, Y., 1988. Semi-synthetic aequorin: an improved test for the measurement of calcium ion concentration. *Biochemical Journal*, **251**, 405-410. [Ch. 18]

Sidelnicov, I.I., 1981. *Light Fishing of Pacific Fish and Squid*. Moscow: Light and Food Industry. [In Russian.] [Ch. 15]

Simpson, J.H. & Bowers, D.G., 1984. The role of tidal stirring in the controlling the seasonal heat cycle in shelf seas. *Annales Geophysicae*, **2**, 411-416. [Ch. 4]

Simpson, J.H., Tett, P.B., Argote-Espinoza, M.L., Edwards, A., Jones, K.J. & Savidge, G., 1982. Mixing and phytoplankton growth around an island in a stratified sea. *Continental Shelf Research*, **1**, 15-31. [Ch. 2]

Simpson, J.J., Koblinsky, C.J., Pelaez, J., Haury, L.R. & Wiesenhahn, D., 1986. Temperature - plant pigment - optical relations in a recurrent offshore mesoscale eddy near Point Conception, California. *Journal of Geophysical Research*, **91**, 12919-12936. [Ch. 2]

Sivak, J.G., 1976. Optics of the eye of the 'four-eyed fish' (*Anableps anableps*). *Vision Research*, **16**, 531-534. [Ch. 9]

Sivak, J.G., 1978. Vertebrate strategies for vision in air and water. In *Sensory Ecology* (ed. M.M. Ali), pp. 503-519. New York: Plenum. [Ch. 9]

Sivak, J.G., Hildebrand, T. & Lebert, C., 1985. Magnitude and rate of accommodation in diving and non-diving birds. *Vision Research*, **25**, 925-933. [Ch. 9]

Sivak, J.G. & Millodot, M., 1977. Optical performance of the penguin eye in air and water. *Journal of Comparative Physiology*, **119**, 241-247. [Ch. 9]

Smith, B.M. & Melis, A., 1987. Photosystem stoichiometry and excitation distribution in chloroplasts from surface and minus 20 meter blades of *Macrocystis pyrifera*, the giant kelp. *Plant Physiology*, **84**, 1325-1330. [Ch. 5]

Smith, E.L., 1936. Photosynthesis in relation to light and carbon dioxide. *Proceedings of the National Academy of Sciences of the United States of America*, **22**, 504-511. [Ch. 1, 4]

Smith, H. (ed.), 1981. *Plants and the Daylight Spectrum*. London: Academic Press. [Ch. 7]

Smith, R.C., 1974. Structure of solar radiation in the upper layers of the sea. In *Optical Aspects of Oceanography* (ed. N.G. Jerlov and E. Steemann Nielsen), pp. 95-119. London: Academic Press. [Ch. 10]

Smith, R.C. & Baker, K.S., 1978a. Optical classification of natural waters. *Limnology and Oceanography*, **23**, 260-267. [Ch. 1, 2]

Smith, R.C. & Baker, K.S., 1978b. The bio-optical state of ocean waters and remote sensing. *Limnology and Oceanography*, **23**, 247-259. [Ch. 2, 4]

Smith, R.C. & Baker, K.S., 1981. Optical properties of the clearest natural waters (200-800 nm). *Applied Optics*, **20**, 177-184. [Ch. 3, 10]

Smith, R.C., Bidigare, R.R., Prezelin, B.B., Baker, K.S. & Brooks, J.M., 1987. Optical characterization of primary productivity across a coastal front. *Marine Biology*, **96**, 575-591. [Ch. 5]

Smith, R.C., Booth, C.R. & Star, J.L., 1984. An oceanographic bio-optical profiling system. *Applied Optics*, **23**, 2791-2797. [Ch. 3]

Smith, R.C., Eppley, R.W. & Baker, K.S., 1982. Correlation of primary production as measured aboard ship in Southern California coastal waters and as estimated from satellite chlorophyll images. *Marine Biology*, **66**, 281-288. [Ch. 1]

Smith, R.C. & Wilson, W.H., 1981. Ship and satellite bio-optical research in the Californian Bight. In *Oceanography from Space* (ed. J.F.R. Gower), pp. 281-294. New York: Plenum Press. [Ch. 2]

Somiya, H., 1976. Functional significance of the yellow lens in the eyes of *Argyropelecus affinis*. *Marine Biology*, **34**, 93-99. [Ch. 8]

Somiya, H., 1982. 'Yellow lens' eyes of the stomiatoid deep-sea fish, *Malacosteus niger*. *Proceedings of the Royal Society* (B), **215**, 481-489. [Ch. 8]

Song, P.-S., 1980. Spectroscopic and photochemical characterization of flavoprotein and carotenoproteins as blue light photoreceptors. In *The Blue Light Syndrome* (ed. H. Senger), pp. 157-175. Berlin: Springer-Verlag. [Ch. 7]

Song, P.-S., 1987. Possible primary photoreceptors. In *Blue Light Responses: Phenomena and Occurrence in Plants and Microorganisms* (ed. H. Senger), pp. 3-18. Florida: CRC Press Inc. [Ch. 7]

Sournia, A., 1982. Is there a shade flora in the marine plankton? *Journal of Plankton Research*, **4**, 391-399. [Ch. 6]

Speksnijder, J.E., Miller, A.L., Weisenseel, M.H., Chen, T.-H. & Jaffe, L.F., 1989. Calcium buffer injections block fucoid egg development by speeding calcium diffusion. *Proceedings of the National Academy of Sciences of the United States of America*, in press. [Ch. 7]

Spinrad, R.W. & Yentsch, C.M., 1987. Observations on the intra- and interspecific single cell optical variability of marine phytoplankton. *Applied Optics*, **26**, 357-362. [Ch. 10]

Starck, W. A. & Davis, W. P., 1966. Night habits of fishes of Alligator Reef, Florida. *Ichthyologica*, **38**, 313-356. [Ch. 14]

Stauber, J.L. & Jeffrey, S.W., 1988. Photosynthetic pigments in fifty-one species of marine diatoms. *Journal of Phycology*, **24**, 158-172. [Ch. 5]

Steele, J.H. & Henderson, E.W., 1977. Plankton patches in the northern North Sea. In *Fisheries Mathematics* (ed. J.H. Steele), pp. 1-19. London: Academic Press. [Ch. 4]

Steemann Nielsen, E., 1952. The use of radioactive carbon (^{14}C) for measuring organic production in the sea. *Journal du Conseil*, **18**, 117-140. [Ch. 4]

Steemann Nielsen, E. & Jorgensen, E.G., 1968. The adaptation of plankton algae. I. General part. *Physiologia Plantarum*, **21**, 401-413. [Ch. 4]

Steemann Nielsen, E. & Park, T.S., 1964. On the time course in adapting to low light intensities in marine phytoplankton. *Journal du Conseil*, **29**, 19-24. [Ch. 4]

Steenstrup, S. & Munk, O., 1980. Optical function of the convexiclivate fovea with particular regard to notosudid deep-sea teleosts. *Optica Acta*, **27**(7), 949-964. [Ch. 8]

Stephenson, R.L. & Power, M.J., 1988. Semi-diel vertical movements in Atlantic herring *Clupea harengus* larvae: a mechanism for larval retention? *Marine Ecology - Progress Series*, **50**, 3-11. [Ch. 13]

Stewart, P.A.M., 1981. An investigation into the reactions of fish to electrified barriers and bubble curtains. *Fisheries Research*, **1**, 3-22. [Ch. 15]

Stewart, R.H., 1985. *Methods of Satellite Oceanography*. Berkeley, California: University of California Press. [Ch. 2]

Stryer, L., 1986. Cyclic GMP cascade of vision. *Annual Review of Neuroscience*, **9**, 87-119. [Ch. 11, 12]

Stryer, L. & Bourne, H.B., 1986. G-proteins: a family of signal transducers. *Annual Review of Cell Biology*, **2**, 391-419. [Ch. 12]

Sturm, B., 1981. The atmospheric correction of remotely sensed data and the quantitative determination of suspended matter in marine water surface layers. In *Remote Sensing in Meteorology, Oceanography and Hydrology* (ed. A.P. Cracknell), pp. 163-197. Chichester: Ellis Horwood. [Ch. 2]

Sturm, B., 1983. Selected topics of Coastal Zone Color Scanner data evaluation. In *Remote Sensing Applications in Marine Science and Technology* (ed. A.P. Cracknell), pp. 137-168. Dordrecht: D. Reidel Publishing Co. [Ch. 2]

Sverdrup, H.U., 1953. On conditions for the vernal blooming of phytoplankton. *Journal du Conseil*, **18**, 287-295. [Ch. 4]

Swinney, G.N., Clarke, M.R. & Maddock, L., 1986. Influence of an electric light on the capture of deep-sea fish in Biscay. *Journal of the Marine Biological Association of the United Kingdom*, **66**, 483-496. [Ch. 15]

Swinney, G.N., Pascoe, P.L., Clarke, M.R. & Maddock, L., in preparation. Influence of electric light on the capture of deep-sea fish near Madeira. [Ch. 15]

Szuts, E., Wood, S., Reid, M. & Fein, A., 1986. Light stimulates the rapid formation of inositol trisphosphate in squid retinas. *Biochemical Journal*, **240**, 929-932. [Ch. 12]

Takahashi, M. & Bienfang, P.K., 1983. Size structure of phytoplankton biomass and photosynthesis in subtropical Hawaiian waters. *Marine Biology*, **76**, 203-211. [Ch. 6]

Talling, J.F., 1957a. Photosynthetic characteristics of some freshwater plankton diatoms in relation to underwater radiation. *New Phytologist*, **56**, 29-50. [Ch. 4, 6]

Talling, J.F., 1957b. The phytoplankton population as a compound photosynthetic system. *New Phytologist*, **56**, 133-149. [Ch. 4]

Terouanne, B., Nicolas, J.C. & Crastes de Paulet, A., 1986. Bioluminescent immunosorbent for rapid immunoassays. *Analytical Biochemistry*, **154**, 118-125. [Ch. 18]

Tett, P., Edwards, A., Grantham, B., Jones, K. & Turner, M., 1988. Microplankton dynamics in an enclosed coastal water column in summer. In *Algae and the Aquatic Environment* (ed. F.E. Round), pp. 339-368. Bristol: Biopress. [Ch. 4]

Tett, P., Gowen, R., Grantham, B., Jones, K. & Miller, B.S., 1986. The phytoplankton ecology of the Firth of Clyde sea-lochs Striven and Fyne. *Proceedings of the Royal Society of Edinburgh* (B), **90**, 223-238. [Ch. 4]

Tett, P.B., 1972. An annual cycle of flash induced luminescence in the euphausiid *Thysanoessa raschii*. *Marine Biology*, **12**, 207-218. [Ch. 16]

Thimann, K.V. & Curry, G.M., 1960. Phototropism and phototaxis. In *Comparative Biochemistry*, vol. 1 (ed. H.S. Mason and M. Florkin), pp. 243-309. New York: Academic Press. [Ch. 7]

Thomas, B., 1981. Specific effects of blue light on plant growth and development. In *Plants and the Daylight Spectrum* (ed. H. Smith), pp. 375-390. London: Academic Press. [Ch. 7]

Thorpe, J. E., 1978. *Rhythmic Activity of Fishes*. London: Academic Press. [Ch. 14]

Tilzer, M.M., 1984. The quantum yield as a fundamental parameter controlling vertical photosynthesis profiles of phytoplankton in Lake Constance. *Archiv für Hydrobiologie*, supplement 69, 169-198. [Ch. 4]

Tobin, E.M. & Silverthorne, J., 1985. Light regulation of gene expression in higher plants. *Annual Review of Plant Physiology*, **36**, 569-593. [Ch. 7]

Toll, R.B., 1983. The lycoteuthid genus *Oregoniateuthis* Voss, 1956, a synonym of *Lycoteuthis* Pfeffer, 1900 (Cephalopoda: Teuthoidea). *Proceedings of the Biological Society of Washington*, **96**, 365-369. [Ch. 16]

Tomita, T., 1968. The electrical response of single photoreceptors. *Proceedings of the Institute of Electrical and Electronic Engineers*, **56**, 1015-1023. [Ch. 12]

Topliss, B.J. & Platt, T., 1986. Passive fluorescence and photosynthesis in the ocean: implications for remote sensing. *Deep-Sea Research*, **33**, 849-864. [Ch. 1]

Torre, V., Matthews, H.R. & Lamb, T.D., 1986. Role of calcium in regulating the cyclic GMP cascade of phototransduction in retinal rods. *Proceedings of the National Academy of Sciences of the United States of America*, **83**, 7109-7113. [Ch. 11]

Tsien, R.Y., 1980. New calcium indicators and buffers with high selectivity against magnesium and protons: design, synthesis and properties of prototype structures. *Biochemistry*, **19**, 2396-2404. [Ch. 11]

Tsuda, M., 1987. Photoreception and phototransduction in invertebrate photoreceptors. *Photochemistry and Photobiology*, **45**, 915-931. [Ch. 12]

Tsuda, M., Tsuda, T., Terayama, Y., Fukada, Y., Akino, Y., Yamanaka, G., Stryer, L., Katada, T., Ui, M. & Ebrey, T., 1986. Kinship of cephalopod photoreceptor G-protein with vertebrate transducin. *FEBS Letters*, **198**, 5-10. [Ch. 12]

Tsuji, F.I. & Hill, E., 1983. Repetitive cycle of bioluminescence and spawning in the polychaete, *Odontosyllis phosphorea*. *Biological Bulletin. Marine Biological Laboratory, Woods Hole, Mass.*, **165**, 444-449. [Ch. 16]

Tsuji, F.I., Lynch, R.V. & Haneda, Y., 1970. Studies on the bioluminescence of the marine ostracod *Cypridina serrata*. *Biological Bulletin. Marine Biological Laboratory, Woods Hole, Mass.*, **139**, 386-401. [Ch. 16]

Tsukita, S., Tsukita, S. & Matsumoto, G., 1988. Light-induced structural changes of cytoskeleton in squid photoreceptor microvilli detected by rapid-freeze method. *Journal of Cell Biology*, **106**, 1151-1160. [Ch. 12]

Turner, J.S., 1973. *Buoyancy Effects in Fluids*. Cambridge: Cambridge University Press. [Ch. 4]

Twigg, J., Patel, R. & Whitaker, M.J., 1988. Translational control of InsP$_3$-induced chromatin condensation during the early cell cycles of sea urchin embryos. *Nature, London*, **332**, 366-369. [Ch. 7]

Tyler, I.D., 1983. *A Carbon Budget for Creran, a Scottish Sea Loch*. PhD Thesis. University of Strathclyde. [Ch. 4]

Tyler, J.E., 1960a. Observed and computed path radiance in the underwater light field. *Journal of Marine Research* **18**, 157-167. [Ch. 10]

Tyler, J.E., 1960b. Radiance distribution as a function of depth in an underwater environment. *Bulletin of the Scripps Institution of Oceanography*, **7**, 363-411. [Ch. 8]

Tyler, J.E., 1968. The Secchi disc. *Limnology and Oceanography*, **13**, 1-6. [Ch. 3]

Tyler, J.E., 1975. The *in situ* quantum efficiency of natural phytoplankton populations. *Limnology and Oceanography*, **20**, 976-980. [Ch. 4]

Tyler, J.E. & Smith, R.C., 1970. *Measurement of Spectral Irradiance Under Water*. New York: Gordon & Breach. [Ch. 10]

Ulitzur, S. & Hastings, J.W., 1978. Myristic acid stimulation of bacterial bioluminescence of aldehyde mutants. *Proceedings of the National Academy of Sciences of the United States of America*, **75**, 266-269. [Ch. 17]

Ulitzur, S. & Kuan, J., 1989. A new approach for the specific determination of bacteria and their antimicrobial susceptibilties. *Journal of Bioluminescence and Chemiluminescence*, in press. [Ch. 18]

UNESCO, 1966. Report of the second meeting of the joint group of experts on photosynthetic radiant energy. *UNESCO Technical Papers in Marine Science*, no. 2, 1-11. [Ch. 4]

Vandenberg, C.A. & Montal, M., 1984a. Light-regulated biochemical events in invertebrate photoreceptors. 1. Light-activated guanosine triphosphatase, guanine nucleotide binding, and cholera toxin catalyzed labelling of squid photoreceptor membranes. *Biochemistry*, **23**, 2339-2347. [Ch. 12]

Vandenberg, C.A. & Montal, M., 1984b. Light-regulated biochemical events in invertebrate photoreceptors. 2. Light regulated phosphorylation of rhodopsin and phosphoinositides in squid photoreceptor membranes. *Biochemistry*, **23**, 2347-2352. [Ch. 12]

Venrick, E.L., McGowan, J.A. & Mantyla, A.W., 1973. Deep maxima of photosynthetic chlorophyll in the Pacific Ocean. *Fishery Bulletin. National Oceanic and Atmospheric Administration of the United States*, **71**, 41-52. [Ch. 4]

Verheijen, F.J., 1956. On a method for collecting and keeping clupeoids for experimental purposes. *Pubblicazioni della Stazione Zoologica di Napoli*, **28**, 225-240. [Ch. 13]

Verheijen, F.J., 1958. The mechanisms of the trapping effect of artificial light sources upon animals. *Archives de Néerlandaises de Zoologie*, **13**, 1-107. [Ch. 15]

Verity, P.G., 1981. Effects of temperature, irradiance and daylength on the marine diatom *Leptocylindrus danicus* Cleve. I. Photosynthesis and cellular composition. *Journal of Experimental Marine Biology and Ecology*, **55**, 79-91. [Ch. 4]

Verity, P.G., 1982a. Effects of temperature, irradiance and daylength on the marine diatom *Leptocylindrus danicus* Cleve. III. Dark respiration. *Journal of Experimental Marine Biology and Ecology*, **60**, 197-207. [Ch. 4]

Verity, P.G., 1982b. Effects of temperature, irradiance and daylength on the marine diatom *Leptocylindrus danicus* Cleve. IV. Growth. *Journal of Experimental Marine Biology and Ecology*, **60**, 209-222. [Ch. 4]

Vesk, M. & Jeffrey, S.W., 1977. Effects of blue-green light on photosynthetic pigments and chloroplast structure in unicellular marine algae from six classes. *Journal of Phycology*, **13**, 280-288. [Ch. 5]

Vesk, M. & Jeffrey, S.W., 1987. Ultrastructure and pigments of two strains of the picoplanktonic alga *Pelagococcus subviridis* (Chrysophyceae). *Journal of Phycology*, **23**, 322-336. [Ch. 5]

Vierstra, R.D. & Poff, K.L., 1981. Role of carotenoids in the phototropic response of corn seedlings. *Plant Physiology*, **68**, 798-801. [Ch. 7]

Vilter, V., 1953. Existence d'une rétine a plusiers mosaïques photoreceptrices chez un poisson abyssal bathypelagique, *Bathylagus benedicti*. *Compte Rendu des Séances de la Société Biologique*, **148**, 1937-1939. [Ch. 8]

Viollier, M. & Sturm, B., 1984. CZCS data analysis in turbid coastal waters. *Journal of Geophysical Research*, **89**, 4977-4985. [Ch. 2]

Viollier, M., Tanre, D. & Deschamps, P.Y., 1980. An algorithm for remote sensing of water color from space. *Boundary-Layer Meteorology*, **18**, 247-267. [Ch. 3]

Vogt, K., 1975. Zur Optik des Flußkrebsauges. *Zeitschrift für Naturforschung*, **30**c, 691. [Ch. 9]

Vogt, K., 1980. Die Spiegeloptik des Flusßkrebsauges: The optical system of the crayfish eye. *Journal of Comparative Physiology*, **135**, 1-19. [Ch. 9]

Wald, G., Brown, P.K. & Brown, P.S., 1957. Visual pigment and depth of habitat of marine fishes. *Nature, London*, **180**, 969-971. [Ch. 8]

Wald, G., Brown, P.K. & Smith, P.H., 1953. Cyanopsin, a new pigment of cone vision. *Science, New York*, **188**, 505-508. [Ch. 10]

Wales, W., 1975. Extra-retinal control of vertical migration in fish larvae. *Nature, London*, **253**, 42-43. [Ch. 13]

Walls, G.L., 1942. *The Vertebrate Eye and its Adaptive Radiation*. Michigan: Cranbrook Institute. [Reprinted (1967) New York: Hafner.] [Ch. 9]

Walsh, C., 1980. Flavin co-enzymes: at the cross roads of biological redox chemistry. *Accounts of Chemical Research*, **13**, 148-155. [Ch. 17]

Wampler, J.E., 1981. Earthworm bioluminescence. In *Bioluminescence and Chemiluminescence, Basic Chemistry and Analytical Aplications* (ed. M. DeLuca and W.D. McElroy), pp. 249-256. New York: Academic Press. [Ch. 17]

Ward, W.W. & Cormier, M.J., 1978. Energy transfer via protein-protein interaction in *Renilla* bioluminescence. *Photochemistry and Photobiology*, **27**, 389-396. [Ch. 17]

Wardle, C.S., 1983. Fish reactions to towed fishing gears. In *Experimental Biology at Sea* (ed. A.G. Macdonald and I.G. Priede), pp. 167-195. London: Academic Press. [Ch. 15]

Wardle, C.S., 1986. Fish behaviour and fishing gear. In *The Behaviour of Teleost Fishes* (ed. T.J. Pitcher), pp. 463-495. London: Croom Helm. [Ch. 15]

Wardle, C.S., 1987. Investigating the behaviour of fish during capture. In *Developments in Fisheries Research in Scotland* (ed. R.S. Bailey and B.B. Parrish), pp. 139-155. Farnham: Fishing News Books. [Ch. 15]

Warner, J., Latz, M.I. & Case, J.F., 1979. Cryptic bioluminescence in a midwater shrimp. *Science, New York*, **203**, 1109-1110. [Ch. 16]

Watanabe, M.M., Takeda, Y., Sasa, T., Inouye, I., Suda, S., Sawaguchi, T. & Chihara, M., 1987. A green dinoflagellate with chlorophylls *a* and *b*: morphology, fine structure of the chloroplast and chlorophyll composition. *Journal of Phycology*, **23**, 382-389. [Ch. 5]

Waterbury, J.B., Watson, S.W., Guillard, R.R.L. & Brand, L.E., 1979. Widespread occurrence of a unicellular marine phytoplanktonic cyanobacterium. *Nature, London*, **277**, 293-294. [Ch. 6]

Waterbury, J.B., Watson, S.W., Valois, F.W. & Franks, D.G., 1986. Biological and ecological characterisation of the marine unicellular cyanobacterium *Synechococcus*. *Canadian Bulletin of Fisheries and Aquatic Sciences*, no. 214, 17-120. [Ch. 6]

Waterman, T.H., 1975. Natural polarized light and e-vector discrimination by vertebrates. In *Light as an Ecological Factor* II (ed. G.C. Evans and R. Bainbridge), pp. 305-335. Oxford: Blackwell. [Ch. 10]

Waterman, T.H., 1981. Polarization sensitivity. In *Handbook of Sensory Physiology*, vol. VII/6B. *Comparative Physiology and Evolution of Vision in Invertebrates* (ed. H. Autrum), pp. 281-469. Berlin: Springer-Verlag. [Ch. 10]

Waterman, T.H., 1984. Natural polarised light and vision. In *Photoreception and Vision in Invertebrates* (ed. M.A.Ali), pp. 63-114. New York: Plenum Press. [Ch. 8]

Waterman, T.H., 1988. Polarisation of marine light fields and animal orientation. *SPIE*, **925**, *Ocean Optics* IX, 431-436. [Ch. 8]

Weeks, I., Beheshti, I., McCapra, F., Campbell, A.K. & Woodhead, J.S., 1983. Acridinium esters as high specific activity chemiluminescent labels. *Clinical Chemistry*, **29**, 1474-1479. [Ch. 18]

Weeks, I. & Woodhead, J.S., 1987. Chemiluminescence immunoassay: an overview. *Clinical Science*, C70, 403-408. [Ch. 18]

Weidermann, A.D. & Bannister, T.T., 1986. Absorption and scattering coefficients in Irondequoit Bay. *Limnology and Oceanography*, **31**, 567-583. [Ch. 3]

Weidermann, A.D., Bannister, T.T., Effler, S. & Johnson, D., 1985. Particulate and optical properties during $CaCO_3$ precipitation in Otisco Lake. *Limnology and Oceanography*, **30**, 1078-1083. [Ch. 3]

Welsby, V.G., Dunn, J., Chapman, C., Sharman, D.P. & Priestley, R., 1964. Further uses of electronically scanned sonar in the investigation of behaviour of fish. *Nature, London*, **203**, 588-589. [Ch. 13]

Welschmeyer, N.A. & Lorenzen, C.J., 1981. Chlorophyll-specific photosynthesis and quantum efficiency at subsaturating light intensities. *Journal of Phycology*, **17**, 283-293. [Ch. 4]

Wheeler, W.N., 1980. Pigment content and photosynthetic rate of the fronds of *Macrocystis pyrifera*. *Marine Biology*, **56**, 97-102. [Ch. 5]

White, H.H., 1979. Effects of dinoflagellate bioluminescence on the ingestion rates of herbivorous zooplankton. *Journal of Experimental Marine Biology and Ecology*, **36**, 217-224. [Ch. 16]

Whitehead, T.P., Thorpe, G.H., Carter, T.J.N., Groucutt, H.C. & Kricka, L.J., 1983. Enhanced luminescence procedure for sensitive determination of peroxidase-labelled conjugates in immunoassay. *Nature, London*, **305**, 158-159. [Ch. 18]

Wickham, D.A., 1970. Collecting coastal pelagic fishes with artificial light and 5 metre lift net. *Commercial Fisheries Review*, **32**(12), 52-57. [Ch. 15]

Wickham, D.A., 1971. Harvesting coastal pelagic fishes with artificial light and purse seine. *Commercial Fisheries Review*, **33**(1), 30-38. [Ch. 15]

Wickham, D.A., 1973. Collecting and controlling coastal pelagic fish with night lights. *Transactions of the American Fisheries Society.* **102**, 816-825. [Ch. 15]

Widder, E.A., Bernstein, S.A., Bracher, D.F., Case, J.F., Reisenbichler, K.R., Torres, J.J. & Robison, B.H., 1989. Bioluminescence in the Monterey Submarine Canyon: image analysis of video recordings from a midwater submersible. *Marine Biology*, **100**, 541-551. [Ch. 16]

Widder, E.A., Latz, M.F. & Case, J.F., 1983. Marine bioluminescence spectra measured with an optical multi-channel detection system. *Biological Bulletin. Marine Biological Laboratory, Woods Hole, Mass.,* **165**, 791-810. [Ch. 8, 10, 16]

Widder, E.A., Latz, M.I., Herring, P.J. & Case, J.F., 1984. Far red bioluminescence from two deep-sea fishes. *Science, New York*, **225**, 512-514. [Ch. 16]

Widell, S., 1987. Membrane-bound blue light receptors - possible connection to blue light photo-morphogenesis. In *Blue Light Responses: Phenomena and Occurrence in Plants and Microorganisms* (ed. H. Senger), pp. 89-98. Florida: CRC Press Inc. [Ch. 7]

Wiebe, P.H., Boyd, S.H., Davis, B.M. & Cox, J.L., 1982. Avoidance of towed nets by the euphausiid *Nematoscelis megalops*. *Fishery Bulletin. National Oceanic and Atmospheric Administration of the United States*, **80**, 75-91. [Ch. 15]

Wilhelm, C., 1987. Purification and identification of chlorophyll c_i from the green alga *Mantoniella squamata*. *Biochimica et Biophysica Acta*, **892**, 23-29. [Ch. 5]

Wilhelm, C., 1988. The existence of chlorophyll c in the Chl b-containing light-harvesting complex of the green alga *Mantoniella squamata* (Prasinophyceae). *Botanica Acta*, **101**, 14-17. [Ch. 5]

Wilhelm, C. & Lenartz-Weiler, I., 1987. Energy transfer and pigment composition in three chlorophyll b-containing light-harvesting complexes isolated from *Mantoniella squamata* (Prasinophyceae), *Chlorella fusca* (Chlorophyceae) and *Sinapis alba*. *Photosynthesis Research*, **13**, 101-111. [Ch. 5]

Wilhelm, C., Lenartz-Weiler, I., Wiedemann, I. & Wild, A., 1986. The light-harvesting system of a *Micromonas* species (Prasinophyceae): the combination of three different chlorophyll species in one single chlorophyll-protein complex. *Phycologia*, **25**, 304-312. [Ch. 5]

Wilkens, L.A. & Wolken, J.J., 1981. Electroretinograms from *Odontosyllis enopla* (Polychaeta; Syllidae): initial observations on the visual system of the bioluminescent fireworm of Bermuda. *Marine Behaviour and Physiology*, **8**, 55-56. [Ch. 16]

Williams, P.J.le B., Raine, R.C.T. & Bryan, J.R., 1979. Agreement between the ^{14}C and oxygen methods of measuring phytoplankton production: reassessment of the photosynthetic quotient. *Oceanologica Acta*, **2**, 411-416. [Ch. 4]

Williams, R. & Robinson, G.A., 1973. Primary production at Ocean Weather Station India (59°00'N, 19°00'W) in the North Atlantic. *Bulletin of Marine Ecology*, **8**, 115-121. [Ch. 4]

Winn, H. E., 1955. Formation of a mucous envelope at night by parrot fishes. *Zoologica, New York*, **40**, 145-148. [Ch. 14]

Winn, H. E. & Bardach, J. E., 1959. Differential food selection by moray eels and a possible role of the mucous envelope of parrotfishes in reduction of predation. *Ecology*, **40**, 296-298. [Ch. 14]

Winslade, P., 1974. Behavioural studies on the lesser sandeel *Ammodytes marinus* (Raitt). II. The effect of light intensity on activity. *Journal of Fish Biology*, **6**, 577-586. [Ch. 14]

Wolf, K.U. & Woods, J.D., 1988. Lagrangian simulation of primary production in the physical environment - the deep chlorophyll maximum and nutricline. In *Towards a Theory of Biological-Physical Interactions in the World Ocean* (ed. B.J. Rothschild), pp. 51-70. Dordrecht: D. Reidel Publishing Co. [Ch. 4]

Wood, A.M., 1985. Adaptation of photosynthetic apparatus of marine ultraphytoplankton to natural light fields. *Nature, London*, **316**, 253-255. [Ch. 5, 6]

Woodhead, P.M.J., 1966. The behaviour of fish in relation to light in the sea. *Oceanography and Marine Biology, an Annual Review*, **4**, 337-403. [Ch. 14, 15]

Woods, J.D. & Barkman, W., 1986. The response of the upper ocean to solar heating. I. The mixed layer. *Quarterly Journal of the Royal Meteorological Society*, **112**, 1-27. [Ch. 4]

Woods, J.D. & Onken, R., 1982. Diurnal variation and primary production in the ocean - preliminary results of a Lagrangian ensemble model. *Journal of Plankton Research*, **4**, 735-756. [Ch. 4]

Woods, N.M., Cuthbertson, K.S.R. & Cobbold, P.H., 1986. Repetitive transient rises in cytoplasmic calcium in hormone stimulated hepatocytes. *Nature, London,* **319**, 600-602. [Ch. 18]

Worthington, C.R., Wang, S.K. & Folzer C.M., 1976. Low angle X-ray diffraction patterns of squid retina. *Nature, London,* **262**, 626-628. [Ch. 12]

Wright, S.W. & Jeffrey, S.W., 1987. Fucoxanthin pigment markers of marine phytoplankton analysed by HPLC and HPTLC. *Marine Ecology - Progress Series,* **38**, 259-266. [Ch. 5]

Wyman, M., Gregory, R.P.F. & Carr, N.G., 1985. Novel role for phycoerythrin in a marine cyanobacterium, *Synechococcus* strain DC2. *Science, New York,* **230**, 818-820. [Ch. 6]

Yau, K.-W. & Baylor, D.A., 1989. Cyclic GMP-activated conductance of retinal photoreceptor cells. *Annual Review of Neuroscience,* **12**, 289-327. [Ch. 11]

Yau, K.-W. & Nakatani, K., 1984a. Cation selectivity of light-sensitive conductance in retinal rods. *Nature, London,* **309**, 352-354. [Ch. 11]

Yau, K.-W. & Nakatani, K., 1984b. Electrogenic Na-Ca exchange in retinal rod outer segment. *Nature, London,* **311**, 661-663. [Ch. 11]

Yau, K.-W. & Nakatani, K., 1985a. Light-induced reduction of cytoplasmic free calcium in retinal rod outer segment. *Nature, London,* **313**, 579-582. [Ch. 11]

Yau, K.-W. & Nakatani, K., 1985b. Light-suppressible, cyclic GMP-sensitive conductance in the plasma membrane of a truncated rod outer segment. *Nature, London,* **317**, 252-255. [Ch. 11]

Yau, K.-W., McNaughton, P.A. & Hodgkin, A.L., 1981. Effect of ions on the light-sensitive current in retinal rods. *Nature, London,* **292**, 502-505. [Ch. 11]

Yentsch, C.M., Cucci, L. & Phinney, D.A., 1984. Flow cytometry and cell sorting: problems and promises for biological ocean science research. In *Marine Phytoplankton and Productivity* (ed. O. Holm-Hansen *et al.*), pp. 141-155. Berlin: Springer-Verlag. [Ch. 1]

Yentsch, C.M & Yentsch, C.S., 1984. Emergence of optical instrumentation for measuring biological properties. *Oceanography and Marine Biology, an Annual Review,* **22**, 55-98. [Ch. 1, 6]

Yentsch, C.S. & Phinney, D.A., 1985. Spectral fluorescence: an ataxonomic tool for studying the structure of phytoplankton populations. *Journal of Plankton Research,* **7**, 617-632. [Ch. 1]

Yoshikami, S. & Hagins, W.A., 1973. Control of the dark current in vertebrate rods and cones. In *Biochemistry and Physiology of Visual Pigments* (ed. H. Langer), pp. 245-255. Berlin: Springer-Verlag. [Ch. 11]

Young, J.Z., 1962. The retina of cephalopods and its degeneration after optic nerve section. *Philosophical Transactions of the Royal Society* (B), **245**, 1-18. [Ch. 12]

Young, R.E., 1977. Ventral bioluminescent countershading in midwater cephalopods. *Symposia of the Zoological Society of London,* no. 38, 161-190. [Ch. 8, 16]

Young, R.E., 1983. Oceanic bioluminescence: an overview of general functions. *Bulletin of Marine Science,* **33**, 829-845. [Ch. 16]

Young, R.E. & Arnold, J.M., 1982. The functional morphology of a ventral photophore from the mesopelagic squid. *Abralia trigonura. Malacologia,* **23**, 135-163. [Ch. 8, 16]

Young, R.E. & Mencher, F.M., 1980. Bioluminescence in mesopelagic squid: diel color change during counterillumination. *Science, New York,* **208**, 1286-1288. [Ch. 16]

Young, R.E. & Roper, C.F.E., 1976. Bioluminescent countershading in midwater animals: evidence from living squid. *Science, New York,* **191**, 1046-1048. [Ch. 16]

Young, R.E. & Roper, C.F.E., 1977. Intensity regulation of bioluminescence during counterillumination in living midwater animals. *Fishery Bulletin. National Oceanic and Atmospheric Administration of the United States,* **75**, 239-252. [Ch. 8, 16]

Young, R.E., Kampa, E.M., Maynard, S.D., Mencher, F.M. & Roper, C.F.E., 1980. Counterillumination and the upper depth limit of midwater animals. *Deep-Sea Research,* 27A, 671-691. [Ch. 8, 16]

Young, R.E., Seapy, R.R., Mangold, K.M. & Hochberg, F.G., 1982. Luminescent flashing in the midwater squids *Pterygioteuthis microlampas* and *P. giardi. Marine Biology,* **69**, 299-308. [Ch. 16]

Zonana, H.V., 1961. Fine structure of the squid retina. *Johns Hopkins Hospital Bulletin,* **109**, 185-205. [Ch. 12]

Index

Page numbers in italics refer to Figures and Tables

Herring, P.J., Campbell, A.K., Whitfield, M. & Maddock, L. (ed.). *Light and Life in the Sea.*
© 1990. Cambridge University Press.

INDEX